Nyquist AD Converters, Sensor Interfaces, and Robustness

Arthur H.M. van Roermund • Andrea Baschirotto • Michiel Steyaert
Editors

Nyquist AD Converters, Sensor Interfaces, and Robustness

Advances in Analog Circuit Design, 2012

Springer

Editors
Arthur H.M. van Roermund
Electrical Engineering
Eindhoven University of Technology
Eindhoven, Netherlands

Andrea Baschirotto
Department of Physics
University of Milan-Bicocca
Milan, Italy

Michiel Steyaert
Department Elektrotechniek ESAT-MICAS
Leuven, Belgium

ISBN 978-1-4614-4586-9 ISBN 978-1-4614-4587-6 (eBook)
DOI 10.1007/978-1-4614-4587-6
Springer New York Heidelberg Dordrecht London

Library of Congress Control Number: 2012950586

Printed on acid-free paper

Springer is part of Springer Science+Business Media (www.springer.com)

Preface

This book is part of the Analog Circuit Design series and contains contributions of 15 speakers of the 21st workshop on Advances in Analog Circuit Design (AACD). The local chair was Kostas Doris from NXP, the sponsor of the workshop this year. The workshop was held at Château St. Gerlach in Valkenburg, The Netherlands, March 27–29, 2012.

The book comprises three Parts, covering advanced analog and mixed-signal circuit design fields that are considered as very important by the circuit design community:

- Nyquist AD Converters
- Sensor Interfaces
- Robustness

The aim of the AACD workshop is to bring together a group of expert designers to discuss new developments and future options. Each workshop is followed by the publication of a book by Springer in their successful series of Analog Circuit Design. This book is number 21 in this series. The book series can be seen as a reference for all people involved in analog and mixed-signal design. The full list of the previous books and topics in the series is given next.

We are confident that this book, like its predecessors, provides a valuable contribution to our analog and mixed-signal circuit-design community.

Eindhoven, Netherlands — Arthur van Roermund

The topics covered before in this series:

2011	Leuven, Belgium	Low-Voltage Low-Power Data Converters
		Short-Range Wireless Front-Ends
		Power Management and DC-DC
2010	Graz, Austria	Robust Design
		Sigma Delta Converters
		RFID
2009	Lund, Sweden	Smart Data Converters
		Filters on Chip
		Multimode Transmitters
2008	Pavia (Italy)	High-Speed Clock and Data Recovery
		High-Performance Amplifiers
		Power Management
2007	Oostende (Belgium)	Sensors, Actuators and Power Drivers for the Automotive and Industrial Environment
		Integrated PAs from Wireline to RF
		Very High Frequency Front Ends
2006	Maastricht (The Netherlands)	High-Speed AD converters
		Automotive Electronics: EMC Issues
		Ultra Low Power Wireless
2005	Limerick (Ireland)	RF Circuits: Wide Band, Front-Ends, DACs
		Design Methodology and Verification of RF and Mixed-Signal Systems
		Low Power and Low Voltage
2004	Montreux (Swiss)	Sensor and Actuator Interface Electronics
		Integrated High-Voltage Electronics and Power Management
		Low-Power and High-Resolution ADCs
2003	Graz (Austria)	Fractional-N Synthesizers
		Design for Robustness
		Line and Bus drivers
2002	Spa (Belgium)	Structured Mixed-Mode Design
		Multi-Bit Sigma-Delta Converters
		Short-Range RF Circuits

(continued)

2001	Noordwijk (The Netherlands)	Scalable Analog Circuits High-Speed D/A Converters RF Power Amplifiers
2000	Munich (Germany)	High-Speed A/D Converters Mixed-Signal Design PLLs and Synthesizers
1999	Nice (France)	XDSL and Other Communication Systems RF-MOST Models and Behavioural Modelling Integrated Filters and Oscillators
1998	Copenhagen (Denmark)	1-Volt Electronics Mixed-Mode Systems LNAs and RF Power Amps for Telecom
1997	Como (Italy)	RF A/D Converters Sensor and Actuator Interfaces Low-Noise Oscillators, PLLs and Synthesizers
1996	Lausanne (Swiss)	RF CMOS Circuit Design Bandpass Sigma Delta and Other Data Converters Translineair Circuits
1995	Villach (Austria)	Low-Noise/Power/Voltage Mixed-Mode with CAD tools Voltage, Current and Time References
1994	Eindhoven (Netherlands)	Low-Power Low-Voltage Integrated Filters Smart Power
1993	Leuven (Belgium)	Mixed-Mode A/D Design Sensor Interfaces Communication Circuits
1992	Scheveningen (The Netherlands)	OpAmps ADC Analog CAD

Contents

Part I
Nyquist AD Converters

This first part of the book is on 'Nyquist AD Converters', which, as the name suggests, refers to AD converters that can convert signals up to (or close to) the Nyquist frequency. Six chapters will focus on this, from different architecture points of view.

The first two Chapters address the pipeline converter: the various steps of the search algorithm are mapped in time in a successive, serial way. The Chapter of Ahmed Ali, from Analog Devices, US, treats the generic aspects of a pipeline converter without sample-and-hold amplifier, and with calibration. A specific converter, made in 180 nm BiCMOS, with 1.8 V supply, 1 W power consumption, 300 MHz input bandwidth, 250 MS/s, >90 dB SFDR, and 76.5 dB SNDR, is described. The second Chapter, of Jan Mulder from Broadcom, The Netherlands, describes a pipeline converter based on the dual-residue principle, which makes the converter insensitive to gain of the residue amplifier, allowing low gain and thus low power, without calibration. Moreover, the converter described uses four pipes in time interleaved. It is made in 40 nm CMOS, with 1/2.5 V supply, 105 mW, 4×200 MS/s $=$ 800 MS/s, 12 bit resolution, and 59 dB SNDR.

Next, we make a step to two other converter types: the successive approximation converter (the SAR) and the slope converter. Pieter Harpe, from imec/Holst and University of Technology, Eindhoven, The Netherlands, discusses the alternative approach that starts from the inherently low power but low speed direction, choosing the SAR or slope converter, and then achieving speed by using time interleaving, so parallellization in time. A comparison between the SAR and the slope is given. This time-interleaving requires accurate timing. In his case this is achieved intrinsically, by using a proper architecture and layout. The Chapter of Erwin Janssen, from NXP, The Netherlands, uses massive parallellization via time interleaving of SARs, combined with a special open loop sampler that uses the same buffer in sampling and conversion phase to compensate non-linearities, a reduced radix in the SAR, and a redundancy in the number of SARs, to achieve a 2.6 GS/s, 10b conversion for DOCSIS type of multi-stream applications. The time interleaving is done in two steps to make a proper tradeoff in parallellization and sampling issues.

The last two speakers discuss high speed converters above 10 GS/s. Jieh-Tsorng Wu, from National Chiao Tung University, Taiwan, describes the use of flash converters, the ultimate high speed architecture, in a time-interleaved way, to further boost the speed up to a 16 GS/s, 1.5 V, 435 mW converter with an SNDR of 38 dB. He uses regenerative latches that are randomly chopped to reduce the input referred offset, as an alternative to the use of power hungry preamplifiers, and a timing skew calibration detection and correction technique based on equal zero-crossing probability, in background, with extra replica samplers in each AD. Finally, Yuriy Greshishchev from Ciena Corporation, Canada, discusses the fastest converters, made with two-step time-interleaved SARs, for optical communication at 40 GB/s with digital signal processing. Their first entry point in this field was a 2 4GS/s converter in BiCMOS for each of the four 10 GB/s channels of the 4 GS/s optical system; now they realised a 40GS/s converter in full CMOS with $16 \times 10 = 160$ parallel channels; the Fujitsu 56 GS/s 320 paths ADC, in full CMOS, is also shortly discussed.

It is clear that the supremacy of full-flash, time-interleaved flash and pipeline for high-speed Nyquist converters is shifting towards time-interleaved SAR, and that even time-interleaved slope can have its place. An interesting situation that makes the reading of the next six Chapters very interesting.

Arthur van Roermund

Chapter 1
High Performance Piplined A/D Converters in CMOS and BiCMOS Processes

Ahmed M.A. Ali

Abstract This paper describes the design approach and trade-offs in designing high-speed and high performance pipelined A/D converters in CMOS and BiCMOS processes. Design techniques to improve the linearity, lower the noise and reduce the power consumption will be discussed. The discussion will be in the context of a 16-bit 250 MS/s ADC fabricated on a 0.18 μm BiCMOS process. The ADC achieves an SNDR of 76.5 dB and consumes 850 mW from a 1.8 V supply, with an input buffer that consumes 150 mW from a 3 V supply. The measured SFDR is greater than 100 dB for input frequencies up to 100 MHz and 90 dB up to 300 MHz input frequency.

1.1 Introduction

The demand for high-resolution A/D converters (ADCs) with ever increasing sample rates has been unabated. Wireless communication applications have been a major driver of the development of this class of ADCs with excellent linearity and IF/RF sampling capability. In addition, other applications, such as instrumentation, medical imaging and military benefit from having ADCs with higher sample rate and higher performance.

An attractive architecture for this class of A/D converters has been the pipeline architecture, especially as speeds and bandwidths continue to increase. Their algorithmic nature, amenability to performance enhancement using digital signal processing and their proven ability to achieve superb linearity (up to 110 dB) ensure their dominance for the near future.

A block diagram of the pipelined A/D converter discussed in this paper is shown in Fig. 1.1 [1, 2]. Each stage operates in two phases: in the first phase it samples its

A.M.A. Ali (✉)
Analog Devices Inc, 7910 Triad Center Drive, Greensboro, NC 27409, USA
e-mail: ahmed.ali@analog.com

A.H.M. van Roermund et al. (eds.), *Nyquist AD Converters, Sensor Interfaces, and Robustness: Advances in Analog Circuit Design, 2012*,
DOI 10.1007/978-1-4614-4587-6_1,

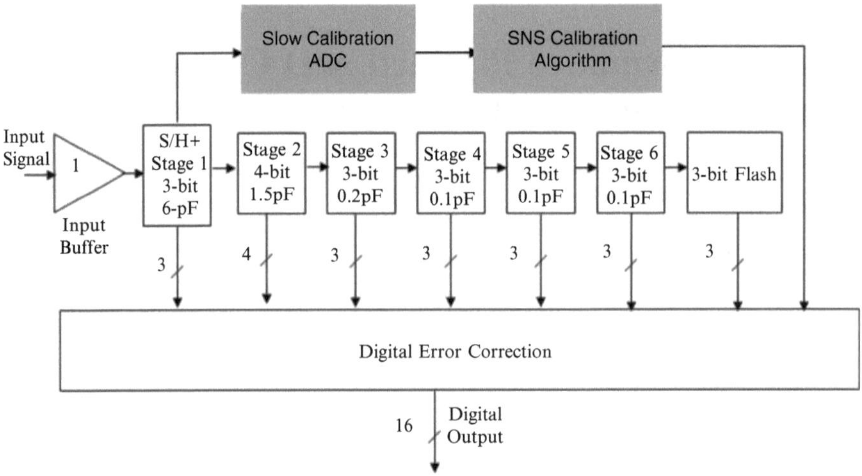

Fig. 1.1 Block diagram of the pipelined ADC

input on the sampling capacitor. In the second phase, it digitizes that input and creates a residue to be sampled by the following stage's sampling capacitance. Each stage consists of a low resolution sub-ADC (usually a flash ADC) that coarsely digitizes the input, and a multiplying DAC (MDAC) that generates and amplifies the residue using a switched capacitor amplifier.

To achieve the desired linearity, the capacitor mismatches of the first two stages need to be calibrated. Since those are supply, temperature and sample rate independent, they can be factory calibrated. This is done by using digital coefficients in the digital correction logic that scale the different sub-ranges appropriately to correct for the errors that are due to the capacitor mismatches.

Since the integrated thermal noise power (kT/C noise) is inversely proportional to the value of the sampling capacitance, achieving a high SNR requires that capacitance to be relatively large. This large capacitance at the ADC front-end is difficult to drive while keeping the power and distortion low. An input buffer using bipolar transistors (BJTs) is needed in order to reduce the charge injection (kick-back) on the ADC driver and help achieve the desired linearity.

In addition, the large sampling capacitance leads to high power consumption in the MDAC's (Multiplying DAC's) residue amplifier (RA). The targeted high performance and sample rate require the RA to have a high gain and a very large bandwidth. Achieving these targets, for reasonable power consumption, represents one of the major design challenges in pipelined ADC design. Background calibration is employed to relax the design targets of the RA and reduce its power consumption without sacrificing performance. This is shown in Fig. 1.1.

Another challenge in implementing such an ADC is preserving the SNR/SFDR performance for high input frequencies. As the input frequency increases, the effect of any non-linearity and mismatches in the input signal path becomes more detrimental. Moreover, the jitter in the clock path degrades the performance at high frequencies significantly. Therefore, the keys to achieving the targeted IF/RF sampling performance is to create a low-noise and low-distortion input front-end capable of handling those high frequency signals, fast and high performance MDACs capable of achieving the desired speed and gain accuracy, together with a very low-jitter low-noise clock path.

1.2 SHA-Less Architecture

In a “SHA-less” architecture, the sample-and-hold (S/H) circuit is integrated in the first MDAC, without a dedicated amplifier. This front-end, shown in Fig. 1.2, provides the same S/H operation to the user, but saves power, improves distortion and reduces noise. It, however, creates two issues that must be addressed for the architecture to work properly.

The first issue with the SHA-less architecture is that it requires matching between the input networks of the flash (sub-ADC) and the MDAC of the first stage. Since the input to the ADC is not a “held” signal, any mismatch between the bandwidths or the sampling instant of these two input networks will result in a mismatch between the sampled values of the MDAC and the flash. This error can be corrected by the digital error correction as long as it does not exceed the redundancy correction range of the first pipeline stage.

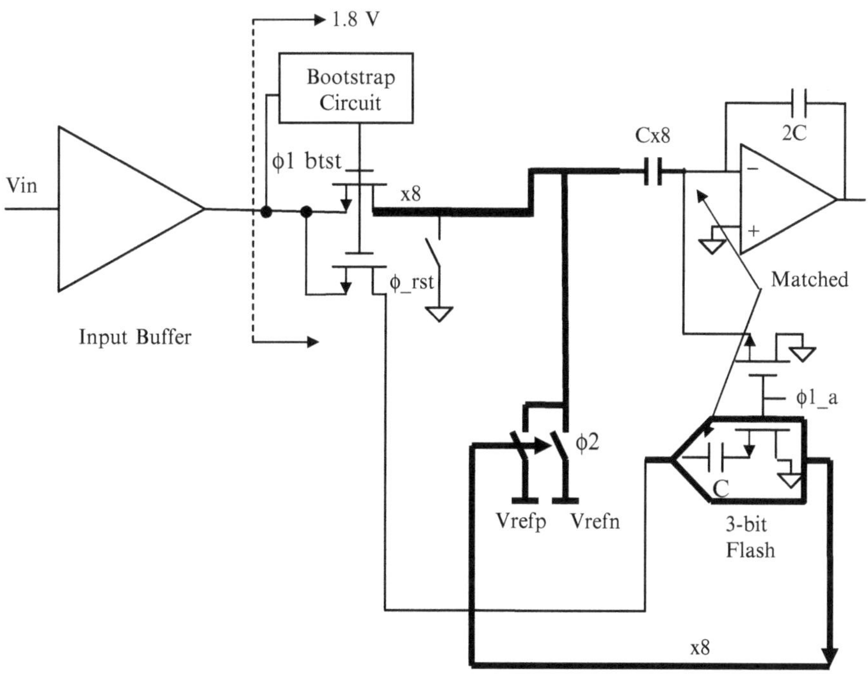

Fig. 1.2 Stage-1 of the ADC showing the input buffer, the first MDAC and the first flash

The second issue with the SHA-less architecture is that it decreases the time available for the settling of the first MDAC, because more time is needed by the flash to make its decision compared to the time it would take had there been a SHA. It also requires a fast comparator to minimize the latching time. A BiCMOS comparator is used, where the high gain-bandwidth product (fT) of the NPN BJTs is utilized to reduce the regeneration/latching time. Nevertheless, the time available for the first MDAC settling is significantly reduced by about 50% from about 2 ns at 250 MS/s to about 1 ns. So the MDAC bandwidth must be increased accordingly. In spite of that, the power saved by removing the SHA is substantial enough to justify using this architecture. In addition, the noise and distortion also improve.

1.3 The Input Buffer

Emitter followers and source followers have traditionally been used as buffers because they have high input impedances and low output impedances. This enables them to isolate the ADC driver from the charge injection (kick-back) of the sampling capacitances and the input switch. In addition, the low output impedance improves the distortion and enables a large acquisition bandwidth. An emitter-follower, shown in Fig. 1.3a, tends to be better than a source follower due to the higher trans-conductance (g_m), higher output impedance and smaller parasitics of the BJT compared to an NMOS device.

One major source of non-linearity in an emitter follower is the current variation through the follower device Q1 in response to the dynamic input signal. Traditionally, this signal current that flows through the sampling capacitance needs to be supplied by the device Q1. This changes the device parameters (such as its g_m) in response to the input signal, which leads to distortion.

One approach to improve the distortion in an emitter- (or source-) follower buffer has been to increase the dc (bias) current of Q1. However, this increases the power consumption and increases the device parasitics, which eventually limit the achievable linearity. Another approach is to use a cascade of two followers, which also increases the power consumption substantially [3].

To overcome these performance limitations and reduce the power consumption, we employ a buffer linearization technique that is shown in Fig. 1.3b. A replica load is used to inject a current into the output node of the follower that is almost equal to the current needed by the load. This reduces the current variation in the emitter-follower device Q1 and hence improves its linearity. It also reduces the power consumption substantially by reducing the DC current needed to achieve the desired linearity.

Figure 1.4 shows a comparison of the performance of this buffer compared to the prior art. This the advantage of the linearized buffer compared to traditional buffers. It also reduces the power consumption by 50–70% [1–3].

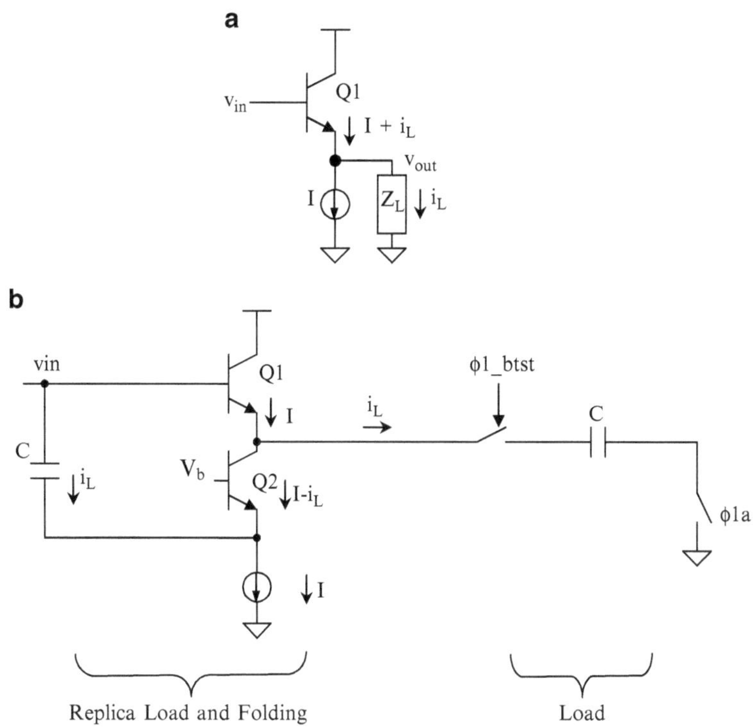

Fig. 1.3 (**a**) A traditional emitter follower. (**b**) A "linearized" emitter follower with current compensation

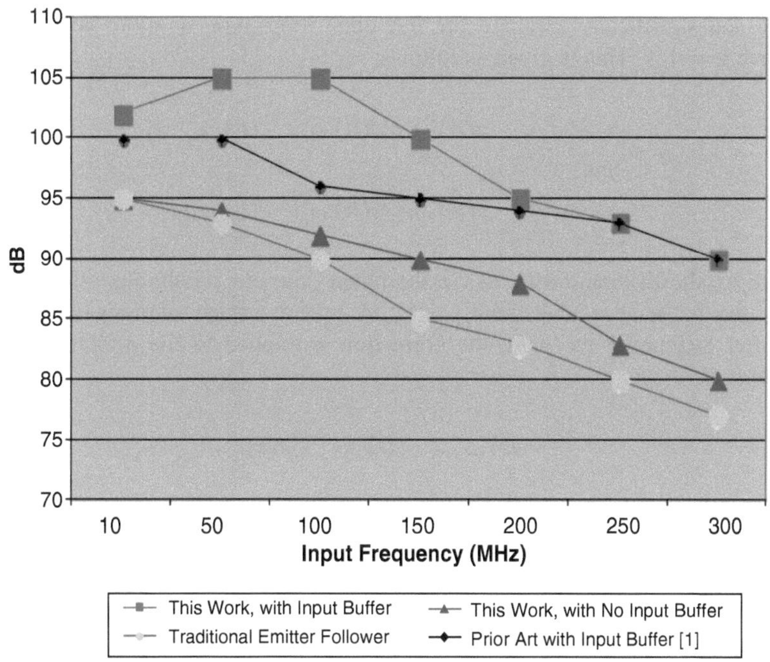

Fig. 1.4 The SFDR measured performance of this work at 250 MS/s compared to the state of the art

1.4 MDAC and Background Calibration

A simplified schematic of the pipeline first stage is shown in Fig. 1.2. During the tracking/sampling phase, the sampling switches are turned on, and the input is sampled on the MDAC and the flash sampling capacitances. During the hold/gain phase, the flash comparators make their decisions. In this design, the flash takes about 1 ns to make its decision, which leaves about 1 ns for the RA output to settle at 250 MS/s to the desired 16-bit accuracy. For this amplifier, this translates into a bandwidth requirement of about 2 GHz and an open loop gain that is larger than 100 dB [1–3]. Relaxing some of these requirements helps reduce the power but causes inter-stage gain errors that tend to be temperature, supply and even sample-rate dependent.

Employing background calibration reduces the power consumption in the first stage by about 50%, while correcting for the resulting errors. Several background calibration techniques exist in the literature [4–8]. In this work, we employ the Summing Node Sampling (SNS) algorithm, which estimates the amplifier open loop gain using its output (i.e. the residue voltage) and the summing node voltage. The summing node voltage is sampled and processed at a much lower sample rate, in combination with the corresponding stage-1 residue sample, to statistically estimate the RA's open loop gain.

After sampling, the summing node voltage is amplified and digitized using an on-chip low-resolution, low-power and low-speed auxiliary ADC. This is shown in Fig. 1.5. The digital values of both voltages are high-pass filtered to remove the offset and processed using the Least-Mean-Square (LMS) algorithm to filter the noise and statistically estimate the RA open loop gain, or more accurately its inverse $\alpha = 1/A$. This is given as follows:

$$\alpha_{i+1} = \alpha_i - \mu D(V_{o1i})[\alpha_i D(V_{o1i}) - D(V_{o1i}/A)]$$

and

$$\alpha_{i+1} = \alpha_i - \mu D(V_{o1i})[\alpha_i D(V_{o1i}) - D(V_{e1i})] \tag{1.1}$$

where α_i is the ith iteration of α, $D(x)$ is the digital value of x, μ is the algorithm step size, V_{o1} is the first stage residue and $V_{e1} = V_{o1}/A$ is the first stage summing node voltage.

After estimating the gain, the correction is applied to the digital residue as follows:

$$DV_{o1\_cal} = DV_{o1}(1 + K\alpha) \tag{1.2}$$

where K is a constant that depends on the feedback factor of the MDAC, DV_{o1} is the digitized stage-1 residue before calibration and DV_{o1_cal} is the digital calibrated value of the stage-1 residue.

Although this algorithm does not depend on the input signal's exact shape or statistics, the input signal needs to be variable (not DC), and its amplitude needs to

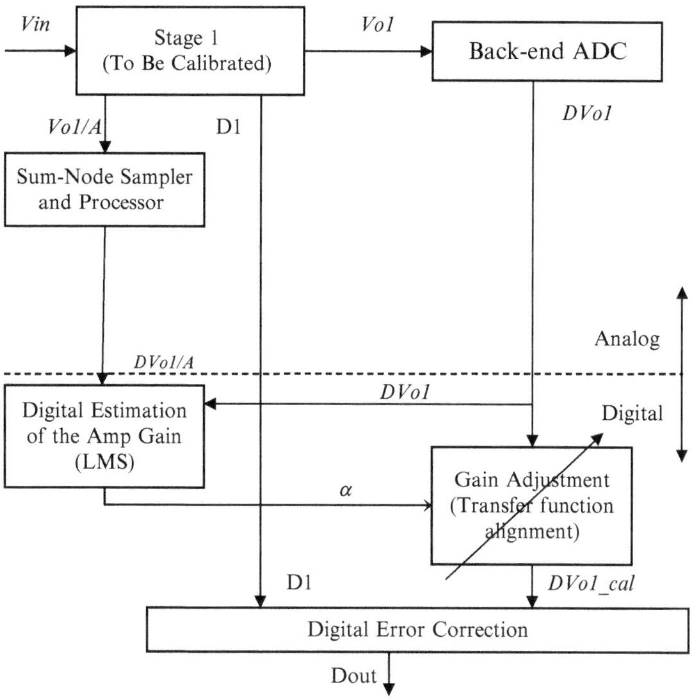

Fig. 1.5 A block diagram illustrating the summing node sampling (SNS) background calibration algorithm of stage-1 amplifier. The summing node voltage is sampled, amplified and digitized using a slow analog processor. The algorithm estimates the open loop gain of the amplifier and its inverse α. This is used to correct the residue voltage (DV_{o1_cal})

be above a certain threshold that is determined by the value of the algorithm's step size μ and the desired convergence time. If this condition is not satisfied, the corresponding samples are ignored and the algorithm is frozen for those samples.

The circuit implementation of the calibration technique is shown in Fig. 1.6. The summing node voltage ($V_{e1} = V_{o1}/A$) is sampled on the capacitance C_{e1}. The value of this capacitance needs to be large enough to achieve a reasonable SNR in the slow path. However, it can't be too large so as to substantially degrade the feedback factor of the main MDAC. The sampled voltage V_{e1} is then amplified by a factor of 64, and digitized using the slow ADC. Given the gain of 64, the resolution of the slow ADC needs to be at least 12 bits, in order to achieve an input-referred linearity of 0.25 LSB at the 16-bit level. The sample rate of the slow ADC is 12.5 MS/s and the power is 10 mW.

In order to reduce the impact of the slow clock on the summing node of the main ADC, a dummy network is used to sample the summing node voltage on the clock cycles where the slow network does not sample the summing node. Therefore, the two networks provide a constant load on the summing node at every clock. In addition, a clock spreading technique is used to randomize the slow clock edge in

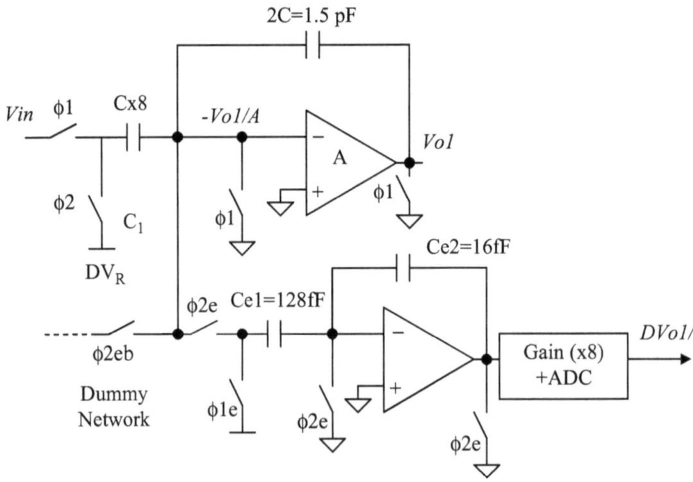

Fig. 1.6 A schematic of the implementation of the summing node voltage sampling for the SNS algorithm

Table 1.1 A comparison of the measured ADC performance with fixed factory calibration only (i.e. without background calibration) compared to using background calibration

Temperature	Parameter	Without background calibration	With background calibration
At 27°C	SNR	76.5 dB	76.5 dB
	SFDR	100 + dB	100 + dB
Worst case over −40°C to −85°C	SNR	74.3 dB	76.2 dB
	SFDR	86 dB	98 dB

order to distribute the energy of any remaining spurs in the noise floor. Silicon results indicate a spur level of better than 105 dB without spreading and better than 120 dB with spreading.

This calibration algorithm is implemented on chip and is shown to improve the SNDR by 5 dB and the SFDR by 10 dB. The measured results are summarized in Table 1.1.

1.5 Clock Jitter

The clock jitter causes degradation in the performance of the ADC that becomes more detrimental as the input frequency increases. To achieve the targeted IF/RF sampling performance, the clock jitter needs to be very small. Coupling from noise sources on the clock path needs to be minimized by layout isolation, shielding and decoupling techniques. In addition, the noise generated by the on-chip devices in the clock signal path itself contributes to the random jitter. This contribution needs to be accurately simulated and minimized in order to reduce the overall jitter.

The ADC jitter noise voltage is given by:

$$N_j = (2\pi f_0 A_{RMS} t_j)^2 \tag{1.3}$$

Where A_{RMS} is the RMS amplitude of the input signal, f_O is the input frequency and t_j is the RMS jitter. The total ADC noise is given by: $N_t = N_{low\_freq} + N_j$, and the resulting total SNR is given by:

$$SNR_t = 10\log(S/N_t) = 10\log[S/(N_{low\_freq} + N_j)] \tag{1.4}$$

Therefore, the RMS clock jitter is given by:

$$t_j(Total) = \frac{\sqrt{N_j}}{2\pi f_0 A_{RMS}} = \frac{\sqrt{\frac{S}{10^{SNRt/10}} - \frac{S}{10^{SNR\_low\_frequency/10}}}}{2\pi f_0 A_{RMS}} \tag{1.5}$$

This represents the total jitter budget, including contributions from the clock source, the on-chip devices in the clock path, and the coupling from substrate, ground and supply noise. Based on that, a budget for the on-chip jitter contribution is calculated and the clock path is designed accordingly using the formula:

$$t_j(ADC) = \sqrt{t_j^2(Total) - t_j^2(Generator)} \tag{1.6}$$

To simulate the random jitter due to the noise generated by the devices in the clock path, strobed (sampled) periodic noise analysis can be used. In this method, the noise analysis is performed around this periodic operating point to account for the periodic time-varying nature of this noise (i.e. cyclostationary noise). Sampling the output clock at the sampling instant, then integrating the resulting noise spectral density, gives the total noise power. The RMS jitter is then obtained by dividing the noise power by the slope of the sampling edge:

$$t_j = \frac{V_{clk\_noise}}{dV_{clk}/dt} = \frac{\sqrt{\int_0^{fs/2} n^2(f)df}}{dV_{clk}/dt} \tag{1.7}$$

Where $n^2(f)$ is the noise power spectral density obtained using the strobed (sampled) periodic noise analysis. This method is used in the current work to optimize the design of the clock path.

To minimize the other sources of jitter, careful substrate isolation is employed. This was done using substrate contacts and N-well isolation. The clock was also carefully shielded, and the supply heavily decoupled.

1.6 ADC Performance Summary

The ADC is fabricated on a 0.18 um BiCMOS process. The SNDR is 76.5 dB and the SFDR is greater than 100 dB. The SNR numbers capture all of the noise sources in the signal and clock paths including the off-chip termination resistors. The ADC consumes 850 mW from a 1.8 V supply, and the input buffer consumes 150 mW from a 3 V supply. The input span is 2.5 Vp-p and the jitter is 60 fs.

1.7 Conclusion

In this paper, we present some of the design techniques used in developing high-speed and high-resolution ADCs in CMOS and BiCMOS processes. This is presented in the context of a 16-bit 250 MS/s pipelined ADC. It employs an input buffer linearization technique that enables an SFDR performance that is 5–10 dB better than the state of the art, while reducing the power consumption by 50–70%. It also employs background calibration algorithm that helps improve the performance and reduce the power consumption of the first stage residue amplifier.

Acknowledgment The author would like to acknowledge Greg Patterson, Paritosh Bhoraskar, Huseyin Dinc, Scott Puckett, Andy Morgan, Chris Dillon, Mike Hensley, Russell Stop, Scott Bardsley, David Lattimore, Jeff Bray, Carroll Speir, Robert Sneed, Val Palmer, Liam Noonan, Cindy Block, John Kornblum, Paul Wilkins, Darren Combs, Robert Shillito, and Chad Shelton at Analog Devices for their contribution to the implementation, evaluation and layout of the ADC described here.

References

1. Ali AMA et al (2010) A 16-bit 250-MS/s IF sampling pipelined ADC with background calibration. In: International solid-state circuits conference, Digest of technical papers, pp 292–293, San Francisco, USA Feb 2010
2. Ali AMA et al (2010) A 16-bit 250-MS/s IF sampling pipelined ADC with background calibration. IEEE J Solid-St Circ 45(12):2602–2612
3. Ali AMA et al (2006) A 14-bit 125 MS/s IF/RF sampling pipelined ADC with 100 dB SFDR and 50 f. jitter. IEEE J Solid-St Circ 41(8):1846–1855
4. Cho TB, Gray PR (1995) A 10 b, 20 Msample/s, 35 mW pipeline A/D converter. IEEE J Solid-St Circ 30(3):166–172
5. Siragusa E, Galton I (2004) A digitally enhanced 1.8-V 15-bit 40-MSample/s CMOS pipelined ADC. IEEE J Solid-St Circ 39:2126–2138
6. Panigada A, Galton I (2009) A 130 mW 100 MS/s pipelined ADC with 69 dB SNDR enabled by digital harmonic distortion correction. In: International solid-state circuits conference, Digest of technical papers, San Francisco, USA pp 162–163, Feb 2009
7. Iroaga E, Murmann B (2007) A 12-Bit 75-MS/s pipelined ADC using incomplete settling. IEEE J Solid-St Circ 42(4):748–756
8. Murmann B, Boser BE (2003) A 12 b 75 MS/s pipelined ADC using open-loop residue amplification. In: International solid-state circuits conference, Digest of technical papers, vol 1, San Francisco, USA pp 328–497

Chapter 2
A 12-bit 800 MS/s Dual-Residue Pipeline ADC

Jan Mulder, Davide Vecchi, Frank M.L. van der Goes, Jan R. Westra, Emre Ayranci, Christopher M. Ward, Jiansong Wan, and Klaas Bult

Abstract This paper presents the design of a pipeline analog-to-digital converter (ADC) based on the dual-residue principle. By applying this technique, the ADC becomes insensitive to the exact gain of the MDAC residue amplifiers. This allows these amplifiers to be designed with a relatively low open-loop gain and low bandwidth, which is favorable for the power consumption of the ADC. The offsets of the residue amplifiers, however, limit the accuracy of the ADC. Therefore, offset calibration is required for the ADC to achieve a high resolution.

A 12-bit 800 MS/s dual-residue ADC was designed and implemented in a standard 40 nm CMOS technology. The high sampling speed was obtained through four times interleaving. The ADC achieves a peak SNDR of 59 dB. It operates from a dual 1 V/2.5 V power supply and consumes 105 mW.

2.1 Introduction

The increasing demands of telecommunication systems are continuously pushing the specifications of integrated circuits. Whereas digital processing can exploit the improvements introduced by deep-submicron technologies, analog interface circuits may be negatively affected by technology scaling. In particular, power supply scaling and intrinsic transistor gain reduction significantly influence the design of high performance analog components.

J. Mulder (✉) • D. Vecchi • F.M.L. van der Goes • J.R. Westra • C.M. Ward • J. Wan • K. Bult
Broadcom Netherlands B.V., Bunnik, The Netherlands
e-mail: jmulder@broadcom.com

E. Ayranci
Now with ClariPhy Communications Inc., Irvine, CA, USA

A.H.M. van Roermund et al. (eds.), *Nyquist AD Converters, Sensor Interfaces, and Robustness: Advances in Analog Circuit Design, 2012*,
DOI 10.1007/978-1-4614-4587-6_2,

High resolution (i.e., > 10 bit) and high conversion rate (> 100 MS/s) are typical requirements today for ADCs. To allow integration on a large System-on-a-Chip (SoC), there are additional limitations on area and power consumption. Pipeline ADCs, given their area and power efficiency, are an excellent architecture to fulfill these requirements [1–9].

Pipeline ADCs, however, suffer from an important limitation: Their linearity is limited by the finite open-loop gain and finite bandwidth of the residue amplifiers. To improve the gain and bandwidth of these amplifiers, design techniques leading to higher power consumption must be implemented. To reduce this power penalty, several calibration algorithms have been proposed [10–12], that can calibrate either for the gain-induced errors (and the incomplete settling due to limited bandwidth) [12] or for the nonlinearity of the residue amplifiers [10] or both [1].

Foreground calibration algorithms cannot track changes caused by temperature, voltage, and aging. On the other hand, background calibrations require long convergence times, making them unsuitable for automatic testing and applications requiring fast start-up. An interesting approach is the "split ADC" technique [13], which shortens the convergence time significantly but for high-accuracy applications requires extra design effort to reduce the mismatch between the two ADCs [14]. Moreover, each component inside the ADC must be duplicated, including the flash ADCs embedded in each MDAC stage. Whereas the power consumption and total chip area required for the residue amplifiers roughly stay the same, it doubles for the flash ADCs.

This paper describes a pipeline ADC based on the dual-residue principle [15–17]. The main benefit of this type of ADC is that it is inherently insensitive to the gain and bandwidth of the residue amplifiers. As a result, calibration for gain and bandwidth is simply not required. This reduces complexity, while allowing for a low power consumption.

Some calibration is still required, as the dual-residue ADC is sensitive to amplifier offset. Calibration of offsets is a very standard procedure, applied regularly in many circuits. In the presented ADC, a simple background calibration technique is used that converges very rapidly.

The paper is organized as follows. The dual-residue operation principle is described in Sect. 2.2. The ADC architecture is treated in Sect. 2.3, together with the calibration of the time-interleaved ADC lanes. The error sources that can impact the performance of a dual-residue ADC are discussed in Sect. 2.4. The design at circuit level is described in Sect. 2.5. The measurements results are presented in Sect. 2.6. Finally, some conclusions are drawn in Sect. 2.7.

2.2 Dual-Residue Operation

Regular pipeline ADCs require the residue amplifiers to have an exact absolute gain. The subsequent MDAC stages share the same reference voltages. Hence, the gain of the residue amplifiers usually has to be exactly equal to some power of two.

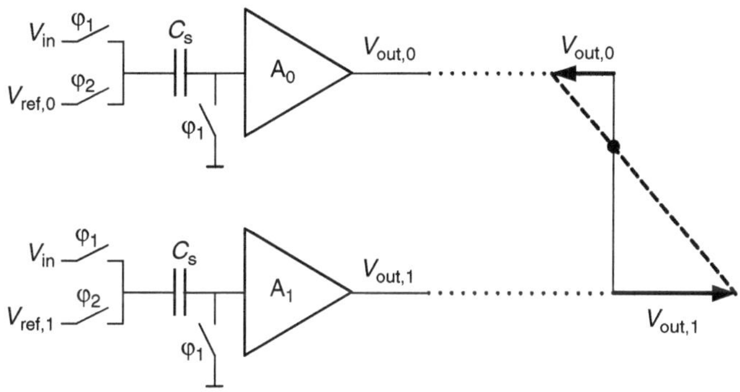

Fig. 2.1 First MDAC stage, based on the dual-residue principle

Achieving an exact absolute gain is challenging in deep-submicron IC processes, where supply voltages and intrinsic transistor voltage gains are low. Alternatively, at the expense of increased complexity, gain calibration can be used to correct for the errors arising from inaccurate gain factors. An interesting alternative is to resort to ADC architectures that are inherently insensitive to the absolute gain of the residue amplifiers. Besides, for example, flash ADCs and successive-approximation ADCs, the dual-residue ADC has this advantageous characteristic.

2.2.1 *Operation of MDAC1*

Instead of processing a single-residue signal throughout the ADC pipeline, in the dual-residue architecture, each MDAC stage processes two residue signals simultaneously. As illustrated in Fig. 2.1 for the first MDAC stage (MDAC1), the input signal V_{in} is sampled on two identical input capacitors C_S.Based on the output of the first Coarse-ADC (CADC1), two DC reference voltages, $V_{\text{ref},0}$ and $V_{\text{ref},1}$, are selected from a reference ladder, which satisfy the inequality:

$$V_{\text{ref},0} < V_{\text{in}} < V_{\text{ref},1}. \tag{2.1}$$

Furthermore, the reference voltages are chosen such that:

$$V_{\text{sub}} = V_{\text{ref},1} - V_{\text{ref},0}, \tag{2.2}$$

where the constant voltage V_{sub} is the size of the MDAC1 subrange interval. By choosing V_{sub} to be greater than the voltage distance between the thresholds of two subsequent CADC1 comparators, overrange can be implemented. The overrange helps to correct errors in the CADC decisions [18], such as comparator offsets.

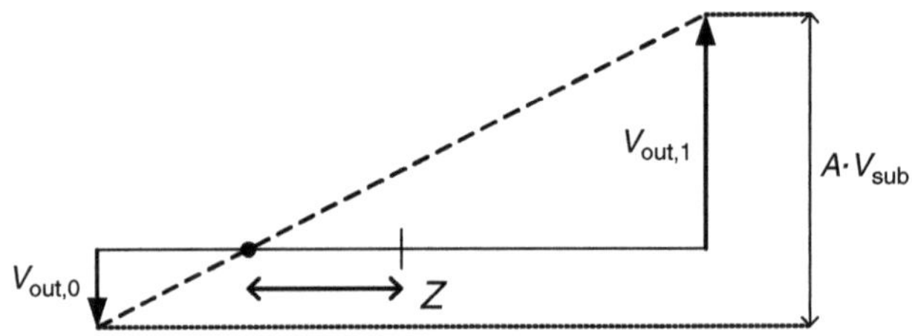

Fig. 2.2 Representation of the MDAC output signal Z in the zero-crossing domain

Two residue signals are obtained by subtracting $V_{ref,0}$ and $V_{ref,1}$ from V_{in}, respectively. After amplification by two residue amplifiers A_0 and A_1, output voltages $V_{out,0}$ and $V_{out,1}$ result:

$$V_{out,0} = A(V_{ref,0} - V_{in}), \tag{2.3}$$

$$V_{out,1} = A(V_{ref,1} - V_{in}), \tag{2.4}$$

where the gain factor A of the residue amplifiers is assumed to be equal. Note that due to the choice of the reference voltages, see Eq. (2.1), one output voltage is always positive, i.e., $V_{out,1} > 0$, while the other is always negative, $V_{out,0} < 0$.

The position of the zero-crossing, denoted by Z, that can be obtained when interpolating between $V_{out,0}$ and $V_{out,1}$, now represents the MDAC residue output signal. This is illustrated in Fig. 2.2. The dimensionless signal Z can be defined by:

$$Z = \frac{V_{out,0} + V_{out,1}}{2(V_{out,0} - V_{out,1})}. \tag{2.5}$$

After substitution of Eqs. (2.2)–(2.4), Z can be written as:

$$Z = \frac{V_{in} - \frac{1}{2}(V_{ref,0} + V_{ref,1})}{V_{sub}}. \tag{2.6}$$

Note that Z varies linearly with the ADC input signal V_{in}.

Equation (2.6) shows that Z does not depend on the *absolute gain* A of the residue amplifiers; only *gain matching* is required. This is a major advantage of this architecture as it relaxes the open-loop gain and settling requirements of the amplifiers, which allows for a low-power implementation.

2.2.2 *Operation of MDAC2*

At the input of the second MDAC (MDAC2) and subsequent MDAC stages, the ADC reference voltages are not required anymore because MDAC1 has transferred the residue signal into the zero-crossing domain. This is illustrated in the block

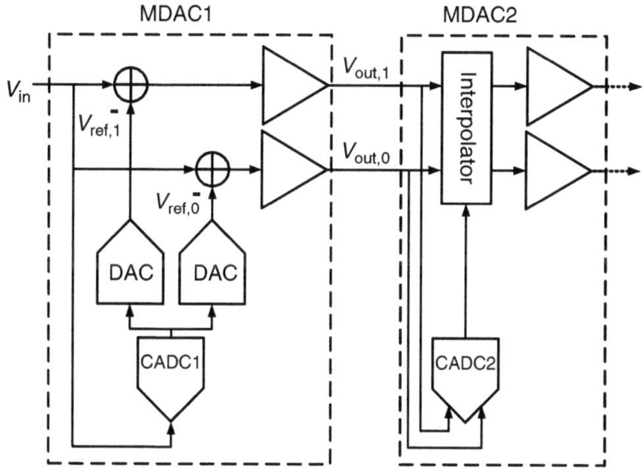

Fig. 2.3 Block diagram of the first and second MDAC stage of a dual-residue ADC

diagram shown in Fig. 2.3. Instead, interpolation between $V_{out,0}$ and $V_{out,1}$ is used to zoom in on the exact position of the zero-crossing Z [19, 20]. The required interpolation factors are determined by the output code of the second Coarse-ADC (CADC2).

CADC2 uses interpolation as well to find the approximate location of the zero-crossing. Figure 2.4 illustrates the operation of a 1.5-bit/stage CADC. The input signals, $V_{comp,0}$ and $V_{comp,1}$, of the two comparators are interpolated values between $V_{out,0}$ and $V_{out,1}$. In a 1.5-bit/stage MDAC, logical choices are:

$$V_{comp,0} = \tfrac{5}{8}V_{out,0} + \tfrac{3}{8}V_{out,1}, \tag{2.7}$$

$$V_{comp,1} = \tfrac{3}{8}V_{out,0} + \tfrac{5}{8}V_{out,1}. \tag{2.8}$$

The input signals, $V_{res,0}$ and $V_{res,1}$, for the residue amplifiers in MDAC2 are obtained through interpolation between $V_{out,0}$ and $V_{out,1}$. For every sample being processed, the interpolation factors to be used are determined by the corresponding digital output value of CADC2. As an example, for the sample being processed in Fig. 2.4, the zero-crossing is located between $V_{comp,0}$ and $V_{comp,1}$. In this case, a suitable choice for $V_{res,0}$ and $V_{res,1}$ is:

$$V_{res,0} = \tfrac{3}{4}V_{out,0} + \tfrac{1}{4}V_{out,1}, \tag{2.9}$$

$$V_{res,1} = \tfrac{1}{4}V_{out,0} + \tfrac{3}{4}V_{out,1}. \tag{2.10}$$

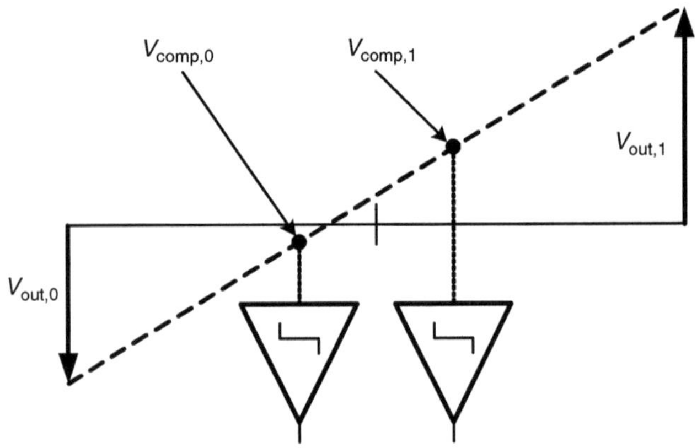

Fig. 2.4 Operation of the CADC in subsequent MDAC stages. Illustrated is a 1.5-bit/stage CADC

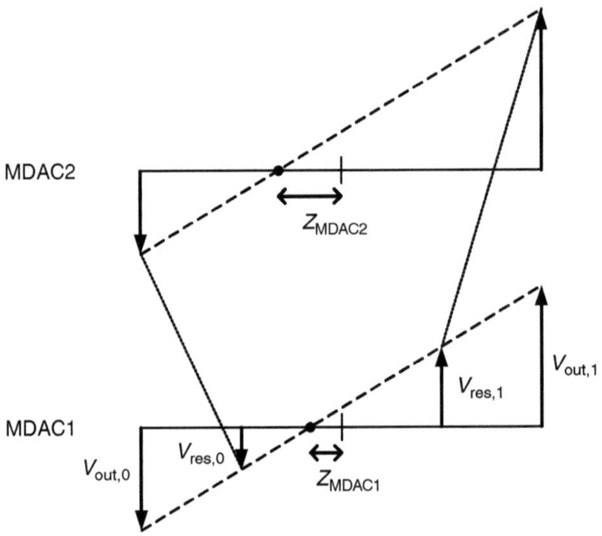

Fig. 2.5 Zooming in through interpolation. Illustrated is a 1.5-bit/stage MDAC stage

This is illustrated in Fig. 2.5. Note that $(V_{res,0} - V_{res,1})$ is chosen to span a wider range than $(V_{comp,0} - V_{comp,1})$. The difference between these two ranges is the implementation of overrange in the zero-crossing domain. The overrange functions exactly in the same way as in a conventional pipeline ADC, where it relaxes the accuracy requirements of the CADC [18].

All subsequent MDAC stages operate exactly like MDAC2, because they all process the residue signal in the zero-crossing domain.

2.3 ADC Architecture

For ADCs used in SoCs, decreasing the quantization noise is often a cost-effective way of decreasing the total ADC noise contribution. In the presented ADC design, shown in Fig. 2.6, the quantization noise has been made negligible. This was accomplished by choosing an overall ADC resolution of 12 bits, implemented by cascading seven stages. The first MDAC stage resolves 5 bits. Choosing such a relatively large number of bits in the first stage has several advantages. First, it allows the two residue amplifiers to have a relatively large gain, of approximately 7 × in this design, which decreases the noise contribution of the subsequent stages. Second, it relaxes the requirements with respect to gain matching, bandwidth matching, and distortion of the residue amplifiers in MDAC1. The second MDAC stage resolves 2.5 bits and has a gain of approximately 2.5 ×. Because of the large gain of MDAC1, aggressive power scaling can be used. The next four MDAC stages all resolve 1.5 bits per stage and have a gain of approximately 2 ×. The last stage is a 2b flash ADC.To achieve an overall sampling frequency F_s of 800 MS/s, four 200 MS/s ADC lanes are interleaved. Because the residue amplifiers in the first ADC stage connect to the reference ladder only in the amplification phase, it is possible to share one reference ladder between two ADC lanes.

The offset and gain errors that exist among the interleaved ADC lanes have been removed by means of digital calibration, which has been implemented on-chip. The offset errors are first calculated by averaging the output data of the ADC lanes and then subtracted digitally from the output data [21]. Subsequently, the gain errors are estimated by comparing the rms-values of the output data of each ADC lane. Digital gain blocks are used next to correct for the differences.

Besides offset and gain errors, timing mismatch is a major issue in high-speed time-interleaved systems [22], especially if the required resolution is high [23]. In this design, standard techniques were implemented to solve this issue. By retiming the 200 MHz clocks controlling the interleaved T/H circuits with a single

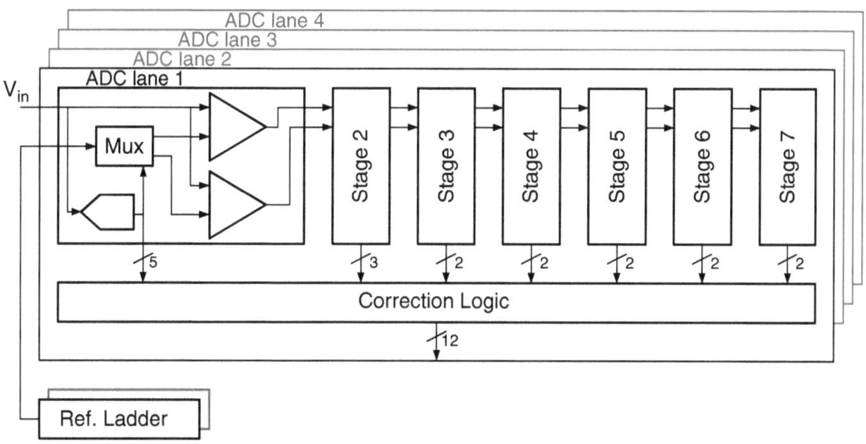

Fig. 2.6 Architecture of the time-interleaved 12-bit 800 MS/s dual-residue pipeline ADC

high-quality 800 MHz clock [23], a significant reduction of the sample time mismatches can be obtained. Special care is required in layout as well. Therefore, tree connections are used for both the ADC input signal and the 800 MHz clock. For the targeted application [24, 25], the required standard deviation for the error due to sampling mismatch was 500 fs. Monte Carlo simulations performed on the designed sampling structure gave a standard deviation of 170 fs, satisfying the required precision. Note that a general-purpose ADC would require a more stringent specification for the timing mismatch.

2.4 Error Sources

The dual-residue principle relies on the matching of two residue amplifiers. Mismatches between these amplifiers will result in errors in the analog-to-digital conversion. Possible mismatches are offset, gain, and bandwidth mismatches.

2.4.1 Amplifier Offset

The INL of a dual-residue ADC is sensitive to the difference in offset voltage of the two residue amplifiers comprising an MDAC stage [16, 17]. In practice, this differential offset is the most significant error source. In high-resolution ADCs, the offset voltage can easily be many LSBs and it might not be possible to reduce the offset sufficiently by scaling of the design; the area penalty would be too large. Hence, some method of offset correction is required [16, 17].

This design uses a simple analog background calibration algorithm, illustrated in Fig. 2.7, which reaches convergence within only a few-thousand clock cycles. During the reset phase, when the inputs of the amplifiers are shorted to a DC

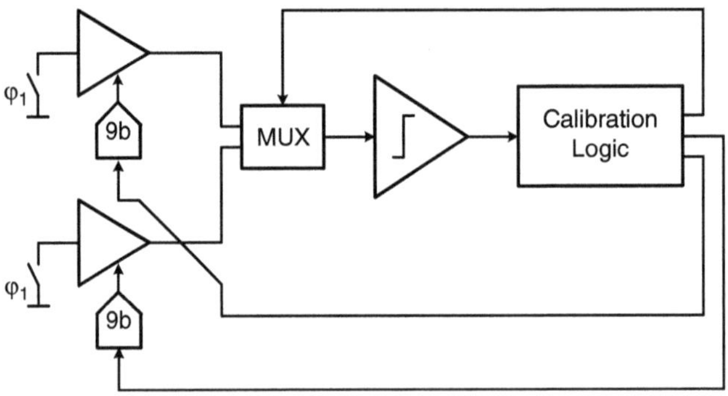

Fig. 2.7 Offset calibration of the residue amplifiers

voltage, the outputs are equal to their respective offset voltages. A comparator is used to determine the sign of the offset of a particular residue amplifier. Depending on the result, the DAC output voltage is increased or decreased. By repeating this measurement, the DAC output voltage converges to the point where the amplifier offset is near zero. To limit the impact of the comparator noise, several subsequent comparator decisions are averaged before updating the DAC input code.

A single comparator is used to subsequently measure the offset of both residue amplifiers comprising an MDAC stage. This way, the offset of the comparator itself is not important because the common offset voltage of the amplifiers does not compromise the ADC performance.

The offset DACs have a 9-bit resolution. To guarantee monotonicity, their design is based on a resistor string [26]. A differential pair is used to convert the DAC voltage into a current that is added to the output current of the amplifier input stage. Although, in principle, the DAC resolution can be scaled down in subsequent MDAC stages, the same DAC was used throughout the ADC to limit design time.

2.4.2 Gain and Bandwidth Mismatch

The dual-residue ADC is not sensitive to the *absolute gain* of the residue amplifiers. It is, however, sensitive to the *matching of the gain* of the two residue amplifiers inside one MDAC stage. In case of mismatch, the residue voltages can be written as:

$$
\begin{aligned}
V_{\text{out},0} &= (1 + a_1)A(V_{\text{ref},0} - V_{\text{in}}),\\
V_{\text{out},1} &= (1 - a_1)A(V_{\text{ref},1} - V_{\text{in}}),
\end{aligned}
\tag{2.11}
$$

where a_1 represents the gain mismatch. Figure 2.8 plots the resulting error in the position of the zero-crossing. The error has a quadratic shape. A maximum error is reached in the middle of the subrange, whereas it is zero at the very edges.

Note that the error has a certain DC offset value associated with it, which does not compromise the linearity of the ADC. The exact value of this DC component depends on the signal distribution across the subrange, which in turn depends on the accuracy of the flash ADC. A more accurate flash ADC design will limit the use of overrange, and, as can be seen from Fig. 2.8, can significantly reduce the nonlinear part of the error.

The amount of gain mismatch can be reduced by increasing the open-loop gain of the residue amplifiers. In this way, the influence of the active devices in the amplifiers is attenuated, and the matching relies more on the passive components in the feedback network. These usually match quite well (properly sized resistors can give a matching level between the closed-loop gains in the order of 0.2%). In the design phase, the tradeoff between matching and power consumption (to obtain higher open-loop gain) has to be carefully evaluated.

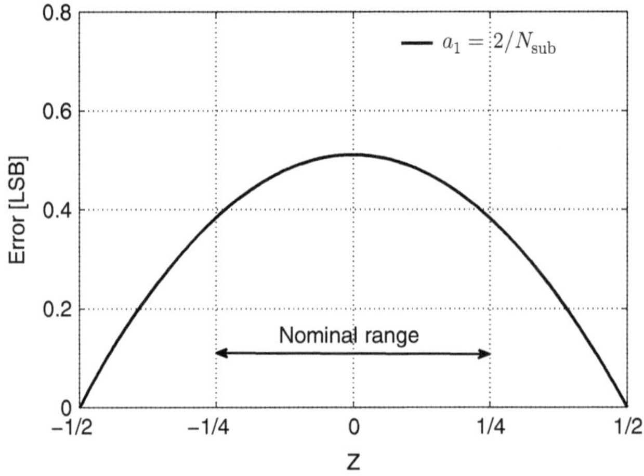

Fig. 2.8 Zero-crossing error due to gain mismatch expressed in LSBs

The effect of a mismatch in bandwidth between two residue amplifiers is identical to the effect of a gain mismatch. The errors can be reduced in two ways, either by improving the matching or by increasing the absolute bandwidth. Increasing the signal bandwidth, however, will also increase the noise bandwidth, ultimately leading to higher power consumption. In practice, sufficient matching can be realized by design, especially if the circuits are designed in such a way that the bandwidth is mainly determined by passive components.

2.5 Circuit Design

This section describes the design of the most important analog blocks comprising the ADC: the residue amplifiers and the reference ladder in the first MDAC stage, and the residue amplifiers in the subsequent MDAC stages.

2.5.1 Residue Amplifier for MDAC1

In contrast to most conventional pipeline ADC designs, this design uses voltage-mode amplifiers in the first MDAC stage to amplify the residue voltages, as shown in Fig. 2.9. Each residue amplifier consists of two operation amplifiers with a resistive feedback network. The sampling capacitor C_{S} is equal to 1 pF (per residue amplifier). The switches connecting C_{S} to V_{in} or V_{ref} are thick-oxide devices, operating on a 2.5 V clock. Fast level shifters were designed to lift the clock signals

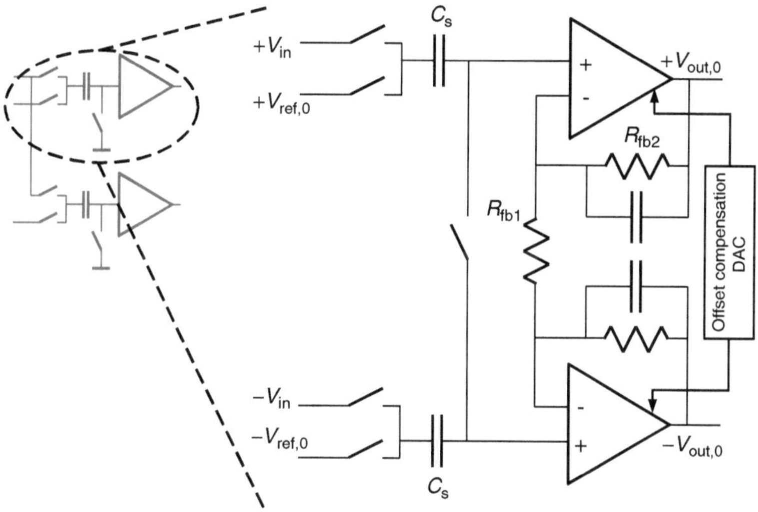

Fig. 2.9 Schematic of one of the residue amplifiers in MDAC1

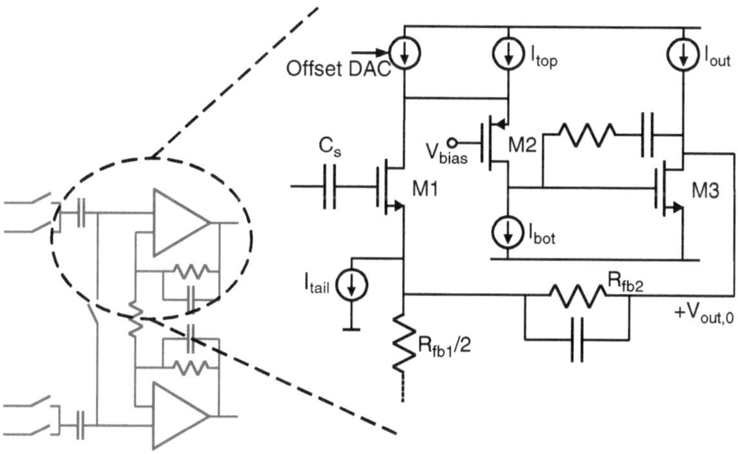

Fig. 2.10 Schematic of the opamp used in MDAC1

from 1 to 2.5 V. An advantage of using voltage-mode amplifiers is that the input capacitors are not discharged during the amplification phase. This condition relaxes the speed requirements for the on-chip reference buffers, described in Sect. 2.5.2, which helps to limit their power consumption.

The implemented op amp is illustrated in Fig. 2.10. A two-stage design is required due to the resistive feedback. The input stage uses a single transistor, M1, instead of a differential pair. In comparison to a differential pair, only half the power is required, while only half the noise is generated. The final noise improvement is, however, partly reduced due to the fact that the feedback resistors

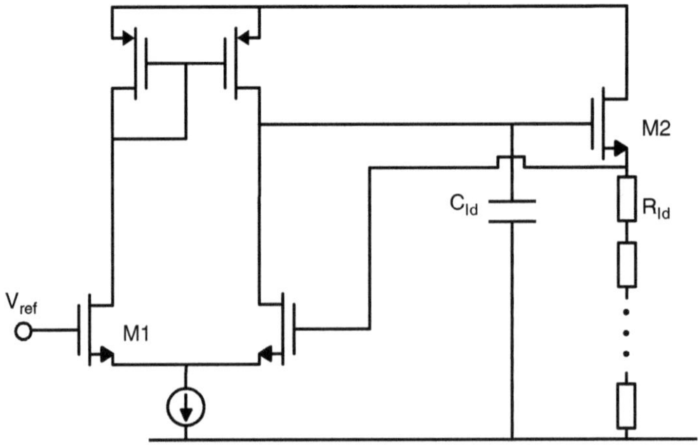

Fig. 2.11 Schematic of the reference ladder, including the driver

degenerate the transconductance g_{m1} of M1. This degeneration increases the equivalent input noise due to the noise generated by I_{top} and I_{bot}. The input stage is followed by a folded-cascode transistor M2 and the output stage transistor M3. A small Miller capacitance around M3 and a feedback capacitor around R_{fb2} provide the amplifier with sufficient phase margin.

Although the open-loop gain of the amplifier typically is only 40 dB, and the closed-loop bandwidth is only $\sim$ 350 MHz, the performance of this pipeline ADC is not compromised, owing to the dual-residue principle.

The power supply for the amplifier is 1 V, and the simulated power consumption per op amp is $\sim$ 3 mW. The simulated worst-case error on the zero-crossing, due to MDAC1, is equal to 0.25 LSBs at the output of MDAC1. This is including gain and bandwidth mismatch.

In Fig. 2.10, the current source used to calibrate out the offset voltage is also shown: the current through M1 is diverted from one-half of the pseudodifferential amplifier to the other and the other way around, depending on the sign of the offset.

2.5.2 *Reference Ladder*

When implemented on-chip, reference voltage drivers for pipeline ADCs are usually very power hungry [27]. They have to simultaneously achieve a low output noise, a low output impedance, and a high bandwidth. This is required to allow for complete recovery from the severe kicks introduced by the capacitive DAC used in most pipeline ADCs.

In this design, the reference voltages are not generated by means of a capacitive DAC. Instead, a resistive ladder is used, as shown in Fig. 2.11. As the accuracy of the reference voltages influences the INL of the ADC, a sufficiently large resistor area has to be chosen.

The settling speed of the resistor ladder is determined by the impedance value of the ladder resistors and by the parasitic capacitance of the multiplexers loading it, see Fig. 2.6. To limit the loading presented by the switches comprising the multiplexers, $4 \times$ capacitive interpolation is used. By splitting the MDAC1 input capacitor into four equal parts, the required number of reference voltages and number of switches is reduced roughly four times, significantly improving the settling speed of the reference voltages. Note that a higher level of capacitive interpolation, as used, for example, in most pipeline ADCs, would further reduce the number of switches, but at the same time would increase the size of the voltage kicks.

The top of the reference ladder is driven by a voltage buffer. The buffer consists of a differential pair input stage, M1, a PMOS current mirror and a source-follower output stage, M2. A large capacitor, connected from the gate of M2 to ground, is used to limit the closed-loop bandwidth of the amplifier to around 40 kHz. As a result, at frequencies significantly above 40 kHz, the output impedance and output noise are determined only by the output transistor M2.

The power consumption of the reference circuit is 7.5 mW, where the amplifier only consumes 25 μW, and the rest of the current flows through the resistive ladder itself.

2.5.3 *Residue Amplifier for MDAC2*

Owing to the bits already resolved in the first MDAC stage, the specifications for the second MDAC stage are significantly reduced. This allows for an aggressive scaling of the sampling capacitors and of the amplifier biasing current. The bandwidth, however, cannot be much smaller than the MDAC1 bandwidth. A minimum level of settling has to be guaranteed to prevent the amplified signal from becoming too small, potentially hurting the SNR. The amplifier, therefore, features a bandwidth of 300 MHz, allowing a settling of about 3τ.

The circuit implementation of the residue amplifier in MDAC2 is depicted in Fig. 2.12. The design uses a single-stage folded-cascode charge amplifier. The output bandwidth is determined by the transconductance of the input stage and the total capacitive loading. The offset of the residue amplifier is calibrated out by changing the currents flowing through the two input devices.

In contrast to the first MDAC stage, MDAC2 does not need any reference voltages (see Sect. 2.3). Instead, interpolation between the two output residue voltages of MDAC1 is required. The input capacitors of MDAC2 are used to implement this interpolation function. The interpolation factors are determined by the flash ADC incorporated in MDAC2. Depending on the flash ADC digital output, in the amplification phase, some capacitors are selected to be connected to the charge amplifier. By using the right mix of capacitors charged to $V_{\text{out},0}$ or $V_{\text{out},1}$, respectively, MDAC2 is able to zoom in on the position of the zero-crossing.

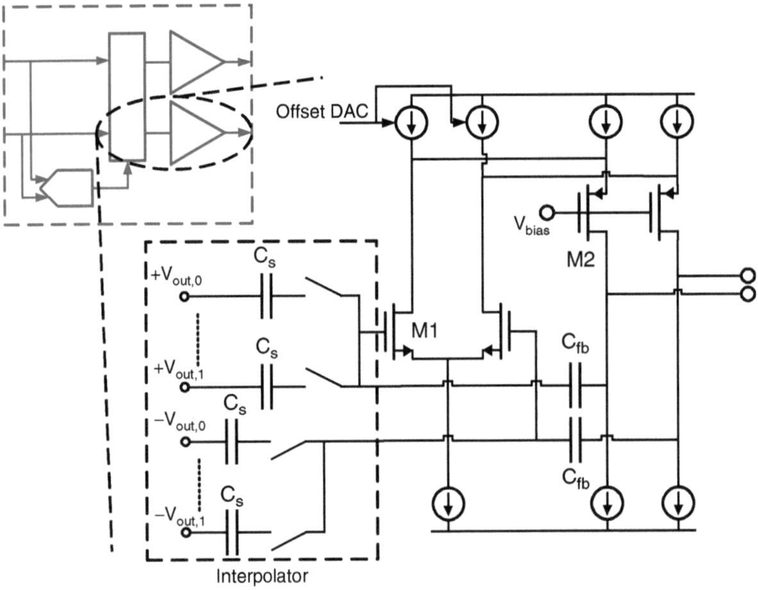

Fig. 2.12 Implementation of the residue amplifier in MDAC2

The power consumption of each residue amplifier is 0.9 mW. The total input capacitance is 160 fF each. This is a scaling factor of almost 7 ×with respect to MDAC1. A larger scaling factor is difficult to achieve due to layout reasons. The interpolation function requires the sampling capacitor to be split in several smaller capacitors.

The third MDAC stage uses the same type of residue amplifier as MDAC2. It is scaled down another 2 ×. The remaining three MDAC stages (4, 5, and 6) are not scaled anymore.

2.6 Experimental Results

The ADC was integrated as part of a 10GBase-T Ethernet [24, 25] transceiver chip, fabricated in a standard 40 nm CMOS process with seven metal layers and one layer of polysilicon. Figure 2.13 shows a die photograph of the ADC, which occupies an area of 0.88 mm^2.

The ADC operates from a dual 1 V/2.5 V power supply. Its power consumption P equals 105 mW, which includes 15 mW for the on-chip ADC reference buffers.

In the measurement setup, the on-chip 10GBase-T transmitter was used as the input signal to the receiver. The ADC is preceded by a Programmable-Gain Amplifier (PGA), which is used to optimize the loading of the ADC. As the PGA could not be bypassed, note that the reported measurement results include the noise and distortion contributions of the PGA.

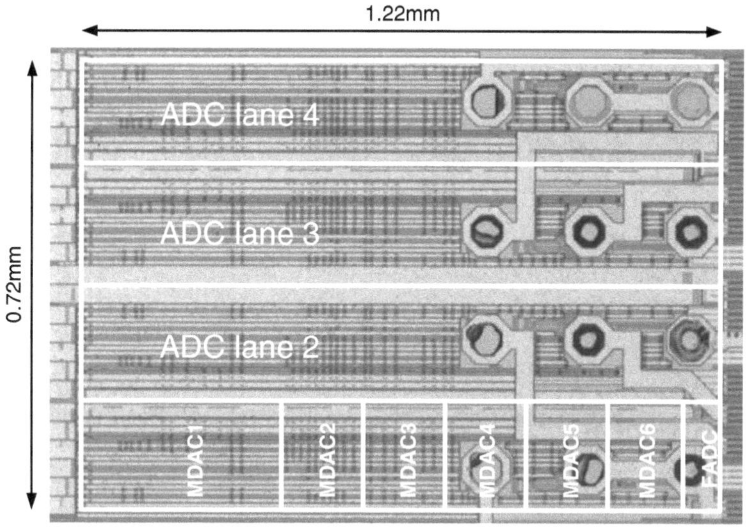

Fig. 2.13 Micrograph of the ADC

Figure 2.14 shows the INL and DNL, measured at $F_s = 800$ MS/s with a two-tone input signal around 37 MHz. The INL equals + 2. 4/−2.5 LSB and is dominated by the third-order distortion caused by the PGA. The DNL equals + 0. 3/−0.4 LSB.

Figure 2.15 shows the ADC output spectrum, measured at $F_{sig} = 360.638$ MHz, and the signal-to-noise-and-distortion-ratio (SNDR) and the spurious-free-dynamic-range (SFDR) as a function of the input frequency F_{sig}. A peak SNDR of 59 dB is achieved. At low frequencies, the SNR drops due to the band-pass characteristic of the PGA, limiting the achievable signal swing at the ADC input. Due to this band-pass characteristic, it was not possible to measure the ADC using input frequencies above Nyquist. The high even harmonics visible in the spectrum are generated by the on-chip 10GBase-T transmitter.

The spurs generated by the sampling time mismatch, located at $F_s/4 \pm F_{sig}$ and $F_s/2 - F_{sig}$, are around −80 dB. The sampling time mismatch of the measured chip is below the simulated standard deviation (1σ) value of 170 fs. For samples with a mismatch closer to the 3σ value, these spurs will be more pronounced. In that case, however, the performance of the ADC is still sufficient for the intended application.

The gain mismatch between the ADC lanes was measured to be in the order of 2% (3σ). The measured offset was 30 LSBs (3σ). Combined, these ADC lane mismatches result in a reduction of the overall ADC dynamic range below 3%.

The energy per conversion step is an often-used figure of merit (FOM) for ADCs. It is defined by:

$$\mathrm{FOM} = \frac{\mathrm{P}}{\min(2 \cdot \mathrm{ERBW}, \mathrm{F_s}) \cdot 2^{\mathrm{ENOB}}}, \tag{2.12}$$

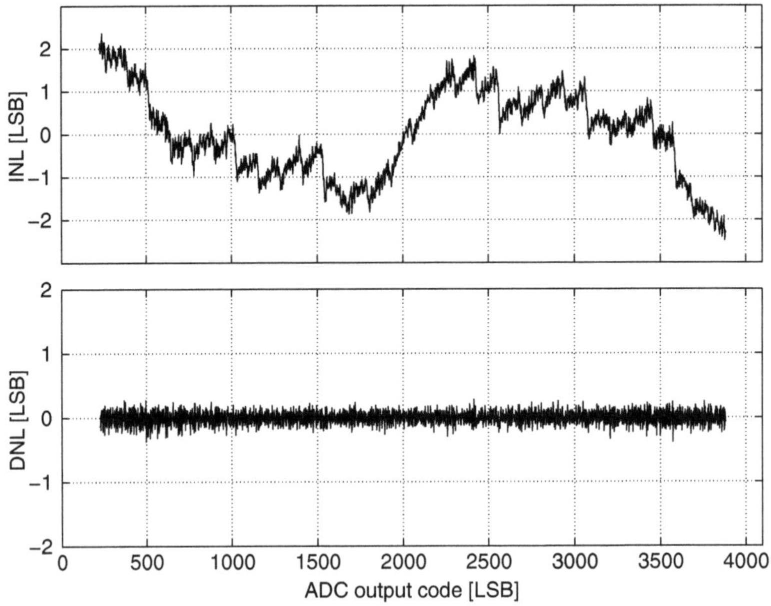

Fig. 2.14 Measured INL and DNL

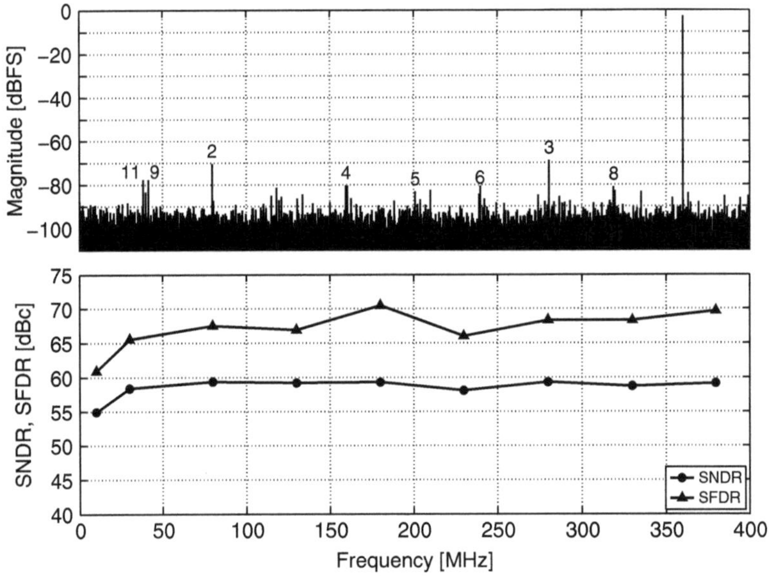

Fig. 2.15 Measured output spectrum at $F_{sig} = 360$ MHz, and SNDR and SNR versus F_{sig}

where ERBW is the effective resolution bandwidth, and ENOB is the effective number of bits. For this design, the FOM equals 0.18 pJ/conversion at 800 MS/s.

2.7 Conclusions

In contrast to the conventional pipeline ADC architecture, in the dual-residue architecture, two residue amplifiers are required in each MDAC stage. All signal processing is carried out in the zero-crossing domain. As a result, absolute precision requirements on gain and bandwidth of the residue amplifiers are eliminated and are replaced by requirements on the matching of offset, gain, and bandwidth. This significantly relaxes the specifications of these amplifiers, allowing for a low-power implementation. By design, a sufficient level of matching of gain and bandwidth can be reached quite easily. The difference in offset, however, requires calibration for the ADC to achieve a high resolution.

Demonstrating the merits of the dual-residue principle, a 12-bit 800 MS/s ADC was designed in a standard 40 nm CMOS process. It achieves a peak SNDR of 59 dB, consuming 105 mW of power.

References

1. Sahoo B, Razavi B (2009) A 12-bit 200-MHz CMOS ADC. IEEE J Solid State Circuits 44(9):2366–2380
2. Verma A, Razavi B (2009) A 10b 500MHz 55mW CMOS ADC. In: ISSCC digest of technical papers. IEEE, New York/Piscataway, pp 84–85
3. Ali A, Morgan A, Dillon C, Patterson G, Puckett S, Hensley M, Stop R, Bhoraskar P, Bardsley S, Lattimore D, Brey J, Speir C, Sneed R (2010) A 16b 250 MS/s IF-sampling pipelined A/D converter with background calibration. In: ISSCC digest of technical papers. IEEE, Piscataway, pp 292–293
4. Hernes B, Briskemyret A, Andersen T, Telsto F, Bonnerud T, Moldsvor O (2004) A 1.2V 220 MS/s 10b pipeline ADC implemented in 0.13 μm digital CMOS. In: ISSCC digest of technical papers. IEEE, Piscataway, p 256
5. Lee S, Jeon Y, Kim K, Kwon J, Kim J, Moon J, Lee W (2007) A 10b 205 MS/s 1 mm2 90 nm CMOS Pipeline ADC for flat-panel display applications. In: ISSCC digest of technical papers. Digital Pub., Lisbon Falls, p 458
6. Hernes B, Bjornsen J, Andersen T, Virnje A, Korsvoll H, Telsto F, Briskemyr A, Holdo C, Moldsvor O (2007) A 92.5 mW 205 MS/s 10b pipeline IF ADC implemented in 1.2 V/3.3 V 0.13 μm CMOS. In: ISSCC digest of technical papers. Digital Pub., Lisbon Falls, p 462
7. Hsu C, Huang F, Shih C, Huang C, Lin Y, Lee C, Razavi B (2007) An 11b 800 MS/s time-interleaved ADC with digital background calibration. In: ISSCC digest of technical papers. Digital Pub., Lisbon Falls, p 464
8. Hsueh K-W, Chou Y, Tu Y, Chen Y, Yang Y, Li H (2008) A 1V 11b 200 MS/s pipelined ADC with digital background calibration in 65 nm CMOS. In: ISSCC digest of technical papers. IEEE, Piscataway, p 546
9. Gupta S, Choi M, Inerfield M, Wang J, (2006) A 1 GS/s 11b time-interleaved ADC in 0.13 μm CMOS. In: ISSCC digest of technical papers. IEEE, Piscataway, pp 2360–2369

10. Murmann B, Boser B, (2003) A 12 bit 75 MS/s pipelined ADC using open-loop residue amplification. In: ISSCC digest of technical papers. IEEE, Piscataway, pp 328–329
11. Galton I (2000) Digital cancellation of D/A converter noise in pipelined A/D converters. IEEE Trans CAS-II 47:185–196
12. Chuang S-Y, Sculley T, (2002) A digitally self-calibrating 14-bit 10-MHz CMOS pipelined A/D converter. IEEE J Solid State Circuits 37(6):778–786
13. McNeill J, Coln M, Larivee B (2005) Split ADC architecture for deterministic digital background calibration of a 16-bit 1-MS/s ADC. IEEE J Solid State Circuits 40(12):2437–2445
14. McNeill J, Coln M, Brown D, Larivee B (2009) Digital background calibration algorithm for "Split ADC" Architecture. IEEE Trans CAS-I 56(2):294–306
15. Mangelsdorf C, Malik H, Lee S, Hisano S, Martin M (1993) A two-residue architecture for multistage ADCs. In: ISSCC digest of technical papers. IEEE, New York, pp 64–65
16. van der Ploeg H, Hoogzaad G, Termeer H, Vertregt M, Roovers R (2001) A 2.5-V 12-b 54-Msample/s 0.25-μm CMOS ADC in 1-mm^2 with mixed-signal chopping and calibration. IEEE J Solid State Circuits 36(12):1859–1867
17. van der Ploeg H, Remmers R (1999) A 3.3-V, 10-b, 250MSample/s two-step ADC in 0.35-μm CMOS. IEEE J Solid State Circuits 34(12):1803–1811
18. Lewis S, Fetterman H, Gross GF Jr, Ramachandran R, Viswanathan T (1992) A 10-b 20-Msample/s analog-to-digital converter. IEEE J Solid State Circuits 27(3):351–358
19. Sandner C, Clara M, Santner A, Hartig T, Kuttner F (2005) A 6-bit 1.2-GS/s low-power flash-ADC in 0.13-μm digital CMOS. IEEE J Solid State Circuits 40(7):1499–1505
20. Mulder J, Ward C, Lin C-H, Kruse D, Westra J, Lugthart M, Arslan E, van de Plassche R, Bult K, van der Goes F (2004) A 21-mW 8-b 125-MSample/s ADC in 0.09-mm^2 0.13-μm CMOS. IEEE J Solid State Circuits 39(12):2116–2125
21. Draxelmayr D (2004) A 6b 600 MHz 10 mW ADC array in digital 90 nm CMOS. In: ISSCC digest of technical papers. IEEE, Piscataway, p 264
22. Louwsma S, van Tuijl A, Vertregt M, Nauta B (2008) A 1.35 GS/s, 10 b, 175 mW time-interleaved ADConverter in 0.13 μm CMOS. IEEE J Solid State Circuits 43(4):778–786
23. Doris K, Janssen E, Nani C, Zanikopoulos A, van der Weide G (2011) A 480 mW 2.6 GS/s 10b 65 nm CMOS time-interleaved ADC with 48.5 dB SNDR up tp Nyquist. In: ISSCC digest of technical papers. IEEE, Piscataway, pp 180–181
24. Spencer R (2003) Analog front ends for Ethernet on copper. Plenary week 10GBASE-T study group meeting. [Online]. Available: http://www.ieee802.org/3/10GBT/public/jul03/spencer_1_0703.pdf
25. CSMA/CD (Ethernet) access method, IEEE Std. 802.3 http://standards.ieee.org/about/get/802/802.3.html
26. van de Plassche R (2003) CMOS integrated analog-to-digital and digital-to-analog converters. Kluwer, Boston
27. Zanchi A, Tsay F (2005) A 16-bit 65-MS/s 3.3-V pipeline ADC core in SiGe BiCMOS with 78-dB SNR and 180-fs jitter. IEEE J Solid State Circuits 40(6):1225–1237

Chapter 3
Time-Interleaved SAR and Slope Converters

Pieter Harpe, Ming Ding, Ben Büsze, Cui Zhou, Kathleen Philips, and Harmke de Groot

Abstract This paper investigates time-interleaved SAR and time-interleaved slope converters, targeting low-power, low-resolution, high-speed applications. Fundamentally, these two architectures can be relatively power-efficient as compared to other architectures. At the same time, complex calibration schemes are not required thanks to their inherent accuracy. The architectures are examined and compared, circuit implementations and measurement results are discussed and an outlook to the future will be given.

3.1 Introduction

Wireless communication standards using Impulse Radio UWB (IR-UWB), like 802.15.4a WPAN and 802.15.6 WBAN, require low-power, low-resolution (3–6 bit), high-speed (0.5–2 GHz) AD converters [1, 2]. Because of the high speed of operation, flash-based converters are often selected for this application [3], but alternatives based on pipelining [4] or SAR [5] are also being developed. When deciding for a suitable ADC architecture, one could either take the high-speed requirement or the low-power aim as a starting point. Both these approaches will be briefly discussed in the following sections.Eventually, the time-interleaved SAR and slope architectures are chosen for further analysis and implementation.

P. Harpe (✉)
Holst Centre and imec, Eindhoven, The Netherlands

Eindhoven University of Technology, Eindhoven, The Netherlands
e-mail: p.j.a.harpe@tue.nl

M. Ding • B. Büsze • C. Zhou • K. Philips • H. de Groot
Holst Centre and imec, Eindhoven, The Netherlands

A.H.M. van Roermund et al. (eds.), *Nyquist AD Converters, Sensor Interfaces, and Robustness: Advances in Analog Circuit Design, 2012*,
DOI 10.1007/978-1-4614-4587-6_3,

3.1.1 High-Speed Approach

When focussing on speed, flash and pipelined architectures offer an advantage because of the parallel operation of hardware: flash converters use a set of parallel comparators to maximize speed, while pipelined ADCs can decide the various bits in parallel by pipelining the operations. While this is advantageous for speed, it also leads to drawbacks in terms of power-efficiency and calibration complexity.

In case of an N-bit flash converter, the number of comparators equals $2^N - 1$. Noting that the power consumption of the comparator is an important contributor to the overall power, the exponentially increasing number of comparators implies that the power-efficiency may be sub-optimal as compared to e.g. an architecture in which the number of comparators grows linear with the resolution N. To improve the power-efficiency of flash converters, the devices in the comparators are usually reduced in size to minimize the power consumption. In this way, excellent power-efficiencies are feasible such as demonstrated in [3]. However, the drawback of these down-scaled devices is the increase of mismatch, which limits the intrinsic linearity that can be achieved. This can be solved by calibration techniques that measure and correct the offset of each individual comparator. Since each comparator in a flash ADC has a different threshold voltage, a large set of precise input signals is required to measure and tune each comparator to the required level, thus increasing the system complexity.

In pipelined ADCs, various analog components such as the amplifier are important contributors to the overall power. As explained in [4], a combination of techniques such as dynamic amplification and a pipelined binary search algorithm (PLBS) can lead to very good power-efficiencies. However, similar to flash converters, this optimization results in an increased calibration complexity since both offsets and non-linearities need to be corrected.

In summary, the high-speed approach achieves high-speed *intrinsically* (by architecture choice), while the combination of power efficiency and accuracy is achieved *extrinsically* (by technology-scaling and calibration).

3.1.2 Low-Power Approach

In contrast to the high-speed approach, the low-power approach starts with selecting architectures that potentially enable a better power-efficiency. Throughout the various ADC architectures, the comparator is typically substantial for the overall power consumption. For that reason, Table 3.1 gives an overview of various ADC architectures in terms of the number of comparisons needed for a complete conversion cycle. Note that the number of comparisons is given rather than the number of comparators, since (assuming a dynamic comparator) it matches better to the expected power consumption. This gives a different result in certain cases, since e.g. the CABS approach [6] has $2^N - 1$ comparators but activates only N of them,

Table 3.1 Number of comparisons for an N-bit conversion for various ADC architectures

Architecture	Comparisons per conversion
Flash	$2^N - 1$
Folding (with factor M)	2^{N-M}
Pipeline	$> N, < 2^N$
SAR	N
Slope	$2^N - 1$ or 1

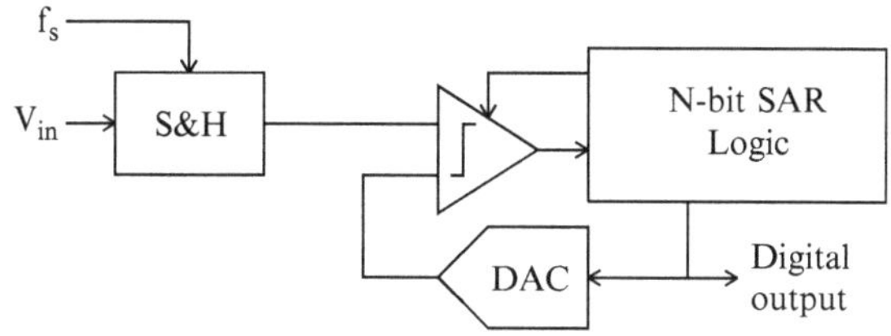

Fig. 3.1 SAR ADC architecture

while a typical SAR converter has only 1 comparator, but it is activated N times per conversion. As shown in the table, some architectures grow exponential in the number of bits N, while other architectures grow linear in N, remain constant, or have a relation in between one of these trends. For slope-type converters, two cases can be distinguished: systems in which the comparator is clocked, and systems in which a continuous-time comparator is used. While a clocked system needs $2^N - 1$ comparisons, the comparator in a continuous-time system would toggle only once. The latter case would potentially enable a highly efficient conversion.

Based on this analysis, one could expect that SAR and slope converters would be amongst the most power-efficient topologies.Partially this is confirmed by the extensive analysis from [7], which summarizes ADC performance over the past 15 years: on average, the SAR converters achieve very good power-efficiencies when compared to other architectures. On the other hand, this overview includes only one ADC based on the slope architecture [8], revealing that this architecture has not been very popular as a stand-alone ADC.

Figures 3.1 and 3.2 illustrate the basic topologies of a SAR and a digital-slope converter: in the SAR converter, the analog input signal is sampled and compared against the output of a DAC, which is tuned towards the sampled signal by means of a digitally implemented binary-search algorithm. A digital-slope ADC operates very similar to a SAR ADC, except that a linear-search is performed instead of a binary-search. The similarity suggests that most components (S&H, comparator, DAC) could be implemented in a similar way. Fundamentally, SAR and slope-based converters can be advantageous in terms of power-efficiency because of the small number of comparisons required. Moreover, the accuracy of these converters is given by the feedback DAC. As opposed to the matching of comparators, which comes at the cost of power or calibration, the matching of e.g. switched-capacitor

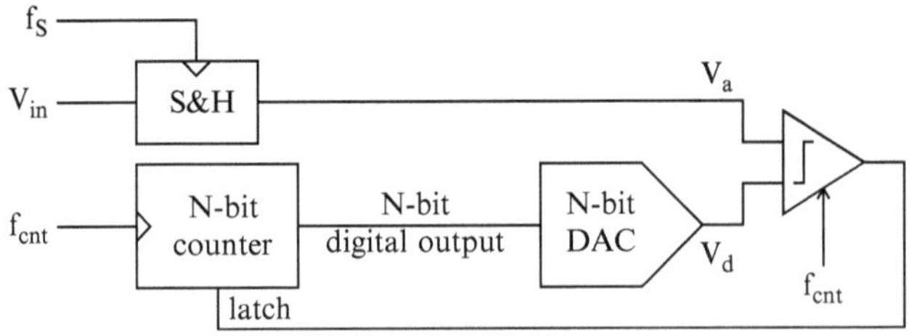

Fig. 3.2 Digital-slope ADC architecture

DACs can be made sufficiently precise without a significant penalty in power or calibration [9]. However, both of these topologies are relatively unsuitable for high-speed operation due to the nature of their search algorithms.In order to achieve the required sampling rates, time-interleaving [10] can be applied.

In summary, the low-power approach achieves accuracy and power efficiency *intrinsically* (by architecture choice), while high-speed needs to be achieved *extrinsically* (by technology-scaling and time-interleaving).

Since both SAR and slope architectures appear to be potential candidates for power-efficient ADCs, these two topologies are further investigated in Sects. 3.2 and 3.3. In both cases, the aim is to design a low-power 5 bit converter for high-speed applications. After the discussion on the separate designs, the architectures will be compared to each other in Sect. 3.4.

3.2 A 5 bit 1 GS/s Time-Interleaved SAR ADC

This section reviews the design of a time-interleaved SAR ADC, as originally published in [11]. Aiming for UWB receiver applications, this design realizes a 5 bit 1 GS/s ADC by interleaving 16 converters, each operating at 62.5 MS/s. First, the design of an individual channel is described, followed by the integration into the time-interleaved structure.

3.2.1 Single Channel: Asynchronous SAR ADC

Each 5 bit 62.5 MS/s SAR ADC is implemented according to the architecture in Fig. 3.1. While a typical SAR ADC requires an oversampled clock to synchronize the various bit-cycles, asynchronous implementations such as in [9] require a sample-rate clock only. Thus, a single 62.5 MHz clock is sufficient in this particular case.

The analog part of the ADC (composed of S&H and DAC) is implemented by a switched-capacitor network (Fig. 3.3), with a total equivalent sampling capacitance C_s of only 60 fF to save power. The analog input is sampled on the complete array of capacitors, while the feedback DAC is implemented by a charge redistribution

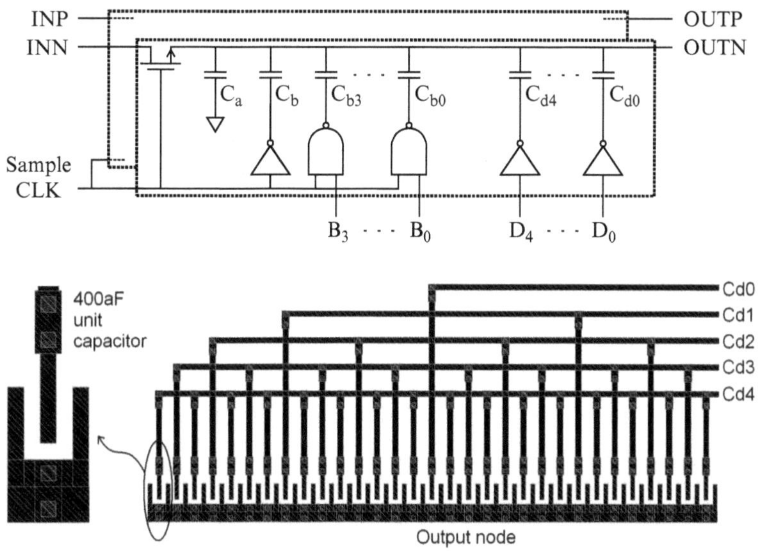

Fig. 3.3 Switched capacitor core, 400 aF unit capacitor implementation and part of the layout of the capacitor array

structure, controlled by bits $D_4 \cdots D_0$. Furthermore, a common-mode shift operation with programmable offset correction is included. The common-mode shift, implemented with capacitor C_b as in [12], increases the common-mode level after sampling from 0.2 to 0.5 V. The low common-mode voltage during sampling allows the use of a single NMOS device of $\frac{1.5\,\mu m}{0.1\,\mu m}$ as sampling switch. The higher common-mode voltage after sampling enables faster operation of the comparator. In parallel to the fixed capacitor C_b, the binary scaled capacitors $C_{b3} \cdots C_{b0}$ can be programmed digitally for each half of the differential structure.In that way, the common-mode shift becomes programmable, such that a differential offset can be added in the range of $\pm$ 7. 5 LSB with 0. 5 LSB steps, which is used to calibrate the ADC offset. The noise and matching requirements are relatively relaxed since the target resolution is only 5 bit. Thus, to save power and to maximize the signal bandwidth, it is preferable to minimize the values of the capacitors as much as possible. Dedicated lateral metal-metal capacitors can be applied in order to achieve this [9, 13]. As detailed in [9], sub-fF capacitors are still able to achieve sufficient matching. For that reason, this work uses capacitor elements of 400 aF (Fig. 3.3). A part of the layout of the capacitor array is also shown, revealing the common-centroid approach used for the binary-scaled DAC elements.

The comparator and logic in this ADC are designed as in [9, 11]. The key aspect in these circuits is to use asynchronous components with a dynamic power consumption. The dynamic power consumption enables inherent scalability of the ADC's power consumption as a function of the sample rate. The asynchronous operation simplifies the clocking scheme, which reduces the power and the frequency of the required clock.

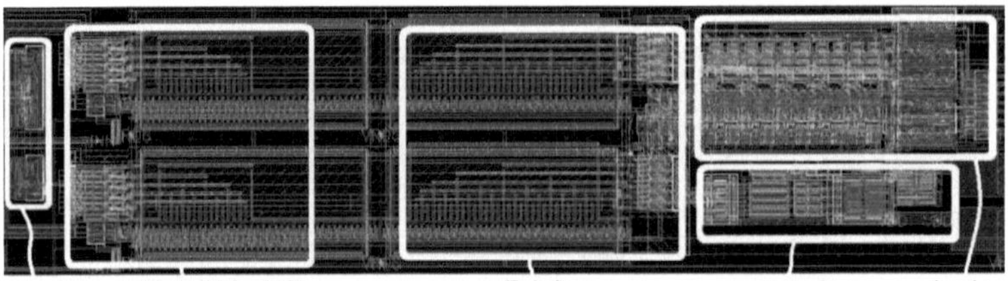

Fig. 3.4 Layout view of a single channel of the ADC, occupying 96 ×24 μm

Figure 3.4 shows the layout of a single 5 bit 62.5 MS/s ADC. The small area of only 96 ×24 μm in 90 nm CMOS is advantageous for various reasons: it minimizes power loss due to layout parasitics, it enables integration of many channels without consuming a large chip area, and it minimizes mismatches between the channels of the time-interleaved ADC. This very small size can be achieved because of the small-area capacitors, the asynchronous dynamic logic (which reduces the number of transistors), by putting the supply decoupling capacitors below the signal capacitors and by careful layout optimization.

3.2.2 16-Channel Time-Interleaved ADC

Time-interleaved converter arrays are widely used to increase the speed of operation while maintaining power-efficiency [10]. However, channel mismatches result in undesirable spurious components. Thus, either sufficient matching needs to be achieved by intrinsic design or otherwise calibration or correction techniques are required to compensate for these errors. The three main types of mismatch are: offset mismatch, gain mismatch and time-skew, as analyzed in a.o. [10]. Gain mismatch can be neglected in this time-interleaved SAR architecture, as the gain is determined by precise capacitor ratio's. On the other hand, the offset mismatch and time-skew need to be taken into account, and are therefore discussed in the following sections.

3.2.2.1 Offset Mismatch and Correction

Calibration of the channel-offset is needed due to the random and systematic mismatch of the comparators in each channel. A Monte-Carlo simulation of the comparator predicts a random offset with a standard deviation of 0.5 LSB. Furthermore, gradients in the layout or systematic layout asymmetries can lead to systematic mismatch between the comparators. As described in Sect. 3.2.1, the offset can be corrected in the analog domain inside every channel within a range of ± 7. 5 LSB with 0. 5 LSB steps. The offset calibration is performed as follows: at startup, the average output code per channel is determined, which corresponds to the channel offset. Then, the offset-correction code is determined for each channel and loaded into the calibration register that controls the programmable capacitors $C_{b3...0}$.

Note that the offset calibration required in this case is more simple as compared to offset calibration in a flash converter: in this case, all comparators need to be tuned to the same decision level, equal to 0 V. This can be done on-chip easily without the need for a precise reference level. In case of a flash converter, a set of $2^N - 1$ reference voltages would be required, which is less trivial for an on-chip implementation.

3.2.2.2 Time-Skew Mismatch and Intrinsic Matching

Time-skew mismatch [10] is related to the inaccuracy of the sampling moments of the various channels, and becomes especially important in high-speed ADCs. Since on-chip detection and correction of time-skew mismatch is not trivial [5], the timing precision is designed to be intrinsically sufficient in this design.

Time-skew mismatch is mostly determined by the sampling networks and the global wiring of the input signals to all channels of the ADC. A first logical method to minimize time-skew is by designing a small form-factor layout: this ensures the global wiring will be shorter and thus it will have less impact on time-skew. This motivates why the single channels are optimized for a small and narrow size (24 μm high). Secondly, the layout of analog and clock input signals should be carefully done to prevent systematic time-skew errors. Usually this is done by designing the global clock and signal paths for equal delay (e.g. a tree structure as in Fig. 3.5a). However, since four signals (differential clock and differential signal) need to be implemented precisely, this results in a complex layout and increased wiring length and chip area. To alleviate these problems, a *matched-delay* approach can be applied to minimize time-skew instead. The layout sketch in Fig. 3.5b shows the distribution of the clock and input signal to the 16 channels: the channels are placed in a single column, while the global connections are implemented with straight vertical wires. The time-skew between the channels is minimized in this situation by matching the clock-delay and signal-delay for each channel: e.g. channel I will be delayed by Δ compared to channel A. However, since both the sampling clock and the input signal are delayed by a similar amount Δ, the delay will have little impact on the sampled values. Post-layout simulations can be performed to analyze the timing precision of the matched-delay layout. In this way, the layout can be iteratively designed until sufficient intrinsic matching is achieved. In the designed chip, the time-skew has been reduced down to $\sigma = 0.4$ ps, which is abundantly sufficient for a 5 bit 1 GS/s converter [10].

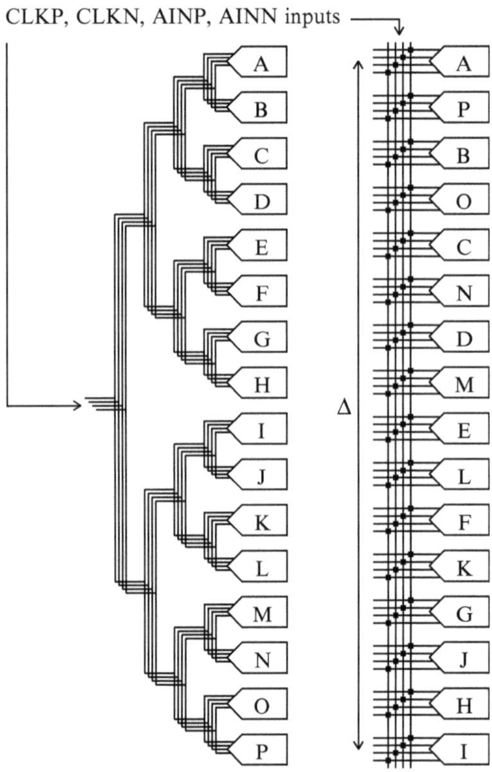

Fig. 3.5 Layout diagram for clock and signal distribution: tree structure (*left*) and proposed matched-delay layout (*right*)

3.2.3 Overall Implementation

Figure 3.6 shows the realized ADC in 90 nm CMOS, which occupies 0.11 mm^2 including decoupling capacitors. An SPI interface controls the offset-calibration register and a multiplexer is added for data-interfacing. The output data is transmitted off-chip to an FPGA board for realtime data capturing.

3.2.4 Measurement Results

The time-interleaved ADC was measured at 1 GS/s with 1.0 V supply. The offset measurement is performed once at startup. It is done off-chip using an external signal generator and a PC for calculating the channel offsets. After the offset measurement, the resulting calibration coefficients are loaded into the chip to perform the offset correction as described previously.

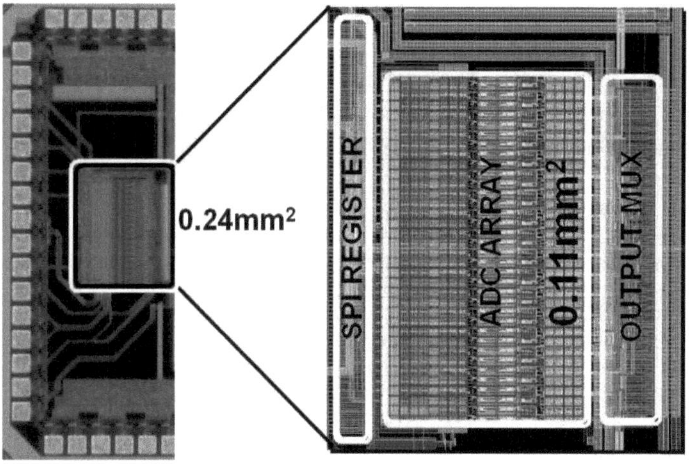

Fig. 3.6 Die photo and layout view of the ADC in 90 nm

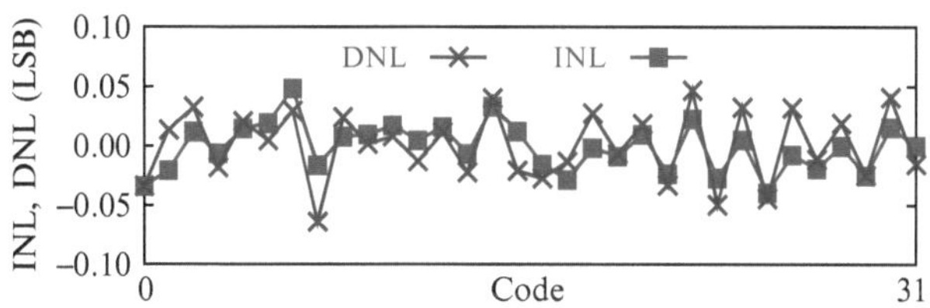

Fig. 3.7 Measured INL and DNL

After offset calibration, the measured INL and DNL (Fig. 3.7) are well below 0.1 LSB. Figure 3.8 shows the output spectrum for a near-Nyquist tone at 1 GS/s both before and after offset calibration. Before calibration, the linearity is limited by offset mismatch. After calibration, the offset-spurs are reduced to – 49.1 dB which is negligible for 5 bit accuracy. The figure also confirms that gain and time-skew errors are negligible in this design (– 49. 0 dB). From the measurements, a time-skew with $\sigma = 1.\ 8$ ps is estimated, which is mainly caused by random variations in the clock divider logic and sampling switches.

Figure 3.9 shows the ENOB before and after offset calibration. After calibration, an ENOB of 4.8 bit is achieved, while the ERBW is far beyond Nyquist. The measured power consumption equals 1.6 mW. Using the FoM:

$$FoM = \frac{Power}{2^{ENOB} \cdot \min(f_s, 2ERBW)},$$

this results in a power efficiency of 57 fJ/conversion-step.

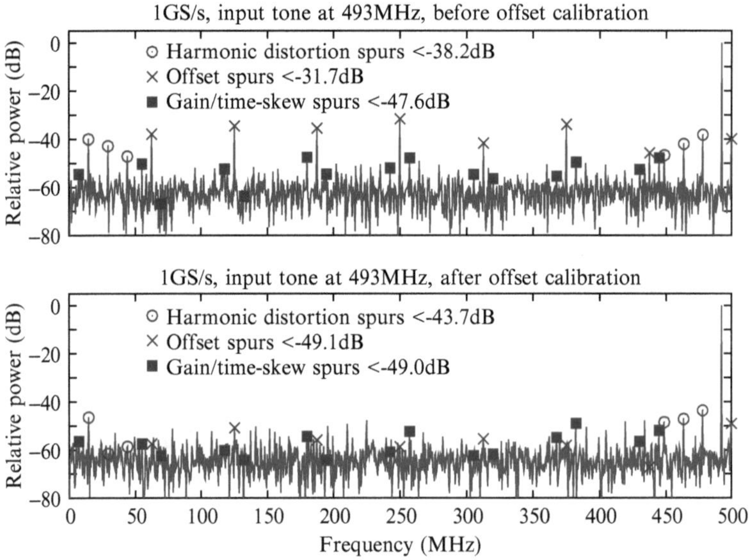

Fig. 3.8 Measured spectrum before and after offset calibration at 1 GS/s with 493 MHz input tone

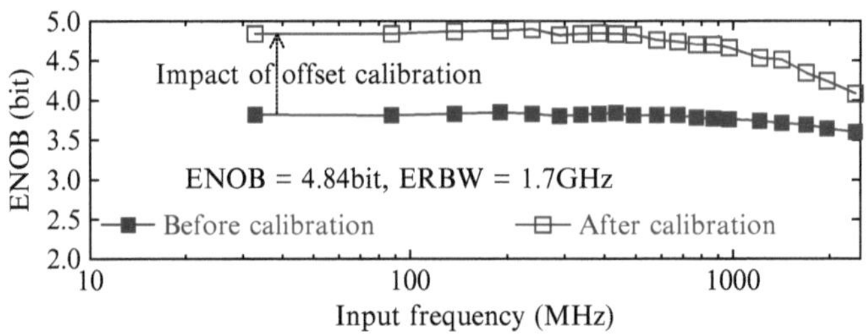

Fig. 3.9 Measured ENOB before and after offset calibration at 1 GS/s and 1.0 V supply

3.3 A 5 bit 250 MS/s Time-Interleaved Asynchronous Digital Slope ADC

This section reviews the design of a time-interleaved asynchronous digital slope ADC, as published in [14]. As a first prototype, this design realizes a 5 bit 250 MS/s ADC by interleaving two converters, each operating at 125 MS/s. First, the architecture of an asynchronous digital slope ADC is explained, followed by the actual design of the circuit.

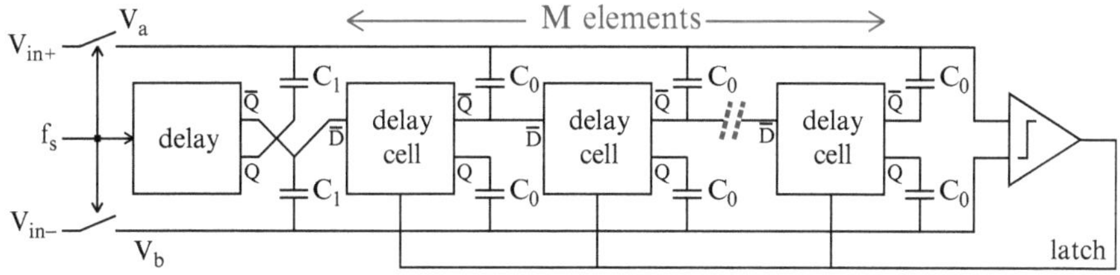

Fig. 3.10 Asynchronous digital slope ADC architecture

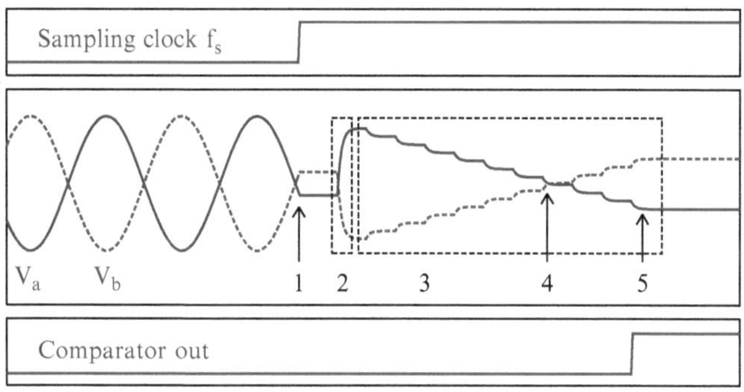

Fig. 3.11 Time-domain behavior of the ADC

3.3.1 Asynchronous Digital Slope ADC Architecture

A diagram of a typical digital-slope ADC was already shown in Fig. 3.2. In terms of speed, this architecture is limited by the fact that the comparator, counter and DAC need to operate at a frequency f_{cnt}, which is given by $f_{cnt} = 2^N \cdot f_s$. Hence, a 5 bit converter operating at e.g. 100 MS/s would already require a 3.2 GHz clock. Due to this property, slope converters are usually applied for low-speed applications, such as [15] and [16], which achieve sampling rates in the order of 100 kS/s. Recently, a few designs have been able to achieve sampling rates from 250 MS/s to 1 GS/s by architectural innovations and time-interleaving [8, 14]. In [14], the speed limitation is resolved by introducing an asynchronous digital slope architecture that eliminates the oversampled clock f_{cnt}. This architecture is illustrated in Fig. 3.10, while its behavior is illustrated in Fig. 3.11: first, the bipolar differential input signal is sampled on the array of capacitors; this is indicated with the number (1) in Fig. 3.11. During phase (2), by means of charge redistribution through capacitors C_1, the bipolar input range $[-V_{max}, +V_{max}]$ is shifted to a unipolar input range $[0, +2V_{max}]$, such that $V_a \geq V_b$ before the actual conversion starts. Then, in phase (3), the digital delay line is activated: like dominos, the cells will switch one by one from a logical 0 to a logical 1. By means of charge redistribution through

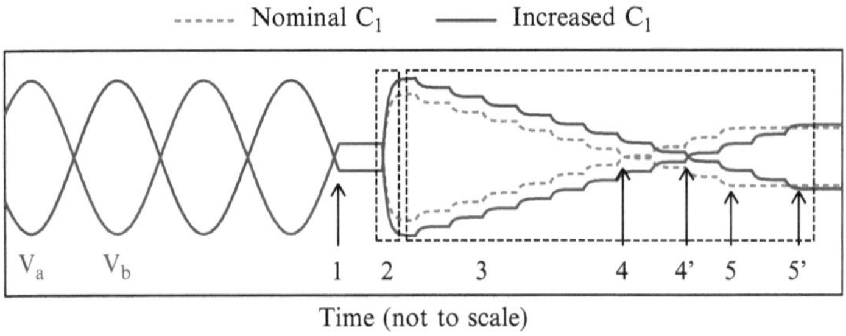

Fig. 3.12 Offset/delay compensation by programming capacitor C_1

capacitors C_0, the delay line produces a differential slope, as illustrated in Fig. 3.11. Note that the delay line operates asynchronously, thus enabling a high-speed digitally generated slope triggered by a low-speed sample rate clock. As soon as $V_a < V_b$ (4), the output of the comparator will toggle and latch the delay line at its current position. The latched thermometer code, stored in the delay line, is proportional to the input voltage and thus provides the required digital output. Because of the latency of the comparator, the moment of latching (5) will be delayed compared to the crossing-point (4). For that reason, the delay line is extended with several additional cells (thus $M > 2^N$) to prevent saturation at the end of the range. By doing so, the constant latency simply translates to a constant code-shift of the AD converter which does not degrade the performance. In a practical implementation (shown later), the overhead in delay cells due to comparator latency is about 12%. Since the comparator operates without a clock and the latency is tolerated by design, the speed of the comparator can be slow compared to the latency of the delay cells, thus enabling a higher overall speed.

A simple single-point calibration can be performed to cancel ADC offset. To implement offset correction, capacitors C_1 (Fig. 3.10) are made digitally programmable: by tuning C_1, the signal-shift during phase (2) can be altered, as shown by the example in Fig. 3.12: when increasing C_1, an additional step is added to $V_a - V_b$, such that the crossing-point ($4'$) will be delayed as well as the moment of latching ($5'$), thus resulting in a shifted output code. By applying a zero-input to the ADC and tuning C_1 until the output code reaches mid-scale, the offset can be calibrated. Since a latency variation of the comparator also manifests itself as a shift of the output code, the calibration method will automatically compensate for the comparator latency error as well.

3.3.2 *Circuit Design and Implementation*

The core of the proposed ADC architecture (Fig. 3.10) is formed by the digital delay line and capacitor network. Figure 3.13 shows the implementation of a delay cell. It is composed of a dynamic memory element and two inverters to create the

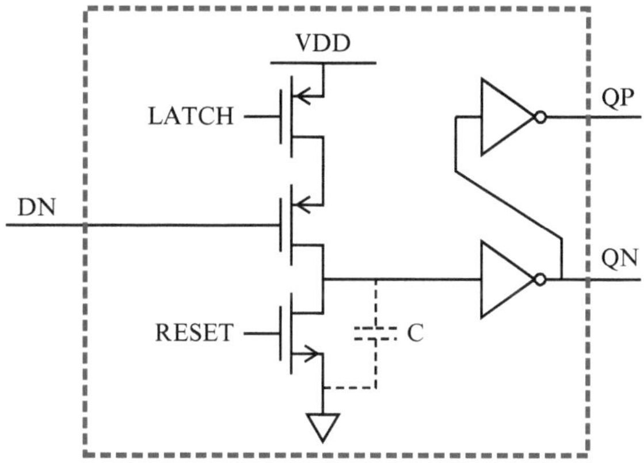

Fig. 3.13 Delay cell using a dynamic memory element

differential output. The 1-bit memory element is implemented by capacitor C, which is simply the parasitic capacitance of the node, caused by the connected transistors plus wiring. Upon RESET, the element is precharged low. The delay cell is activated by a falling edge on input DN, which is generated by the preceding element in the delay line. At that moment, DN charges C to a high-level and the output will change state. This causes a step of the analog ramp and it activates the next element of the delay line. As soon as the comparator latches, the LATCH input will go high, thus disabling input DN, and freezing all delay cells at their present state. The simple implementation of the memory element (with only three transistors) results in a low power consumption and it enables high speed operation. In total, 44 delay cells are implemented to accommodate for comparator latency and to provide additional range for calibration. In this prototype, the delay of each cell is 110 ps, leading to a duration of almost 5 ns for the complete slope. Though the cells could be made faster in the given technology, the achieved speed is sufficient for a sampling rate of 125 MS/s.

Similar to the case of the SAR ADC discussed in Sect. 3.2, the capacitors creating the slope function should be minimized in value to minimize the power consumption. Thus, dedicated 0.5 fF lateral metal-metal capacitors are constructed, comparable to the approach used for the SAR ADC. Figure 3.14 shows part of the layout of the capacitor array, and its connection to the delay cells. Finally, a standard complementary switch is added to sample the analog input signal onto the capacitor arrays.

The two-stage clockless comparator is implemented as shown in Fig. 3.15: two inverters act as pre-amplifier, while a cross-coupled inverter pair is used for the second stage. After each conversion, the comparator is disabled to reduce the power consumption by turning of the first stage. Also, the second stage is reset to remove possible memory effects, thereby ensuring that the comparator latency remains constant from one conversion to the other. The simulated latency of the comparator is 420 ps, corresponding to the latency of approximately four delay cells.

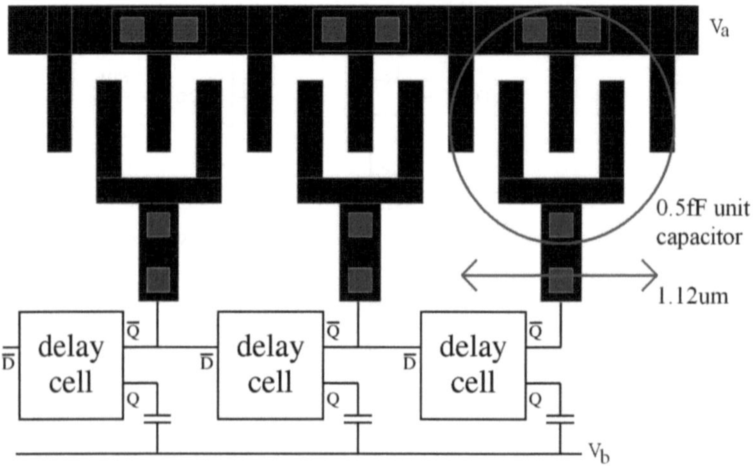

Fig. 3.14 Part of the delay line and capacitor implementation

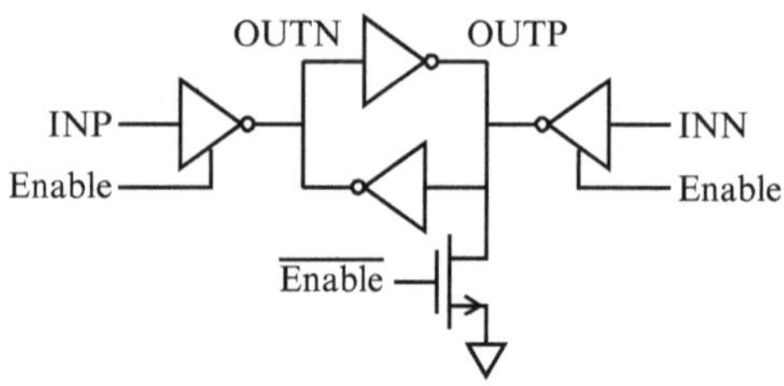

Fig. 3.15 High-speed digital-style comparator

3.3.3 *Overall Implementation*

A photo of the implemented prototype in 90 nm CMOS is shown in Fig. 3.16. The test-chip contains a 250 MS/s two-channel time-interleaved ADC, which occupies $160 \times 300\ \mu m^2$ excluding decoupling capacitors. Each channel uses a 5 bit asynchronous digital slope ADC, operating at 125 MS/s. A clock divider is included as well as thermometer to gray encoders and a digital register that allows control of the C_1 capacitors for offset/delay calibration. The output data of the ADC is transmitted off-chip to a logic analyzer for realtime data capturing.

3.3.4 *Measurement Results*

The performance of the ADC was measured at 250 MS/s with a 1.0 V supply. For all measurements, the offset compensation register was loaded with the default value, thus no calibration is used. The measured INL and DNL are shown in

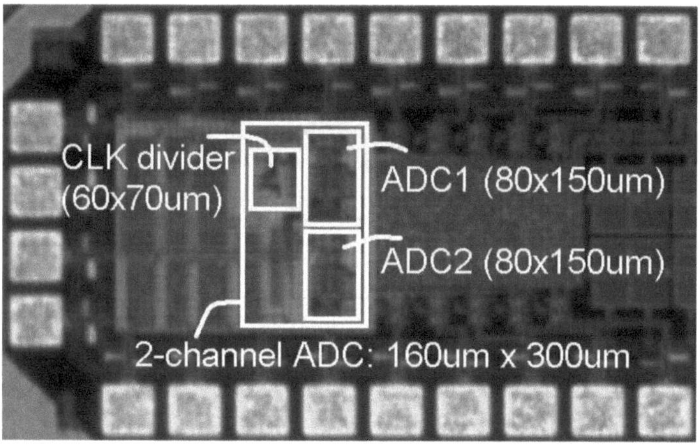

Fig. 3.16 Die photo of the time-interleaved ADC in 90 nm CMOS

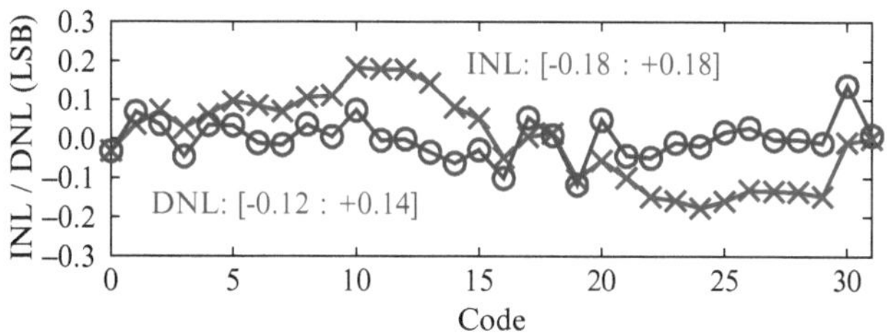

Fig. 3.17 Measured INL and DNL

Fig. 3.17, reaching a maximum error of 0.18 LSB. The dynamic performance was verified by applying single-tone inputs. Figure 3.18 shows two measured spectra. No tones due to time-skew or gain-mismatch can be observed, but a modulated spur due to offset can be seen. However, this tone is negligible for 5 bit performance, thus revealing that the two channels are sufficiently matched and no calibration is required to achieve the target accuracy. Figure 3.19 shows the measured ENOB as a function of the applied input frequency. The low-frequency ENOB equals 4.6 bit, while the ERBW is beyond Nyquist. The measured power consumption (including clock divider and thermometer to gray encoders) is 0.8 mW, yielding a FoM of 130 fJ/conversion-step.

3.4 SAR and Slope Converter Performance Comparison

Table 3.2 gives an overview of the measured performance of the SAR and slope converter, as well as a comparison to existing solutions. First, the SAR and slope architecture will be compared to each other. From the table, it appears that

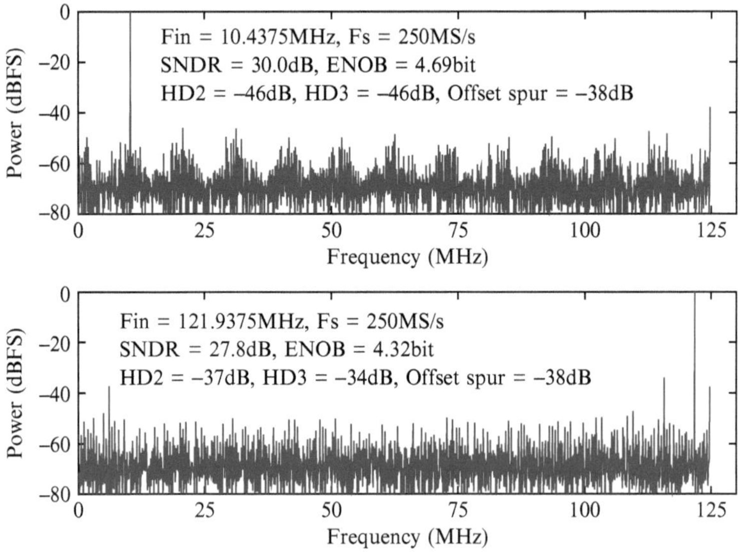

Fig. 3.18 Measured spectra for a low-frequency tone and a near-Nyquist tone

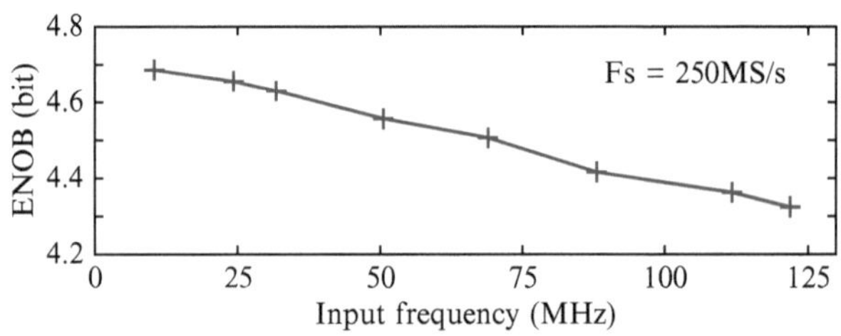

Fig. 3.19 Measured ENOB as a function of the input frequency

Table 3.2 Measured performance summary and comparison

	[3]	[4]	[5]	[17]	This work	
Architecture	Flash	Pipeline	SAR	Binary-search	SAR	Slope
Resolution (bit)	5	6	5	5	5	5
Sample frequency (GS/s)	1.75	2.2	0.25	0.8	1.0	0.25
# Interleaved channels	1	4	36	1	16	2
Power supply (V)	1.0	1.1	0.8, 1.2	1.0	1.0	1.0
Power consumption (mW)	2.2	2.6	1.2	2.0	1.6	0.8
ENOB (bit)	4.7	4.9	4.6	4.4	4.8	4.6
ERBW (GHz)	0.9	2	≈ 0.2	0.7	1.7	≈ 0.15
FoM (fJ/conversion-step)	50	40	240	116	57	130
Technology (nm)	90	40	65	65	90	90

Table 3.3 Distribution of energy per conversion for SAR and slope ADC

Architecture	SAR (pJ)	Slope (pJ)
Logic	1.07	1.47
Comparator	0.37	1.38
DAC	0.16	0.35
Total	**1.6**	**3.2**

the current implementation of the slope ADC is not as good as the SAR converter in terms of speed, bandwidth and power-efficiency: in terms of speed, the SAR ADC operates at 1 GS/s while the slope ADC operates at 250 MS/s. However, this is due to the fact that the SAR design interleaves 16 channels and the slope design only 2. Looking to the speed of the individual channels, the slope ADC is actually 2 × faster than the SAR ADC.Logically, the slope ADC could be extended to 1 GS/s as well by interleaving more channels. For example, a second generation slope converter achieves 1 GS/s by interleaving four channels of 250 MS/s each [18].

In terms of static accuracy, both designs approximate an ENOB of 5 bit. For dynamic accuracy, the SAR implementation is showing a much better performance by reaching an ERBW of 1.7 GHz, while the slope ADC reaches an ERBW around 150 MHz. This is caused by the different sampling switches being used: in case of the SAR design, the small NMOS switch with reduced CM-level yields a large bandwidth. The bandwidth of the complementary switch in the slope ADC is much smaller since the threshold voltage of the devices is relatively large, causing a higher on-resistance. However, this limitation can be solved easily by adopting the same sampling switch as used in the SAR design, as shown by an improved version of the slope converter [18].

Finally, the power consumption of both designs can be compared when observing the energy per conversion. Table 3.3 shows this information for both designs, and also separates the contributions from the logic, comparator and DAC. Overall, the current implementation of the slope ADC has a 2 ×higher consumption, mostly caused by the logic and comparator. While the SAR logic has been optimized for efficiency (by minimizing layout parasitics and by selecting a moderate speed of 62.5 MS/s per channel), the slope logic was optimized for higher speed (in order to achieve 125 MS/s per channel). Most likely, the logic in the slope converter would be more energy-efficient when optimized for a lower speed, making the difference between the two architectures small. Furthermore, the largest difference in efficiency is given by the comparator, which consumes 0.37 pJ for the SAR architecture and 1.38 pJ for the slope design.The higher consumption of the comparator in the slope ADC is for two reasons: first of all, it has been overdesigned in terms of noise and offset. Based on simulations, it is predicted that the comparator can be scaled down by 2-to-4 ×without penalty in performance. Secondly, the comparator in the slope ADC is continuously comparing. As soon as the comparator toggles state, the delayline is latched and the analog value at the input of the comparator will also remain in its present state. With the implemented comparator, this implies that *after* the decision is already taken, the input amplifier will remain in the metastable state, causing a large leakage current through the inverters. Simulations show that the energy consumption due to leakage after the decision-moment is around 50% of the total consumption. Thus, by simply disabling the comparator after the decision moment rather than after the complete conversion cycle, another factor of 2 can be gained. With these modifications, it is expected that the comparator in the slope ADC could actually be more efficient than the one in the SAR ADC.

In summary, the differences between SAR and slope converter are mostly caused by implementation aspects, and not by the architecture itself. After proper optimization of the comparator in the slope ADC, both designs should achieve a similar power-efficiency and both would be limited by the power consumed in the logic. Thus, both architectures will benefit from scaled technologies. Since the logic of a SAR converter grows linear with the resolution N, while the logic in a slope converter grows exponential with the resolution N, it can be expected that SAR converters will be more suitable for higher resolutions, whereas slope converters will be more suitable for lower resolutions.

When the presented designs are compared against alternative architectures with similar resolution (Table 3.2), it appears the power-efficiency is in a similar range, despite the relatively old 90 nm technology. Furthermore, from practical point of view, the presented designs benefit from the fact that they only need offset calibration as opposed to the complicated techniques used in most of the other works.

3.5 Conclusion

In this work, low-power SAR and slope converter architectures for high-speed applications have been discussed.As an alternative to the more commonly used architectures such as flash and pipelined ADCs, the proposed topologies can achieve similar performance with a lower calibration complexity. Moreover, the mainly digital design style allows easy migration while technology down-scaling and circuit optimization will further improve the power-efficiency.

References

1. Zheng YJ et al (2010) A 0.92/5.3 nJ/b UWB impulse radio SoC for communication and localization. In: ISSCC digest of technical papers. IEEE, Piscataway, pp 230–231 .
2. de Nil M, Busze B, Young A, Neirynck D, Pflug H, Philips K, Huisken J, Stuyt J, de Groot H (2010) Low power IEEE 802.15.4a UWB digital Rx baseband architecture. In: 2010 I.E. international conference on ultra-wideband (ICUWB). IEEE, Piscataway, pp 1–4
3. Verbruggen B, Craninckx J, Kuijk M, Wambacq P, van der Plas G (2008) A 2.2 mW 5b 1.75 GS/s folding flash ADC in 90 nm digital CMOS. In: ISSCC digest of technical papers. IEEE, Piscataway, pp 252–253
4. Verbruggen B, Craninckx J, Kuijk M, Wambacq P, van der Plas G (2010) A 2.6 mW 6b 2.2 GS/s 4-times interleaved fully dynamic pipelined ADC in 40 nm digital CMOS. In: ISSCC digest of technical papers.IEEE, Piscataway, pp 296–297

5. Ginsburg BP, Chandrakasan AP (2008) Highly interleaved 5-bit, 250-MSample/s, 1.2-mW ADC with redundant channels in 65-nm CMOS. IEEE J Solid State Circuits 43(12):2641–2650
6. van der Plas G, Verbruggen B (2008) A 150 MS/s 133 μW 7b ADC in 90 nm digital CMOS using a comparator-based asynchronous binary-search sub-ADC. In: ISSCC digest of technical papers. IEEE, Piscataway, pp 242–243
7. Murmann B (2010) ADC performance survey 1997–2010. [Online]. Available: http://www.stanford.edu/\~murmann/adcsurvey.html
8. Danesh S, Hurwitz J, Findlater K, Renshaw D, Henderson R (2011) A reconfigurable 1GSps to 250MSps, 7-bit to 9-bit highly time-interleaved counter ADC in 0.13 μm CMOS. In: IEEE symposium on VLSI circuits. IEEE, Los Alamitos, pp 268–269
9. Harpe P, Zhou C, Bi Y, van der Meijs NP, Wang X, Philips K, Dolmans G, de Groot H (2011) A 26 μW 8 bit 10 MS/s asynchronous SAR ADC for low energy radios. IEEE J Solid State Circuits 46(7):1585–1595
10. Black WC, Hodges DA (1980) Time interleaved converter arrays. IEEE J Solid State Circuits 15:1022–1029
11. Harpe P, Busze B, Philips K, de Groot H, (2011) A 0.47–1.6 mW 5bit 0.5–1 GS/s time-interleaved SAR ADC for low-power UWB radios. In: Proceedings of ESSCIRC, pp 147–150, accepted in extended form for IEEE J. Solid State Circuits 47(7):1594–1602, July 2012
12. Harpe P, Zhou C, Wang X, Dolmans G, de Groot H (2010) A 30 fJ/Conversion-Step 8b 0-to-10 MS/s asynchronous SAR ADC in 90 nm CMOS. In: ISSCC digest of technical papers. IEEE, Piscataway, pp 388–389
13. Bach J (2011) Capacitive array. US Patent 7,873,191, Jan 2011
14. Harpe P, Zhou C, Philips K, de Groot H (2011) A 0.8-mW 5-bit 250-MS/s time-interleaved asynchronous digital slope ADC. IEEE J Solid State Circuits 46(11):2450–2457
15. Nitta Y et al (2006) High-speed digital double sampling with analog CDS on column parallel ADC architecture for low-noise active pixel sensor In: ISSCC digest of technical papers. IEEE, Piscataway, pp 500–501
16. Snoeij MF, Donegan P, Theuwissen AJP, Makinwa KAA, Huijsing JH (2007) A CMOS image sensor with a column-level multiple-ramp single-slope ADC. In: ISSCC digest of technical papers. Digital Pub., Lisbon Falls, pp 506–507
17. Lin Y-Z, Chang S-J, Liu Y-T, Liu C-C, Huang G-Y, (2009) A 5b 800 MS/s 2 mW asynchronous binary-search ADC in 65 nm CMOS. In: ISSCC digest of technical papers. IEEE, New York/Piscataway, pp 80–81
18. Ding M, Harpe P, Hegt H, Philips K, de Groot H, van Roermund A (2012) A 5 bit 1 GS/s 2.7 mW 0.05 mm^2 asynchronous digital slope ADC in 90 nm CMOS for IR UWB radio. In: Proceedings of the IEEE RFIC

Chapter 4
GS/s AD Conversion for Broadband Multi-stream Reception

Erwin Janssen, Athon Zanikopoulos, Kostas Doris, Claudio Nani, and Gerard van der Weide

Abstract In this paper we present a fully integrated solution for broadband multi-stream reception, based on the direct sampling receiver architecture. The key enabler of such a solution is a 64-times interleaved 2.6 GS/s 10 b Successive-Approximation-Register ADC. The ADC combines interleaving hierarchy with an open-loop buffer array operated in feedforward-sampling and feedback-SAR mode. It is used in a fully integrated direct sampling receiver for DOCSIS 3.0 including a digital multi-channel selection filter and a PLL. The ADC achieves an SNDR of 48.5 dB and a THD of less than − 58 dB at Nyquist with an input signal of $1.4V_{pp\ -\ diff}$. It consumes 480 mW from 1.2/1.3/1.6 V supplies and occupies an area of 5.1 mm^2 in 65 nm CMOS.

4.1 Introduction

During the last years we are witnessing an increasing demand for higher data throughput rates over cable networks. The introduction of the DOCSIS 3.0 standard enabled this increase by means of channel bonding, realizing a total throughput rate of 152 Mb/s [1]. A straightforward DOCSIS 3.0 compliant receiver implementation, which uses a traditional structure, results in excess power consumption (1.6 W) [2]. Another solution [3] presents a limited dual tuner solution, since it operates on two 32 MHz bands instead of receiving a single block of 64 MHz. This fully analog dual tuner approach allows for more flexibility in the cable frequency planning, but it is not easily scalable for higher number of channels. For every additional channel it requires the complete RF processing chain, including the LO

E. Janssen (✉) • A. Zanikopoulos • K. Doris • G. van der Weide
NXP Semiconductors, High Speed Data Acquisition - office 2.20, High Tech Campus 32, 5656AE Eindhoven, NL, The Netherlands
e-mail: erwin.e.janssen@nxp.com

C. Nani
Marvell, Pavia, Italy

A.H.M. van Roermund et al. (eds.), *Nyquist AD Converters, Sensor Interfaces, and Robustness: Advances in Analog Circuit Design, 2012*,
DOI 10.1007/978-1-4614-4587-6_4, © Springer Science+Business Media New York 2013

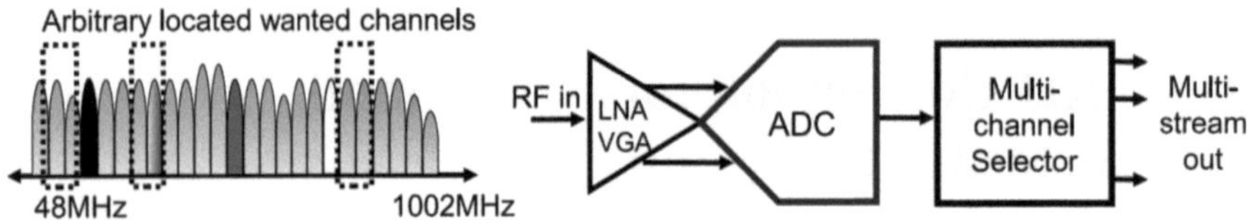

Fig. 4.1 Direct sampling receiver for cable applications

generation, to be duplicated. The use of this architecture to fulfill the market trend (full frequency flexibility with 16+ channels) is therefore considered challenging.

The solution presented in this work is a full spectrum receiver (FSR) that digitizes the complete cable band, spanning from 48 MHz to 1,002 MHz (Fig. 4.1). This direct sampling approach is scalable to down convert many RF channels, e.g. 16–32, in a power and area efficient way by only expanding the digital part of the chip while keeping the same analog frontend.

An essential advantage of the direct RF sampling receiver is that it doesn't suffer from local oscillator (LO) harmonics and image problems, typical issues for receivers involving mixers. Moreover, having captured the entire cable band, it is easy to scale the number of channels by modifying only the digital part.

The output signals are directly available in digital formats (no need for an ADC in the channel demodulator), enabling multiplexing according to reduce the number of IC pins. This is particular important when multiple channels are considered.

In order to efficiently implement a direct sampling receiver that supports multi-stream reception, we should be able to integrate the ADC on the same chip with the channel selection filter. Implementing the ADC and channel selection functions in different technologies, essentially a multi-chip approach, is not an option because high-speed data (data captured by the ADC, representing the entire cable band) should be driven out of the ADC chip and received by the channel selection chip. This results in a very power hungry solution, because multiple LVDS buffers should be employed operating at high speed.

Therefore, it is clear that both functions should be implemented in the same technology. The obvious choice is the deep sub-micron CMOS technology, where the channel selection function benefits the most. However, this technology choice poses a number of challenges for the ADC realization.

The remainder of this paper is organized as follows. In Chap. 2 the architecture of the direct sampling receiver is presented. The architecture of the time-interleaved ADC is discussed in Chap. 3. Chapter 4 deals with the implementation of the ADC. Measurement results are given in Chap. 5, followed by conclusions in Chap. 6.

4.2 Receiver Architecture

The architecture of the complete FSR solution is shown in Fig. 4.2.

The cable is connected to an external low-noise amplifier/variable gain amplifier (LNA/VGA , VGLNA) that drives the FSR. The ADC samples the cable signal and

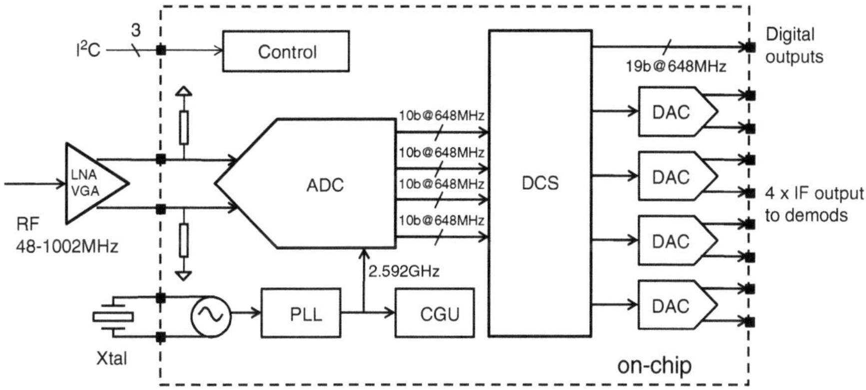

Fig. 4.2 The full spectrum receiver block diagram

provides data to the digital channel selection (DCS) filter that performs down conversion to baseband. In the current implementation the DCS is able to simultaneously output four channels (6 or 8 MHz) in a digital 13.5 MS/s IQ format (multiplexed over the digital output) and as analog IF signals by means of upsampling DACs. External to the single-chip solution, the receiver only requires a VGLNA and a crystal to drive the integrated PLL.

4.2.1 Digital Channel Selection Filter

It is clear that having an efficient high-speed ADC is not enough to effectively implement a direct sampling receiver. An efficient way to deal with all captured data (representing essentially the complete cable band) is needed. This is the role of the DCS filter, by which output streams are selected by a combination of hierarchical band splitting and fine channel selection, as illustrated in Fig. 4.3.

As the figure illustrates, the hierarchical band splitting reduces the clock frequency and employs a frequency overlap technique such that no information is lost [4]. The filter operates on sub-multiples of 648 MHz and the number of output channels can be simply increased by just adding more fine selection blocks. For its realization the standard digital design flow is used, resulting in an area efficient and low power implementation.

4.3 ADC Architecture

In this chapter first the requirements on the performance of the ADC, necessary for the realization of a DOCSIS receiver, are discussed. Next, in Sect. 4.3.2, the difficulties of interleaving are shortly discussed. To overcome these difficulties,

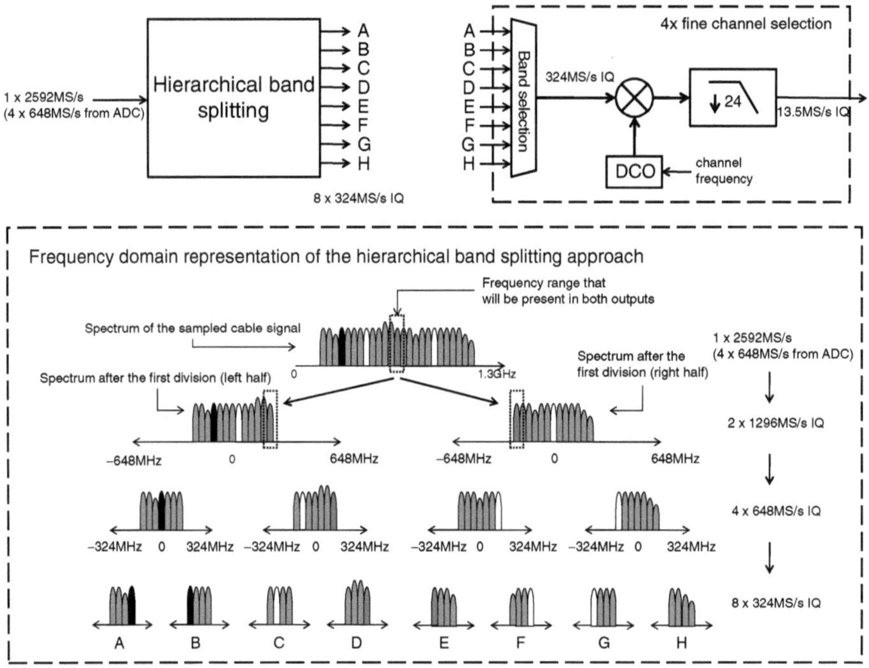

Fig. 4.3 Hierarchical channel selection process

time-interleaving hierarchy is used. In Sect. 4.3.3 the selected approach and the sub-ADC choice is explained. The complete ADC architecture is presented in Sect. 4.3.4.

4.3.1 Performance Requirements

To reach DOCSIS 3.0 performance a sampling speed of 2.6 GS/s with a resolution of 10b is required. The total integrated thermal noise power should be 55 dB below the power of a full scale sinusoidal carrier. The sampling bandwidth should be larger than the cable band upper limit (1 GHz) to avoid signal attenuation close to Nyquist. The Total Harmonic Distortion (THD) should be below −60 dB to avoid performance degradation due to inter-modulation effects. Additionally, low clock jitter is a necessity since clock jitter results in broadband noise. The jitter noise power is strongly dependent on the spectral power allocation at higher frequencies [5, 6], resulting in jitter specifications below 0.5 ps rms.

Offset and gain mismatches due to time-interleaving (see Sect. 4.3.2) should be kept low to avoid creating spurious in channel locations. Timing mismatches [7, 8] have to be limited below 1 ps levels; otherwise interference tones reduce the performance of the receiver.

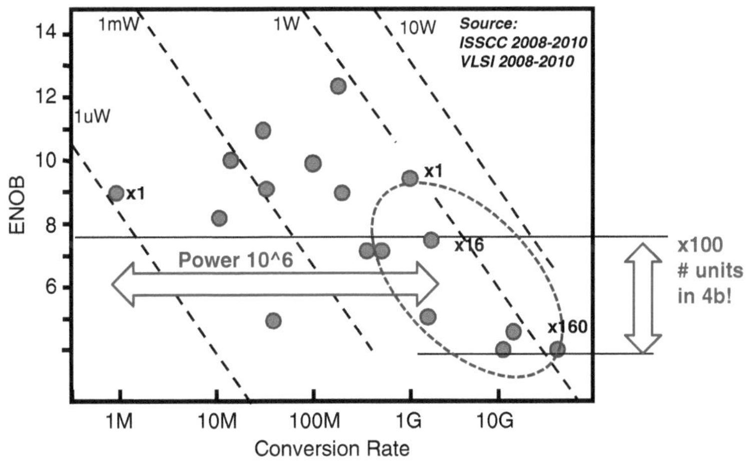

Fig. 4.4 ENOB versus conversion speed for last years ADCs

4.3.2 Time-Interleaving

An efficient approach to reach the required ADC speed and accuracy is to employ parallelism in the time domain, namely time-interleaving [9]. However, even with the use of parallelism it is difficult to meet the performance requirements listed in the previous section.

For low resolution ADCs it is common to employ massive parallelism [10–12]. This is possible since a large ADC array in combination with small sampling capacitors leads to an input load dominated by interconnects, and only limited clocking accuracy is required to achieve high enough signal conversion quality.

Moving to higher resolutions, the sampling capacitance, as it is sized to fulfill kT/C requirements, becomes the dominant limitation [11, 13–16]. This leads to a significant reduction of the efficiently feasible number of interleaved channels.

Furthermore, large time-interleaved arrays counteract the time accuracy offered by nanometer CMOS technologies, due to interconnect and buffer bandwidth limitations and the need for clock buffering. This amplifies the impact of device mismatch on timing skew, so that reaching the 0.1–1 ps level becomes very challenging, either with or without calibrations [10–12, 14, 17].

To put these problems into perspective, we use data from ADCs published the last years in ISSCC and VLSI and plot in Fig. 4.4 the Effective Number of Bits (ENOB) versus Conversion rate.

From this figure, two facts become apparent: First, there is a large reduction in the interleaving factor (number of units) as we move from 4b to 8b and, second, there is a six orders of magnitude increase in power consumption for a three orders of magnitude increase in conversion speed. Both facts support that implementing time-interleaving in ADCs efficiently is more than just placing many ADCs in parallel.

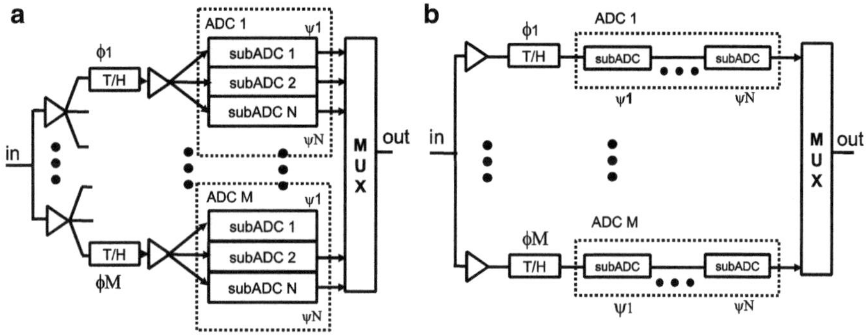

Fig. 4.5 Interleaved architectures

4.3.3 *Time-Interleaving Hierarchy and Sub-ADC Choice*

One way to manage the challenges raised by employing extensive time-interleaving is the use of T/H hierarchy [11]. The basic idea is to allow multiple ADCs to be connected to the same T/H, as Fig. 4.5a depicts. This approach has the benefit of simplifying the distribution of signal and clock, which is especially beneficial at higher frequencies. On the other hand, it increases the performance requirements for the T/H (noise and non-linearity) and dictates buffering.

Another form of hierarchy is the usage of pipelining, as shown in Fig. 4.5b. It requires a smaller number of parallel sub-ADCs per unit ADC, because the throughput is increased through pipelining stages (in contrast to more sub-ADCs in parallel as in Fig. 4.5a). This results in a reduction of the input interconnect.

Recent designs of GS/s pipelined ADCs [16, 18] use few time-interleaved ADC units without hierarchy. They aim at 6b operation with the help of open-loop amplifiers and calibrations. Reference [19] achieves 12b operation at a GS/s rate using multiple internally generated supply/ground rails in combination with blocks designed with a mixture of thin and thick oxide transistors.

A weakness of the pipelined architecture is that it is not suitable for T/H hierarchy using more than four time-interleaved ADCs [20], otherwise noise and power consumption increases. A recent implementation of a four times interleaved pipelined ADC can be found in [21].

The pipelined ADC architecture has been extensively used as the architecture of choice for medium number of bits at medium/high sampling speed applications, with some designs achieving very good power efficiency [22–24].

However, recent literature proves that the Successive-Approximation-Register (SAR) ADC architecture also offers an excellent choice to realize energy efficient converters [25–28]. The SAR ADC architecture is intrinsically slower than the pipeline, because it realizes the binary search algorithm (a serial algorithm) using the same hardware operating in different time slots. In contrast, the pipelined ADC implements the binary search algorithm using parallel simultaneous operating hardware in different time slots.

A key difference between SAR and pipelined ADC architecture is the use of residue amplification. The SAR ADC does not employ a residue amplifier, which is typically the most power-hungry block of a pipelined ADC. Instead it uses a comparator detecting zero-crossings. The absence of amplification means that the residue is not amplified, leading to an increased impact of the SAR loop noise. However, in future nanometer CMOS technologies, it will be easier to design high performance comparators than high performance amplifiers [29].

Using a SAR ADC without T/H hierarchy [30, 31] has the advantage of omitting resampling and therefore reducing noise. On the other hand, it suffers from all time-interleaving limitations mentioned above. Design [13] uses a two-stage pipelined SAR as a way to increase throughput at the expense of an additional resampling phase and a buffer between each T/H and ADC.

Combining T/H hierarchy with the SAR ADC architecture has the advantage of effectively using one sampling operation less compared to a pipelined ADC. The reason is that usually all pipelined stages after the first are designed to contribute, all together, the same amount of noise as the first. Additionally, as mentioned before, we exchange the typically power-hungry residue amplification function with an increase of noise due to SAR loop noise.

A SAR hierarchical interleaved architecture has been selected in this work because of its higher potential for parallelism and the excellent low-power capabilities of the architecture in deep sub-micron CMOS technologies.

4.3.4 Architecture

Figure 4.6 depicts the proposed hierarchical time-interleaved architecture based on SAR ADCs [32].

It employs four front-end T/Hs, each one driving a Quarter ADC (QADC) consisting of 16 reduced-radix SAR ADCs, resulting to a total number of 64 ADC units. Each T/H drives its QADC array with a feedforward-feedback multiplexed open loop buffer interface, which will be detailed in a later section. An on-chip digital calibration engine per QADC removes gain and offset mismatches, corrects the DAC nonlinearity within each SAR ADC, and realizes the non-binary to binary mapping on the output data.

The timing diagram (Fig. 4.7) illustrates the operation of the ADC.

The ADC is clocked from a single $f_s = 2.6$ GHz clock. Each T/H operates at 650 MHz with a duty cycle of 50%, meaning that tracking and hold use two periods of $T_s = 1/f_s$ each. Note that there are always two T/Hs connected to the input, decreasing the load and providing constant input impedance. The SAR ADCs of each QADC re-sample the data provided by the T/H and operate according to the SAR algorithm, with an internal clock cycle of 650 MHz. A SAR ADC unit outputs 10b data after 12 cycles, one for re-sampling and 11 for quantization. Only 12 SAR ADCs per QADC are needed for interleaving at 2.6 GS/s, the rest are implemented

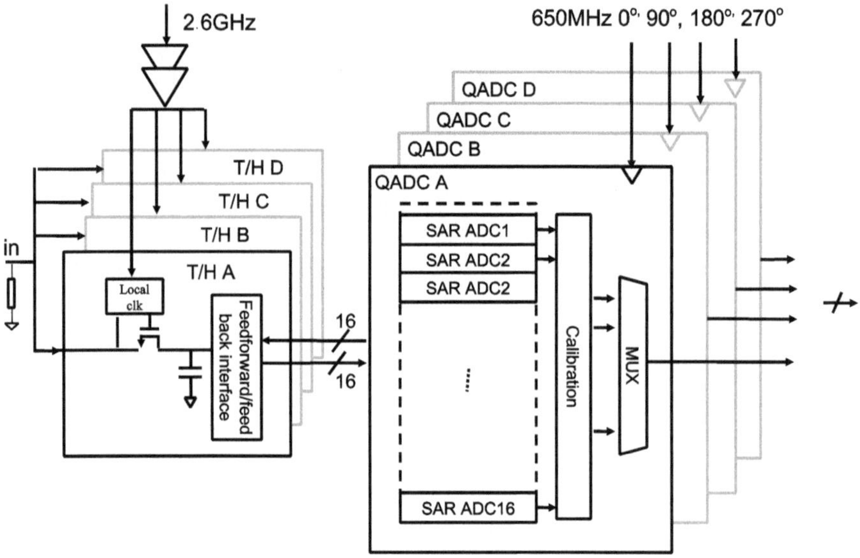

Fig. 4.6 Proposed ADC architecture

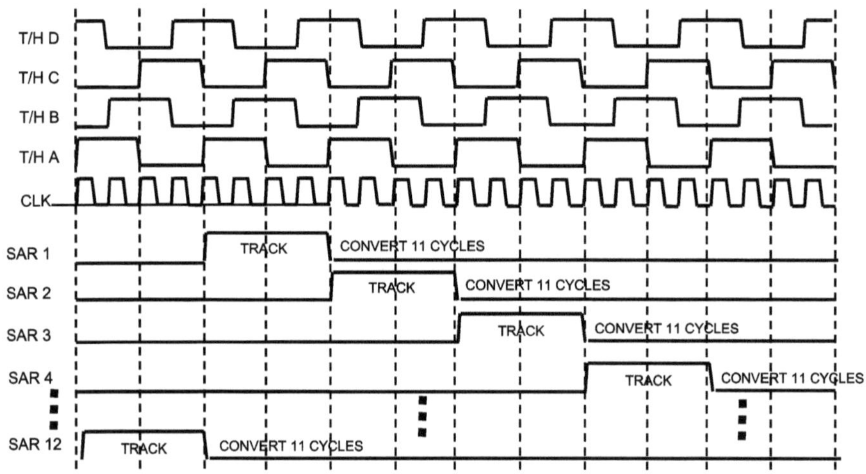

Fig. 4.7 ADC's timing diagram

for redundancy. The total of 16 SAR ADCs can be used in any pre-selected or random order during operation. A data combiner synchronizes the four data streams from the QADCs prior sending them to the ADC output.

The main advantage of the proposed ADC architecture is the partitioning of it in two domains, namely T/H and SAR ADC arrays, which they can then be optimized separately.

The limited number of T/H's simplifies the signal and clock distribution, reducing interconnects. This has as a direct consequence intrinsically high timing accuracy, making unnecessary any calibration (see Sect. 4.4.1).

Furthermore, the feedforward-feedback multiplexed open-loop interface leads to certain advantages enhancing the performance while keeping power consumption low. The load of each T/H remains low, since only a fraction of SAR ADCs are directly connected to it each time. This enables many ADC units to be driven by one T/H without linearity or speed penalty. Moreover, the feedforward-feedback interface eliminates the linearity requirements of the open-loop buffers. This allows the use of a high-swing signal, which directly increases the SNR.

The SAR ADC used in this design splits the sampling and DAC functions, allowing the dimensioning of sampling capacitor for kT/C noise and not for matching, which is preferable given that the targeted accuracy is 10b. The use of a current steering DAC simplifies the reference distributions (particularly challenging in large ADC arrays) and facilitates straightforward gain and offset mismatch calibrations as addition/subtraction of currents.

4.4 ADC Implementation

This chapter discusses a number of implementation aspects of the track and hold (Sect. 4.4.1), the feedforward-feedback interface (Sect. 4.4.2), and the SAR sub-ADC (Sect. 4.4.3).

4.4.1 Track/Hold Front-End

Figure 4.8 shows the clock architecture that has been employed in the ADC.

The clock buffering and distribution is realized using Current-Mode Logic (CML). CML clocking has superior behavior concerning clock spurious generation, supply/substrate noise and Power-Voltage-Temperature (PVT) variations.

Two buffers (in series) receive a low-swing differential clock signal and distribute it to four T/Hs via a shielding H-tree. The use of only four T/Hs in combination with the small height of the ADC array results in a clock interconnect of less than 200 fF.

The circuitry of the local clocking and the sampling switches and capacitors are shown in Fig. 4.9.

The local clocking is done by a small clocking unit, which receives a differential clock and with the help of control signals and bootstrapping generates the CMOS sampling pulse, similar to [13]. The sampling switch is differential and use cross-coupling to cancel feedthrough. Reference voltages necessary for calibration are supplied directly to the sampling capacitors [33].

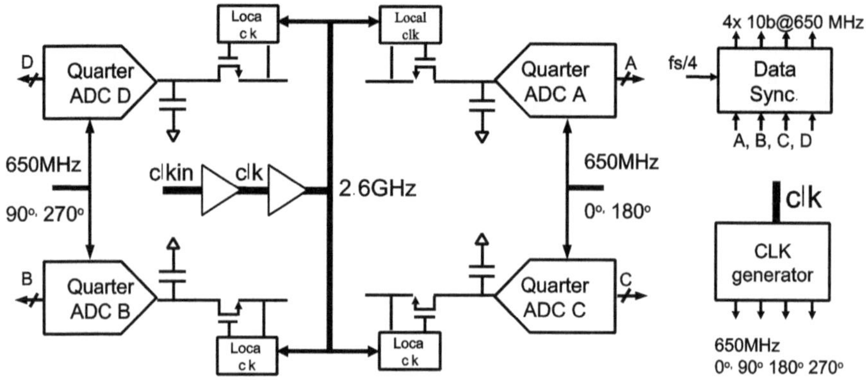

Fig. 4.8 Clock hierarchy and architecture

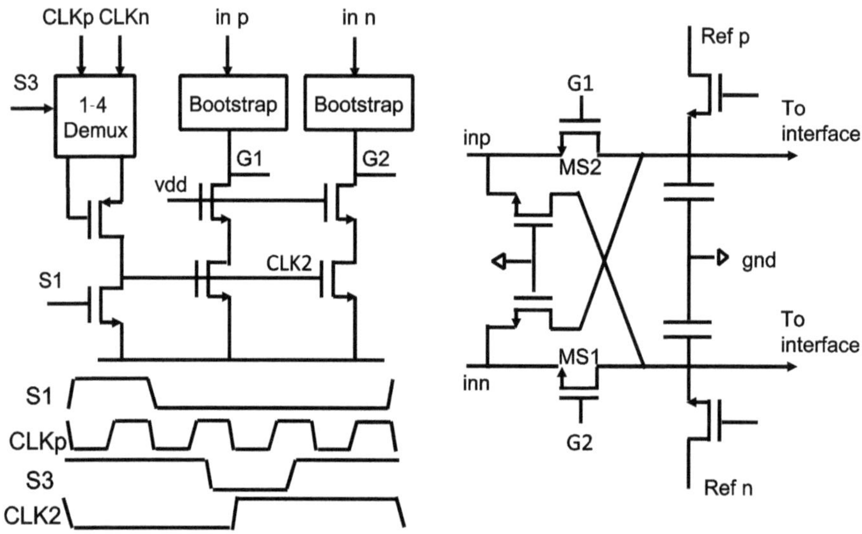

Fig. 4.9 Local clocking and sampling circuitry

Bootstrapping the switch is essential for reaching the targeted performance. This implies limiting the impact of bandwidth mismatch at the sampling node and achieving a sampling linearity of more than 60 dB without calibration. An extensive discussion on the bootstrapping circuit and its operation can be found in [32].

4.4.2 Feedforward: Feedback Interface

The connection between the front-end T/H and the ADC array as depicted in Fig. 4.5a dictates the use of buffers. The source follower topology is known for its

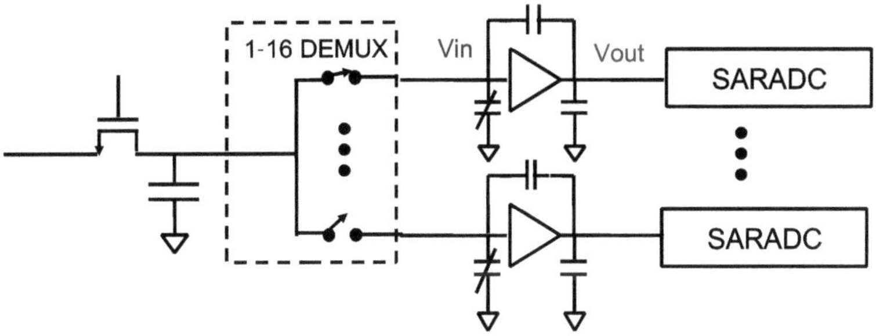

Fig. 4.10 Buffer partitioning and demultiplexer introduction

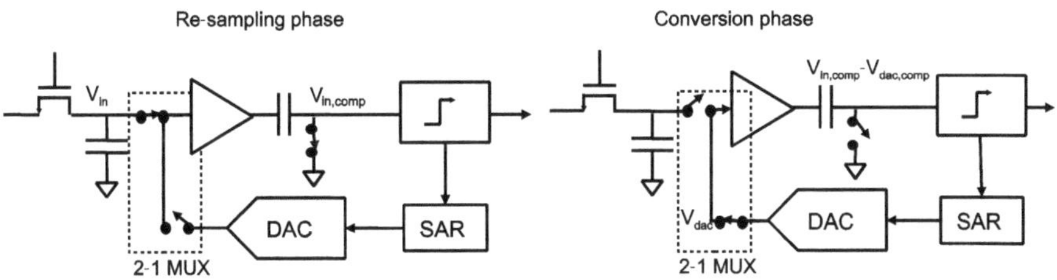

Fig. 4.11 Buffer introduction in the SAR loop; resampling and conversion SAR phase

speed and low-power features [18, 34–36]. The main disadvantage is that it demonstrates strong nonlinearity especially in modern CMOS technologies. This necessitates the use of small signal swing resulting in SNR loss. Furthermore, the large nonlinear gate load limits the sampling linearity especially at high frequencies.

The proposed feedforward-feedback interface alleviates these limitations by changing the circuit topology, which can be described as a two step process. In the first step, the buffer is partitioned in 16 smaller ones, one for each sub-ADC, as shown in Fig. 4.10.

Additionally, a demultiplexer is placed between the front-end T/H and the buffer, assuring that only one buffer is connected to the T/H at any given moment. This greatly reduces the load of the T/H, a major limitation of the hierarchical T/H architectures as shown in Fig. 4.5a.

In the second step, the buffer is introduced in the SAR loop, eliminating linearity limitations intrinsic to open-loop buffers. As Fig. 4.11 illustrates this becomes possible only if the SAR sampling and DAC functions are split from each other.

The multiplexer allows the buffer to be part of the SAR loop. In the feedforward (re-sampling) phase the input signal is compressed by the buffer and resampled on the resampling capacitor. In the feedback (conversion) phase the DAC signal is passed through the same buffer experiencing the same compression. Since the SAR operation is based on zero-crossing detection and the input and DAC signals are equally distorted, the difference signal will still results in the correct decision.

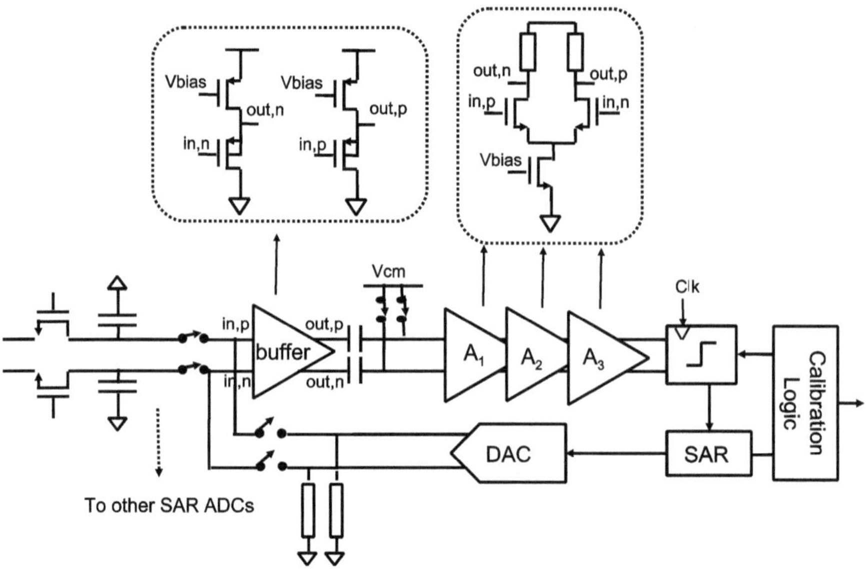

Fig. 4.12 Interface and SAR architecture

4.4.3 SAR ADC

The interface and SAR architecture used in the ADC array is shown in Fig. 4.12.

The choice of the SAR architecture is strongly affected by the usage of the feed-forward feedback structure; it is pseudo-differential and consists of a sampling capacitor, a comparator preceded by a fully differential preamplifier, a SAR digital controller and a current steering main DAC. The SAR loop between the main DAC and the re-sampling SAR capacitor is closed via the front-end multiplexer and the interfacing buffer. Two calibration DACs are used to tune the offset and the gain of the ADC.

The non-binary successive approximation algorithm has been implemented in order to achieve high conversion speed with relaxed settling requirements. Moreover, the redundancy of the non-binary algorithm has been exploited to reduce the main DAC area using a DAC linearity calibration technique similar to [34].

Due to the reduced radix, the complete SAR conversion process requires 12 clock cycles (one for tracking, 11 for conversion) at 650 MHz to achieve a nominal resolution of 10b. While tracking, the input sampled by the front-end is passed via the interfacing buffer to the SAR sampling capacitor where it is sampled using a bottom plate sampling scheme. During the conversion mode the main DAC is connected via the buffer to the SAR sampling capacitor that acts as subtraction point between the DAC and the sampled signal. This difference is then amplified by a preamplifier before being latched.

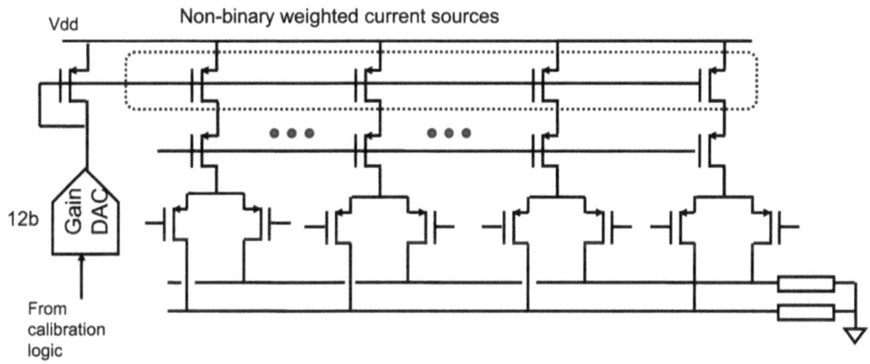

Fig. 4.13 Current steering DAC

The use of a chain of preamplifiers is necessary for a number of reasons. It amplifies the difference signal helping the comparator to a faster decision, it reduces the input referred noise and kickback from the comparator and it limits the noise bandwidth at the input of the comparator, reducing the total noise that is been integrated by the comparator. The preamplifiers are open-loop stages [32] with 28 dB total DC gain and they drive a dynamic regenerative comparator [37].

The offset correction is realized by injecting a differential current at the output of the first preamplifier, by means of a calibration DAC.

Figure 4.13 shows the current-steering non-binary main DAC used in SAR operation, together with the calibration DAC correcting for gain errors.

The choice of using a current steering DAC has certain advantages. It is easy to drive the long interconnect between the DAC's output and the sampling front-end (a switched capacitor DAC would need buffering). It exhibits excellent PVT stability and supply noise rejection. Of equal importance is the simplification of the generation, distribution and correction of the references, defining the gain of the 64 SAR ADCs. The full scale of each main DAC is easily controlled by tuning its biasing current. The use of currents instead of voltages increases the robustness and reduces the sensitivity to supply/substrate noise, PVT variations, parasitic coupling, etc. Furthermore, the size of each main DAC is minimized utilizing the build-in redundancy of the non-binary SAR algorithm [38]. During the startup phase, the main DAC weights are measured and used during the operation to convert the non-binary data to binary. In this way, the mismatch errors between current sources are corrected, without using calibration or trimming.

In order to reach the targeted accuracy, correction of offset and gain interleaving errors is necessary. Figure 4.14 shows the SAR implementation including the calibration DACs.

The actuation of the calibrations takes place in the analog domain, while the measurements are done in the digital domain. The rationale behind the choice for current DACs as actuation means of the offset and gain calibrations are

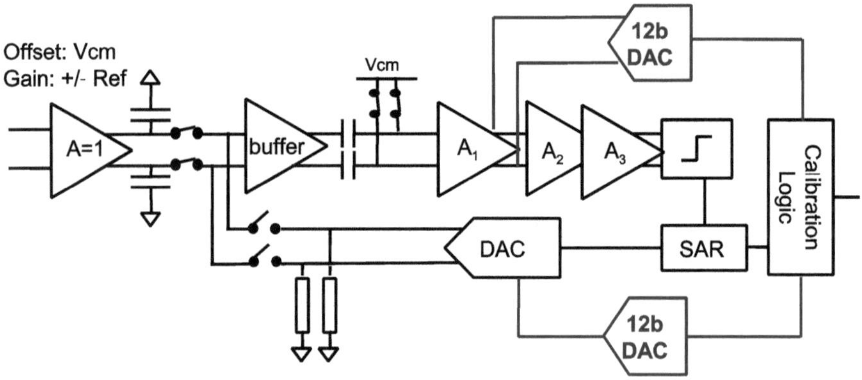

Fig. 4.14 Offset and gain calibration

similar to the choice of the main DAC. During the startup calibration, using externally generated and on-chip buffered reference voltages, the digital calibration logic decides upon the proper control words that minimize the offset and gain error.

4.5 Measurements

The ADC is integrated as part of a DOSCIS 3.0 direct sampling receiver prototype [4] (see Fig. 4.2) using a baseline 7-metal 65 nm CMOS process, packaged in an LGA132 package.

Figure 4.15 shows the die photo of the ADC, which occupies 5.1 mm^2.

On the photo the different parts of the converter have been indicated. The central part is occupied by the T/H and clocking circuitry surrounded by the four QADCs. The digital calibration logic occupies approximately the 50% of the total ADC area and it is placed around the QADCs. The ADC data are synchronized by the data sync unit placed at the top of the converter.

The T/H and CML clock generator operate from a 1.3 V supply. Interface buffers and DACs use a 1.6 V supply, while the SAR ADCs and calibration logic use 1.2 V.

The total power consumption is 480 mW at 2.6 GS/s (the calibration logic consumes only 36 mW), excluding the LVDS-like buffers and external generated references.

The ADC measurements were performed using an external clock generator bypassing the on-chip PLL. The ADC output data were internally decimated by five, according to overcome data acquisition speed limitations, and sent directly to a logic analyzer, without going through the on-chip DSP. The decimation by five operation folds the complete sampled spectrum to one-fifth of the Nyquist frequency, allowing measurement of all spectral artifacts.

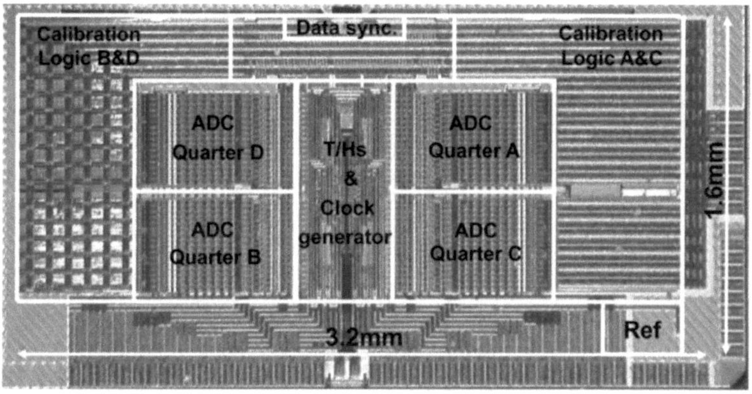

Fig. 4.15 ADC die photo

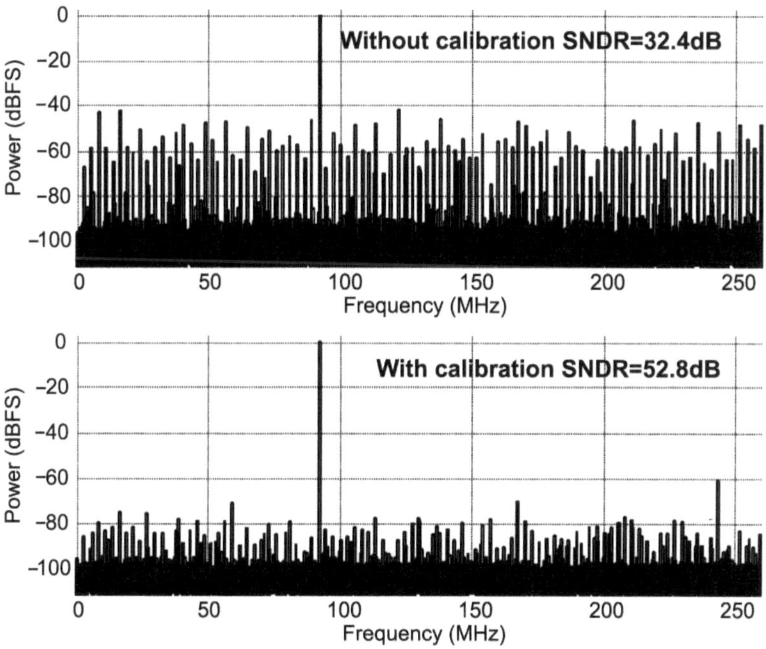

Fig. 4.16 ADC spectrum plots ($f_{in} = 92$ MHz, $f_s = 2.6$ GS/s) with and without calibration

Figure 4.16 shows the effect of the calibration on improving the SNDR of the converter, applying an input signal of $1.4V_{pp-diff}$ and $f_{in} = 92$ MHz at $f_s = 2.6$ GS/s.

Without calibration the SNDR is equal to 32.4 dB, which by activating calibration improves to 52 dB. In this case the total offset and gain mismatch tone power is 10 dB below the total thermal noise.

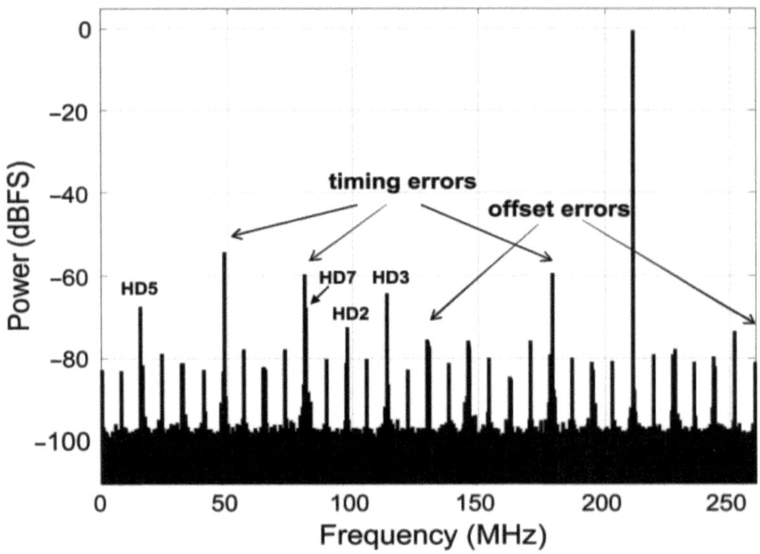

Fig. 4.17 ADC spectrum with $f_{in} = 1.25$ GHz and $f_s = 2.6$ GS/s (decimated by five)

Figure 4.17 depicts the output power spectrum for $f_{in} = 1.25$ GHz at $f_s = 2.6$ GS/s.

The offset tones remain at −75.4 dBFS, similar to the case of the low frequency input signal (Fig. 4.16). On the other hand, spurious tones resulting from timing mismatches (the three dominant tones) have increased, due to their linear dependency with the input signal frequency. The most dominant one reaches −54 dBFS. The HD3 remains at −64.2 dBFS, while HD5 and HD7 remain below −66 dBFS.

Figure 4.18 illustrates the effect of randomization using the redundancy implemented in the number of SAR ADCs (every QADC has 16 units, while only 12 are needed for 10b operation, see Sect. 4.3.4).

The top plot shows the output power spectrum when the SAR ADC units are used in a pre-determined sequence. This leads to a fixed offset and gain tone pattern. However, activating randomization (meaning not fixed sequence) all offset and gain tones are translated to noise, lowering further the spurious tones. The penalty is a minor increase (approximately 0.1 dB) of the noise floor. These measurement results are in agreement with theoretical studies [39].

Figure 4.19 presents how the performance merits SNDR, SFDR, THD and SNR depend on the input signal frequency. The signal amplitude is 1.4 $V_{pp-diff}$ sampled at $f_s = 2.6$ GS/s, while using a fixed SAR ADC sequence (randomization off).

The THD remains better than −58 dB up to Nyquist, while a degradation is observed at higher frequencies, leading to −55 dB at 2 GHz. The degradation is caused by the bandwidth limitations of the buffer during the tracking phase. The SFDR curve profile can be divided in three parts, depending on the dominant mechanism that limits it. From DC up to approximately 200 MHz the SFDR is

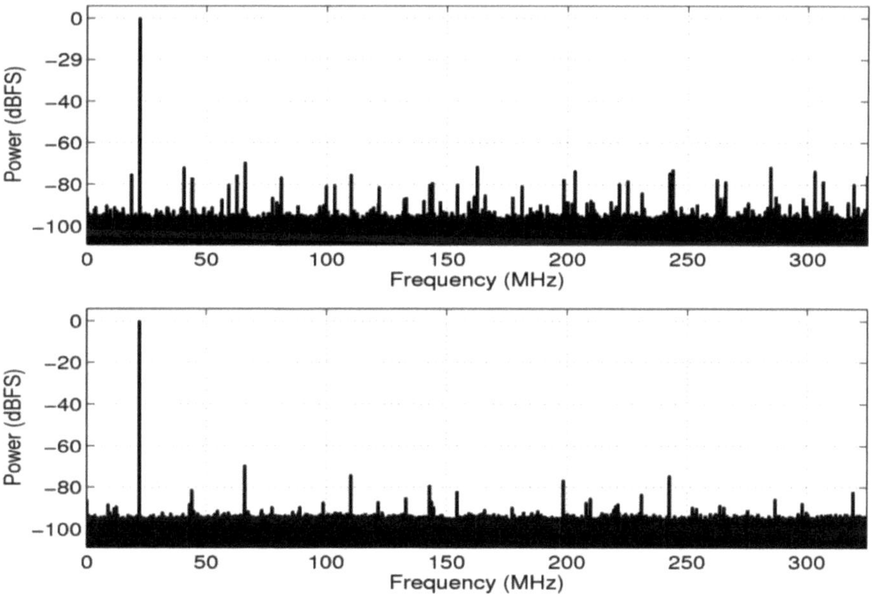

Fig. 4.18 ADC spectrum plots ($f_{in} = 22$ MHz, $f_s = 2.6$ GS/s) without (*top*) and with (*bottom*) randomization mode for one QADC

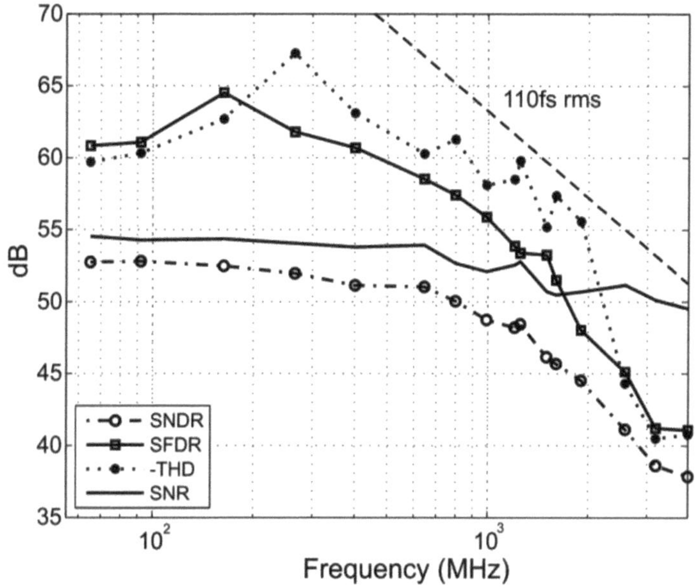

Fig. 4.19 Performance versus input frequency sweep

Table 4.1 Performance summary

Process	65 nm
Resolution	10 b
Active area	5.1 mm^2
Supply	1.2/1.3/1.6 V
Input	1.4 Vpp-diff
Power consumption	480 m W
Sampling rate	2.6 GS/s
Input termination	100 Ohm Diff.
3dB input bandwidth	>5 GHz
SNDR @ Nyq.	48.5 dB
SFDR @ Nyq.	53.8 dB
THD @ Nyq.	<−58 dB
SNR @ Nyq.	>52 dB
Jitter	<110 fs

limited by the harmonic distortion of the signal source. In the decade between 200 MHz and 2 GHz, the SFDR is determined by spurious tones due to timing skew, while beyond 2 GHz harmonic distortion dominates. The timing skew spread is estimated to be 400 fs rms at room temperature.

The SNR curve shown in Fig. 4.19 results from the combined effects of thermal noise, quantization noise (including DNL mismatch) and clock jitter. The SNR at DC is 54.5 dB, predominantly limited by thermal noise. The SNR remains flat over a broad frequency range. At Nyquist, the SNR is 52 dB and remains above 49 dB up to $f_{in} = 4$ GHz. This is accomplished due to low signal attenuation at high frequencies and jitter of less than 110 fs rms. The jitter value calculation is identical to [33] and includes contributions from clock and input signal sources.

Finally, the SNDR, representing the combined effect of all types of noise, harmonic and interleaving spurs, is 52.8 dB at low frequencies dominated by thermal noise. Low clock skew and high linearity help the SNDR to stays above 48.5 dB up to Nyquist.

Furthermore, the performance of this ADC has been evaluated by carrying out measurements in the complete receiver (Fig. 4.2), operating with multi-stream internet and TV functionality. Measurements in such conditions, with a fully loaded cable plant with 158 equal power 6 MHz channels, show no noticeable degradation of the system performance [4].

Table 4.1 summarizes the performance of the ADC.

4.6 Conclusions

This paper presents a hierarchical time-interleaved SAR ADC architecture enabling a direct sampling receiver solution for DOCSIS 3.0. The implemented hierarchy enables separate optimization of the front-end T/H and SAR ADC. The four front-end T/Hs are optimized for speed, linearity and time accuracy, achieving intrinsic accuracy without any time calibration/correction. The SAR ADC uses an open-loop buffer array, operating in a feedforward-sampling feedback-SAR mode. This is enabled by the separation of the sampling and DAC functions and essentially eliminates the buffers linearity requirements. Furthermore, the ADC employs simple differential stages that don't require intrinsic matching, but instead relies on digital calibration techniques.

The implemented ADC architecture demonstrates that massive interleaving of ADCs in advanced CMOS processes is possible, alleviating problems of traditional interleaving approaches, and paves the way for a new generation of direct sampling based receivers.

References

1. Data over cable service interface specifications (2009) DOCSIS3.0, Physical layer specification, CM-SP-PHYv3.0-I08-090121, Cable Labs, 21 Jan 2009
2. CSR (former Microtune) MT2170 datasheet, single-chip DOCSIS 3.0 wideband tuner, available through http://http://www.csr.com/products/107/mt217
3. Gatta F et al (2009) An embedded 65 nm CMOS baseband IQ 48 MHz–1 GHz dual tuner for DOCSIS 3.0. IEEE J Solid-St Circ 44(12):3511–3525
4. Janssen E, Doris K, Zanikopoulos A, van der Weide G, Vertregt M, Jamin O, Courtois F, Blard N, Kristen M, Bertrand S, Riviere F, Deforeit F, Blanc G, Penning Y, Lefebvre F, Viguier D, Dubois M, Vrignaud V, Cazettes C, Schaller L, Jenvrin G (2011) A direct sampling multichannel receiver for DOCSIS 3.0 in 65 nm CMOS. In: Symposium on VLSI circuits (VLSIC), pp 292–293, 15–17 June 2011
5. Balakrishnan A (1962) On the problem of time jitter in sampling. IEEE Trans Inf Theory 8(3):226–236
6. Doris K (2004) High-speed D/A converters: from analysis and synthesis concepts to IC implementation. Ph.D. thesis, Technical University Eindhoven, The Netherlands, 2004
7. Kurosawa N et al (2001) Explicit analysis of channel mismatch effects in time-interleaved ADC systems. IEEE Trans Circuits Syst I, Fundam Theory Appl 48(3):261–271
8. Vogel C (2005) The impact of combined channel mismatch effects in time-interleaved ADCs. IEEE Trans Instrum Meas 54(1):415–427
9. Black WC, Hodges DA (1980) Time interleaved converter arrays. IEEE J Solid-St Circ 15 (6):1022–1029
10. Poulton K et al (2003) A 20GS/s 8b ADC with a 1 MB Memory in 0.18 um CMOS. In: IEEE international solid-state circuits conference, Digest of technical papers, pp 318–319
11. Schvan P et al (2008) A 24 GS/s 6b ADC in 90 nm CMOS. In: IEEE international solid-state circuits conference, Digest of technical papers, pp 544–545
12. Greshishchev D et al (2010) A 40GS/s 6b ADC in 65 nm CMOS. In: IEEE international solid-state circuits conference, Digest of technical papers, pp 390–391
13. Louwsma SM, van Tuijl AJM, Vertregt M, Nauta B (2008) A 1.35 GS/s, 10 b, 175 mW time-interleaved AD converter in 0.13 μm CMOS. IEEE J Solid-St Circ 43(4):778–786
14. Payne R et al (2011) A 12b 1GS/s SiGe BiCMOS two-way time-interleaved pipeline ADC. In: IEEE international solid-state circuits conference, Digest of technical papers, pp 182–184
15. Van de Vel H, Buter BAJ, van der Ploeg H, Vertregt M, Geelen GJGM, Paulus EJF (2009) A 1.2-V 250-mW 14-b 100-MS/s digitally calibrated pipeline ADC in 90-nm CMOS. IEEE J Solid-St Circ 44(4):1047–1056

16. Ali AMA, Dillon C, Sneed R, Morgan AS, Bardsley S, Kornblum J, Wu L (2006) A 14-bit 125 MS/s IF/RF sampling pipelined ADC With 100 dB SFDR and 50 fs Jitter. IEEE J Solid-St Circ 41(8):1846–1855
17. Verbruggen B et al (2010) A 2.6 mW 6b 2.2GS/s 4-times interleaved fully dynamic pipelined ADC in 40 nm digital CMOS. In: IEEE international solid-state circuits conference, Digest of technical papers, pp 296–297
18. Nazemi A, Grace C, Lewyn L, Kobeissy B, Agazzi O, Voois P, Abidin C, Eaton G, Kargar M, Marquez C, Ramprasad S, Bollo F, Posse VA, Wang S, Asmanis G (2008) A 10.3GS/s 6bit (5.1 ENOB at Nyquist) time-interleaved/pipelined ADC using open-loop amplifiers and digital calibration in 90 nm CMOS. In: IEEE symposium on VLSI circuits, pp 18–19, 18–20 June 2008
19. Chen C-Y, Wu J (2011) A 12b 3GS/s pipeline ADC with 500 mW and 0.4 mm^2 in 40 nm digital CMOS. In: Proceedings of the IEEE symposium on VLSI circuits
20. Murmann B (2011) Low-power pipelined A/D conversion. In: Proceedings of the 20th workshop on advances in analog circuit design (AACD), April 2011
21. Vecchi D, Mulder J, van der Goes FML, Westra JR, Ayranci E, Ward CM, Wan J, Bult K (2011) An 800 MS/s dual-residue pipeline ADC in 40 nm CMOS. IEEE J Solid-St Circ 46 (12):2834–2844
22. Ahmed I, Mulder J, Johns DA (2009) A 50MS/s 9.9 mW pipelined ADC with 58 dB SNDR in 0.18 μm CMOS using capacitive charge-pumps. In: IEEE international solid-state circuits conference – Digest of technical papers. ISSCC 2009, vol 165a, pp 164–165, 8–12 Feb 2009
23. Brooks L, Lee H-S (2009) A 12b, 50 MS/s, fully differential zero-crossing based pipelined ADC. IEEE J Solid-St Circ 44(12):3329–3343
24. Huang Y-C, Lee T-C (2011) A 10-bit 100-MS/s 4.5-mW pipelined ADC with a time-sharing technique. IEEE Trans Circuits Syst I, Reg Papers 58(6):1157–1166
25. van Elzakker M, van Tuijl E, Geraedts P, Schinkel D, Klumperink E, Nauta B (2010) A 10-bit charge-redistribution ADC consuming 1.9 μW at 1 MS/s. IEEE J Solid-St Circ 45 (5):1007–1015
26. Harpe P et al (2010) A 30fJ/conversion-step 8b 0-to-10MS/s asynchronous SAR ADC in 90 nm CMOS. In: IEEE international solid-state circuits conference, Digest of technical papers, pp 387–389
27. Liu W et al (2010) A 12b 22.5/45MS/s 3.0 mW 0.059 mm2 CMOS SAR ADC achieving over 90 dB SFDR. In: IEEE international solid-state circuits conference, Digest of technical papers, pp 380–381
28. Liu CC et al. (2010) A 10b 100MS/s 1.13 mW SAR ADC with binary-scaled error compensation. In: IEEE international solid-state circuits conference, Digest of technical papers, pp 386–387
29. Drexelmayr D (2006) Concepts and improvements in pipelined and SAR ADCS. In: Proceedings of the 15th workshop on advances in analog circuit design (AACD), April 2006
30. Drexelmayr D (2004) A 6b 600 MHz 10 mW ADC array in digital 90 nm CMOS. In: IEEE international solid-state circuits conference, Digest of technical papers, pp 264–265
31. Alpman E et al (2009) A 1.1 V 50 mW 2.5GS/s 7b time-interleaved C-2C SAR ADC in 45 nm LP digital CMOS. In: IEEE international solid-state circuits conference, Digest of technical papers, vol 77a, pp 76–77
32. Doris K, Janssen E, Nani C, Zanikopoulos A, van der Weide G (2011) A 480 mW 2.6 GS/s 10b time-interleaved ADC with 48.5 dB SNDR up to Nyquist in 65 nm CMOS. IEEE J Solid-St Circ 46(12):2821–2833
33. Taft RC et al (2009) A 1.8 V 1.0GS/s 10b self-calibrating unified-folding-interpolating ADC with 9.1 ENOB at Nyquist frequency. In: IEEE international solid-state circuits conference, Digest of technical papers, vol 79a, pp 77–78
34. Liu W et al (2009) A 600MS/s 30 mW 0.13um CMOS ADC array achieving over 60 dB SFDR with adaptive digital equalization. In: IEEE international solid-state circuits conference, Digest of technical papers, pp 82–83

35. Hsu C-C (2007) An 11b 800 MS/s time-interleaved ADC with digital background calibration. In: IEEE international solid-state circuits conference, Digest of technical papers, pp 464–465
36. Hsu C-C (2007) A 7b 1.1 GS/s reconfigurable time-interleaved ADC in 90 nm CMOS. In: Proceedings of IEEE symposium on VLSI circuits, pp 66–67
37. Kobayashi T, Nogami K, Shirotori T, Fujimoto Y (1993) A current-controlled latch sense amplifier and a static power-saving input buffer for low-power architecture. IEEE J Solid-St Circ 28(4):523–527
38. Kuttner F (2002) A 1.2 V 10b 20MSample/s non-binary successive approximation ADC in 0.13 μm CMOS. In: IEEE international solid-state circuits conference, Digest of technical papers, vol 1, pp 176–177
39. Elbornsson J, Gustafsson F, Eklund J-E (2005) Analysis of mismatch effects in a randomly interleaved A/D converter system. IEEE Trans Circuits Syst I, Reg Papers 52(3):465–476

Chapter 5
CMOS Ultra-High-Speed Time-Interleaved ADCs

Jieh-Tsorng Wu, Chun-Cheng Huang, and Chung-Yi Wang

Abstract CMOS technologies have been able to fabricate ultra-high-speed time-interleaved (TI) ADCs that achieve a sampling rate over 10 GS/s. The TI architecture relaxes the speed requirement for each A/D channel. It also introduces inter-channel mismatches that cause conversion errors. These errors can be reduced by calibration. An 8-channel 6-bit 16-GS/s TI ADC is presented to illustrate several circuit design and calibration techniques. Each A/D channel is a 6-bit flash ADC. The low-power comparators in the flash ADC are latches with offset calibration. A delay-locked loop generates the 8-phase sampling clocks for the TI ADC. Timing-skew calibration is used to ensure uniform sampling intervals. Both the offset calibration and the timing-skew calibration run continuously in the background. This TI ADC was fabricated using a 65 nm CMOS technology. At 16 GS/s sampling rate, this chip consumes 435 mW from a 1.5 V supply. It achieves a signal-to-distortion-plus-noise ratio (SNDR) of 30.8 dB. The ADC active area is $0.93 \times 1.58\ \text{mm}^2$

5.1 Introduction

CMOS technologies have been able to fabricate ultra-high-speed analog-to-digital converters (ADCs) that provide sampling rates beyond 10 GS/s [1–6]. These are time-interleaved (TI) ADCs with multiple analog-to-digital (A/D) channels, such as 80 current-mode pipelined ADCs [1], 160 SAR ADCs [2, 4], 8 pipelined dual-path ADCs [3], and 8 flash ADCs [5, 6]. Advanced CMOS technologies provide two crucial circuits for ultra-high-speed TI ADCs: (1) input samplers using MOST switches and (2) multi-phase clock generators with a fine timing resolution.

J.-T. Wu (✉) • C.-C. Huang • C.-Y. Wang
Department of Electronics Engineering, National Chiao-Tung University,
1001 Ta-Hsueh Road, Hsin-Chu 300, Taiwan
e-mail: jt.wu@g2.nctu.edu.tw

A.H.M. van Roermund et al. (eds.), *Nyquist AD Converters, Sensor Interfaces, and Robustness: Advances in Analog Circuit Design, 2012*,
DOI 10.1007/978-1-4614-4587-6_5, © Springer Science+Business Media New York 2013

An ADC periodically samples an analog input and digitizes the magnitude of each sampled analog signal into a digital code. Both the input sampler and the magnitude digitizer must meet the speed and resolution requirements. The TI architecture relaxes the speed requirement for the magnitude digitizer. However, the requirements for the input sampler get harsher. The TI architecture also introduces inter-channel mismatches that cause analog-to-digital (A/D) conversion errors.

The remainder of this paper is organized as follows. Section 5.2 discusses the design issues for ultra-high-speed TI ADCs, which include input sampler, inter-channel mismatches and calibrations. Section 5.3 describes an 8-channel 6-bit 16-GS/s TI ADC. It is presented to illustrate several circuit design and calibration techniques. This ADC contains an offset calibration scheme and a timing-skew calibration scheme that can run continuously in the background. Experimental results will be presented. Finally, Sect. 5.4 draws conclusions.

5.2 Design Considerations for Time-Interleaved ADCs

Figure 5.1 shows the architecture of a TI ADC. It consists of M A/D channels. Each A/D channel comprises a MOST input sampler followed by a magnitude digitizer. In the m-th channel, the input $V_i(t)$ is sampled by clock ϕ_{m}, yielding $V_m[k]$. A digitizer converts the voltage $V_m[k]$ into a digital code $s_m[k]$. A multiplexer collects digital streams from all A/D channels, $s_1[k]$ to $s_M[k]$, and reconstructs the TI ADC digital output $s[l]$. The TI ADC requires a multi-phase clock generator that generates sampling clocks ϕ_1 to ϕ_{M}. For the TI ADC shown in Fig. 5.1, the effective sampling rate is $f_s = 1/T_s$, the single-channel conversion time is $T_c = M \times T_s$, and the sampling-mode time of the samplers is T_{sm}.

Design considerations specific to the TI architecture are described in the following subsections.

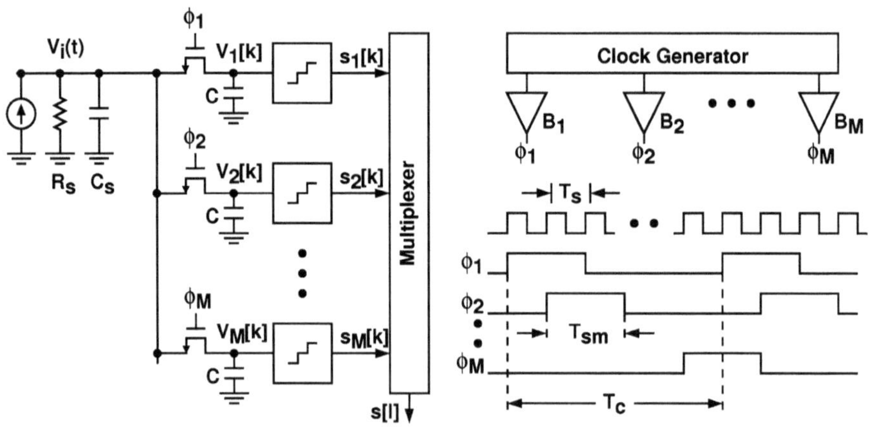

Fig. 5.1 Time-interleaved A/D architecture

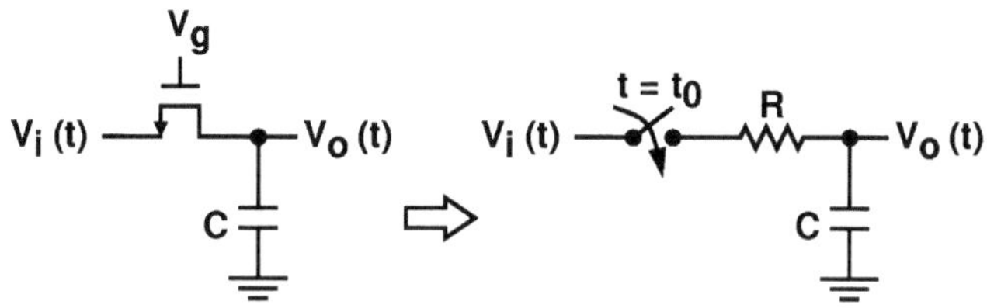

Fig. 5.2 A MOST sampler

5.2.1 MOST Input Sampler

Figure 5.2 shows a MOST sampler in the sampling mode. The MOST is modeled as a switch in series with a turned-on resistor R. Define $\tau_b = RC$. The signal bandwidth of the sampler is $1/\tau_b$.

If the input $V_i(t)$ is a sine wave of frequency ω_i and the MOST switch is turned on at $t = t_0$, then its output $V_o(t)$can be expressed as

$$V_o(t) = \frac{A\sin(\omega_i t - \theta)}{\sqrt{1+\omega_i^2\tau_b^2}} + \left[V_o(t_0) - \frac{A\sin(\omega_i t_0 - \theta)}{\sqrt{1+\omega_i^2\tau_b^2}}\right] \times e^{-(t-t_0)/\tau_b} \qquad \theta = \tan^{-1}(\omega_i\tau_b) \tag{5.1}$$

It consists of a sine wave shaped by the magnitude and phase response of the RC low-pass filter, and a residue exponentially decreasing with a time constant of τ_b. This residue must be smaller than the ADC magnitude resolution at the end of the sampling mode. To achieve B-bit resolution within t_{sm} sampling-mode time, we want

$$t_{sm} > \tau_b \times B \ln 2 = \tau_b \times 0.7B \tag{5.2}$$

Equation 5.2 specifies the minimum t_{sm}. A short t_{sm} is preferred, which reduces the total input loading of the TI ADC. For a minimum input loading, and for $t_{sm} = T_s = 1/f_s$, the sampler bandwidth $1/\tau_b$ must be larger than $0.7B \times f_s$. For a given technology, there is an achievable sampling rate and resolution. If a technology has a transition frequency of f_T and can provide a minimum τ_b of $10/(2\pi f_T)$, then $f_T \approx B \times f_s$. For example, a 65 nm CMOS technology with a $f_T = 200$ GHz can fabricate a 6-bit 33-GS/s TI ADC.

In Fig. 5.2, if the gate voltage V_g is a constant, then the turned-on resistor R varies with $V_i(t)$, resulting in distortion in $V_o(t)$ [7]. Gate-voltage bootstrapping techniques are often used to reduce R and keep R constant at the same time [8, 9]. The sampler is entered the hold mode by turning off the MOST switch. The accompanied clock feedthrough and charge injection should be minimized. The input sampler in an ADC is controlled by a periodic sampling clock. Jitter in the sampling clock yields sampling errors.

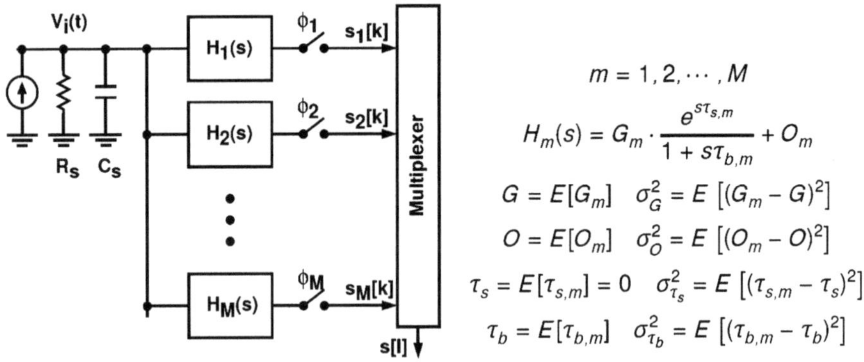

Fig. 5.3 TI ADC mismatch model

5.2.2 Inter-Channel Mismatches

A single A/D channel exhibits its own analog-to-digital transfer function, including conversion offset, conversion gain, and bandwidth. The channel bandwidth is determined by the τ_b of the input sampler and by the bandwidth of the buffer placed before or after the input sampler. In a TI ADC, mismatches of offset, gain, and bandwidth among the A/D channel introduce conversion errors [10–13]. Furthermore, the consecutive input sampling in a TI ADC is carried out by different A/D channels. All time intervals between any two consecutive sampling must be equal to T_s. The sampling switches in the input samplers are controlled by the multi-phase sampling clocks. The signal routes from the clock generator to the input samplers are usually long. Lots of mismatches can happen, resulting in different clock delay called timing skew. The timing skew causes uneven sampling intervals which result in sampling errors. These sampling errors reveal themselves as distortion. Assume a sine wave of frequency $f_i = \omega_i/(2\pi)$ is applied to a TI ADC input. Then, in the output spectrum of the ADC digital output, the inter-channel offset mismatch is manifested as spurious tones at frequencies $k \times f_c$, where $k = 1, 2, \cdots, M-1$. The timing skew and the inter-channel gain and bandwidth mismatches are manifested as spurious tones at frequencies $k \times f_c \pm f_i$, where $k = 1, 2, \cdots, M-1$.

Figure 5.3 shows a mismatch model for a TI ADC. For the m-th channel, its analog-domain characteristic is modeled by a s-domain transfer function $H_m(s)$,

$$H_m(s) = G_m \times \frac{e^{s\tau_{s,m}}}{1+s\tau_{b,m}} + O_m \tag{5.3}$$

where G_m is the channel conversion gain, O_m is the channel conversion offset, $1/\tau_{b,m}$ is the channel bandwidth, and $\tau_{s,m}$ is the timing skew. The average and standard deviation of the above parameters are also defined in Fig. 5.3. For a single-tone test, a sine wave of amplitude A and frequency ω_i is applied to the TI ADC input. In the corresponding output spectrum, the signal power at ω_i is denoted as P_{Signal} and the total power of the spurious tones is denoted as P_{Tones}. The ratio P_{Tones}/P_{Signal} can be found as

$$\frac{P_{\text{Tones}}}{P_{\text{Signal}}} = \left[\frac{\sigma_G^2}{G^2} + \omega_i^2\sigma_{\tau_s}^2 + \frac{\omega_i^2\sigma_{\tau_b}^2}{1+\omega_i^2\tau_b^2} + \frac{\sigma_O^2}{\frac{A^2}{2}\times\frac{G^2}{1+\omega_i^2\tau_b^2}}\right] \times \frac{M-1}{M} \tag{5.4}$$

The above equation quantifies the effect of inter-channel mismatches. The effects of both timing skew and bandwidth mismatch are input-frequency dependent. They can be treated as the same effect if $\omega_i\tau_b \ll 1$, i.e., the input frequency is much less than the signal-path bandwidth.

5.2.3 *Mismatch Calibration*

TI ADCs commonly employ calibration to suppress effects caused by crucial inter-channel mismatches. A calibration scheme contains two basic functions: (1) correction and (2) detection.

Correction involves modeling the mismatches with numerical parameters. Once those parameters are acquired, they are used to correct the errors caused by the mismatches. The correction can be done in the analog domain or digital domain. Most ultra-high-speed TI ADCs correct the mismatches in the analog domain through digital controls [1, 2, 4–6]. For example, timing skew can be corrected by adjusting the phases of the sampling clocks. The A/D conversion errors caused by mismatches can be also corrected by manipulating the digital output data from the A/D channels. Corrections for offset and gain mismatches are straightforward and simple. However, corrections for timing skew and bandwidth mismatches are complex [14–20]. The digital hardware requires large power when operating at high speed.

Detection is to measure the inter-channel mismatches and acquire the necessary parameters for correction. Detection involves applying calibration signals to the input of the ADC and comparing the digital outputs from all A/D channels. For foreground detection, normal A/D operation is stopped, then, a calibration signal of known characteristic is applied [1–4, 16]. Detection is straightforward with known calibration signal.

For background detection without interrupting the ADC's normal operation, the regular ADC input usually serves as the calibration signal [5, 14, 18, 19, 21–24]. However, robust background detection requires specific input conditions. For example, timing-skew detection will fail if the input is a dc signal. Some designs add additional hardware to facilitate detection [5, 6, 25].

Criteria to evaluate a calibration scheme include:

1. Foreground or background. Can the application provide the ADC with spare time to make foreground detection so that the ADC is adaptable to environmental variations?
2. Hardware overhead. What analog and digital circuits are added? Is ADC performance, such as speed and power, deteriorated by the overhead?
3. Signal range overhead. How much extra signal range is required if an additional calibration signal is added to the signal path? This is crucial for circuits already operating at low-voltage supplies.
4. Component matching requirement. Does the calibration scheme hint any matching requirement for circuit components? Is the calibration sensitive to offset or other effects due to mismatches?
5. Detection robustness. Does the detection require specific input condition?
6. Calibration agility. How long does it take for the calibration process to converge? Background detection usually needs to collect huge amount of data to remove unwanted interferences, resulting in slow calibration.

5.3 A 6-Bit 16-GS/s 8-Channel TI ADC

In this section, we use a 6-bit 16-GS/s TI ADC to illustrate several design and calibration techniques. Section 5.3.1 describes the TI ADC architecture. Section 5.3.2 describes the flash A/D channel, and presents the proposed offset calibration. Section 5.3.3 describes the multi-phase clock generator, and presents the proposed timing-skew calibration. This ADC was fabricated using a 65 nm CMOS technology. Section 5.3.4 shows the experimental results.

5.3.1 TI ADC Architecture

Figure 5.4 shows the architecture of the TI ADC. It comprises 8 A/D channels, ADC_1 to ADC_8. The A/D channels are driven respectively by 8-phase clocks, ϕ_1 to ϕ_8. For this TI ADC, number of channels $M = 8$, clock frequency $f_c = 2$ GHz, clock period $T_c = 1/f_c = 500$ ps, sampling rate $f_s = 8f_c = 16$ GS/s, and sampling interval $T_s = 1/f_s = 62.5$ ps.

All A/D channels are identical flash ADCs. Each flash ADC includes a resistor string to generate reference voltages to be compared with the analog input. In fully-differential configuration, the top and bottom references are $V_{RT} = +400$ mV and $V_{RB} = -400$ mV respectively. Thus, the ADC quantization step size is $\Delta V_R = 800/62 \approx 12.9$ mV. Since all flash A/D channels share the same reference voltages, there is no inter-channel offset and gain mismatches.

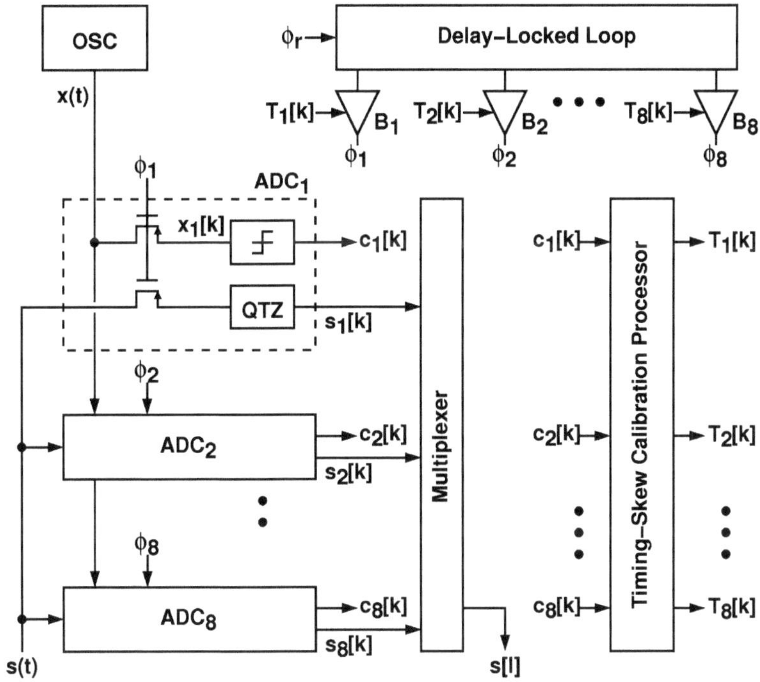

Fig. 5.4 A 6-bit TI ADC

In this TI ADC, a delay-locked loop (DLL) receives a reference clock ϕ_r of frequency f_c and generates 8-phase sampling clocks, ϕ_1 to ϕ_8, of the same frequency. The 8 equally phase-spaced sampling clocks are delivered to the analog samplers in the A/D channels through clock buffers, B_1 to B_8. Due to the device variations in the DLL and the clock buffers, and also due to the mismatches among the clock distribution routes, the clocks may reach their respective samplers with different delays. This phenomenon is called timing skew. Because of timing skew, the phases of the sampling clocks received by the samplers in the A/D channels may no longer be equally spaced. As a result, the TI ADC experiences a periodic variation of sampling intervals. This TI ADC includes a timing-skew calibration processor (TSCP) that automatically adjusts the delay of the 8 clock buffers to ensure sampling interval uniformity.

Analogous to the ADC magnitude resolution ΔV_R, we define the ADC timing resolution as $\Delta T_R = T_s/2^6 \approx 1$ psec. Let the timing skew of a TI ADC be minimized by adjusting the delay of the clock buffers with a control step size of ΔT_R. If the ADC input is a sine wave with a full-scale $2^6 \Delta V_R/2$ amplitude and $f_s/2$ frequency, the sampling error power due to timing skew is similar to the quantization noise power due to magnitude digitization.

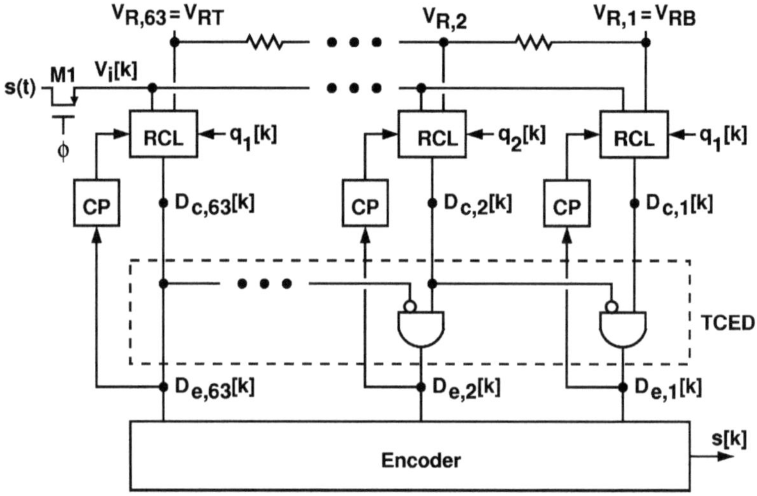

Fig. 5.5 A 6-bit flash A/D channel

5.3.2 *Flash A/D Channel*

Figure 5.5 shows the block diagram of a single A/D channel. It is a flash ADC consisting of 63 comparators with offset calibration. Each comparator comprises a random-chopping latch (RCL) and a calibration processor (CP). Preceding the comparators is a p-channel MOST M1 that functions as an analog input sampler for the ADC.

The reference voltages $V_{R,1}$ to $V_{R,63}$ are generated by using a resistor string. The comparator outputs, $D_{c,1}[k]$ to $D_{c,63}[k]$, are fed into a thermometer-code edge detector (TCED) to generate an edge code, $D_{e,1}[k]$ to $D_{e,63}[k]$. The edge code indicates the location of the 1-to-0 transition edge in $D_{c,1}[k]$ to $D_{c,63}[k]$ of thermometer coding. The TCED is followed by an encoder, which converts the edge code into a Gray code and then converts the Gray code into a binary code.

Figure 5.6 shows the RCL schematic. This comparator employs only regenerative latches for the comparison function. A conventional high-speed comparator usually comprises a regenerative latch preceded by a preamplifier. The latch is power efficient for the comparison function. However, it exhibits an input-referred offset due to device mismatch. The gain of the preamplifier relaxes the offset requirement for the latch. Sometimes offset cancellation techniques, such as capacitor offset storage [26] or spacial filtering [27], are used to reduce the preamplifier offsets. In our design, preamplifiers are removed to save power. As shown in Fig. 5.6, the comparator is a cascade of three latches. The offset of the first latch is adjustable. Its variable-offset control is achieved by changing the loading and pulling strength on nodes V_{a1} and V_{a2} [28]. There are 16 equally-weighted n-channel MOSFET varactor

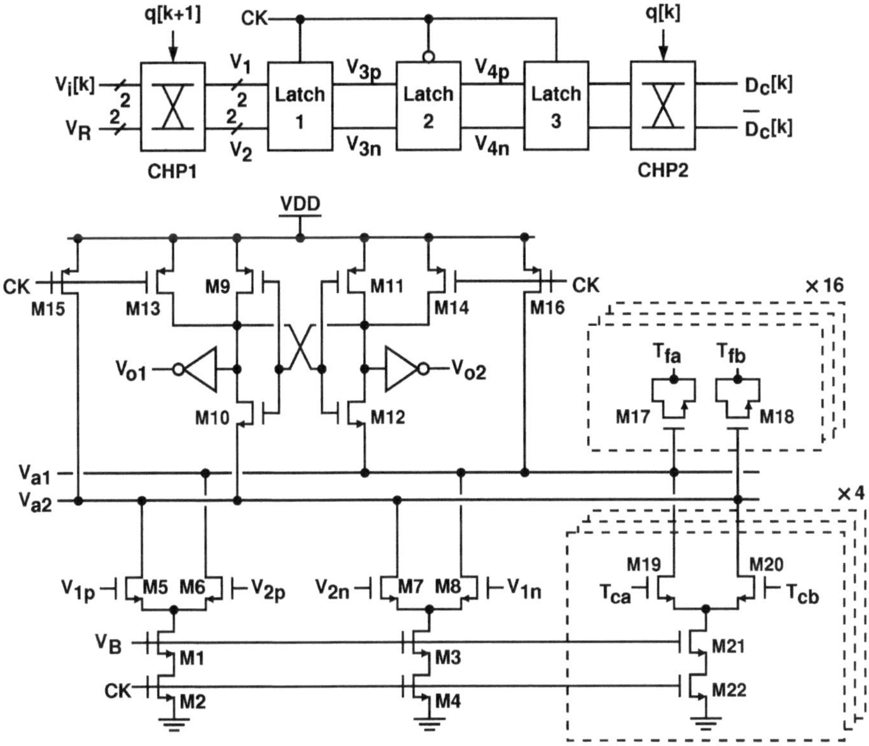

Fig. 5.6 Random-chopping latch (RCL) schematic

pairs (M17–M18) for fine control and 4 equally-weighted pulling current sources (M19–M22) for coarse control. The capacitance of the varactors is varied by switching the voltage on the source and drain nodes. The offset control signals are T_{ca} and T_{cb} for each coarse control current source and T_{fa} and T_{fb} for each fine control varactor pair. They are all full-swing digital signals. From Monte Carlo simulations, the offset variation of the latch has a standard variation of 28.5 mV. The offset coarse control can change $\pm$ 4steps with a step size of 32 mV. The offset fine control can change $\pm$ 16steps with a step size of $\Delta V_{OS} = 3.2$ mV $= (1/4)\Delta V_R$.

A statistics-based offset calibration technique is used to digitally adjust the offset of the first latch in the RCL to eliminate the offset of the overall comparator [29]. Figure 5.7 shows the calibration configuration. The RCL contains two random choppers, CHP1 and CHP2. They are added to detect the offset. They are controlled by the same binary random sequence $q[k] \in \{+1, -1\}$. When $q[k] = +1$, the chopper passes its two inputs to its two corresponding outputs directly. When $q[k] = -1$, the signal paths to the two outputs are interchanged. The chopper CHP1 consists of four analog switches. The chopper CHP2 consists of digital logic gates. Regardless of the q[k] value, the RCL works like a normal comparator, detecting the polarity of $V_i - V_R$ and generating a digital output $D_c[k] \in \{0, 1\}$ accordingly.

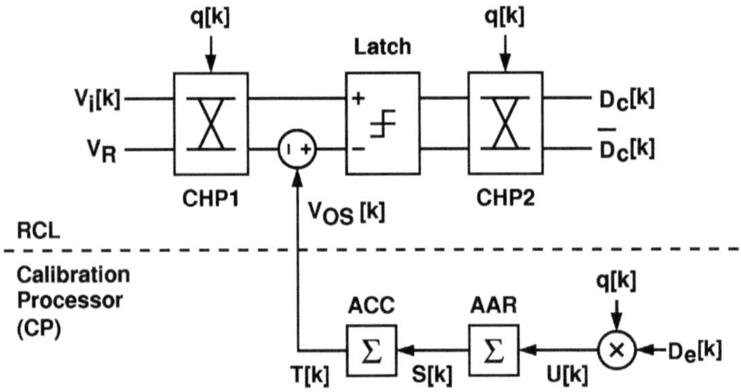

Fig. 5.7 Offset calibration configuration

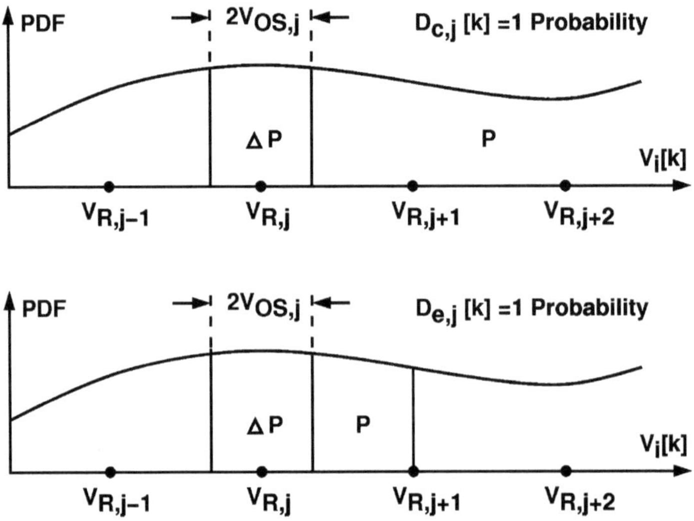

Fig. 5.8 Principle of offset detection

In Fig. 5.7, the RCL is accompanied by a calibration processor (CP). The CP digital output $T[k]$ controls the comparator offset $V_{OS}[k]$ as $V_{OS}[k] = V_{OS,0} + \Delta V_{OS} \times T[k]$, where $V_{OS,0}$ is the comparator offset before calibration, and ΔV_{OS} is the offset control step size. The CP detects the polarity $V_{OS}[k]$ and adjusts $T[k]$ to minimize $|V_{OS}[k]|$.

Figure 5.8 shows the principle of background offset detection. The magnitude of the sampled input $V_i[k]$ is assumed to have stationary probability density function (PDF). For the j-th comparator connected to reference $V_{R,j}$, it exhibits an offset of $V_{OS,j}$. Figure 5.8 assumes $V_{OS,j}>0$. The probability of the comparator output $D_{c,j}[k] = 1$ is illustrated in the upper plot of Fig. 5.8. When $q[k] = +1$, the probability is area P. When $q[k] = -1$, the probability becomes area $P + \Delta P$.

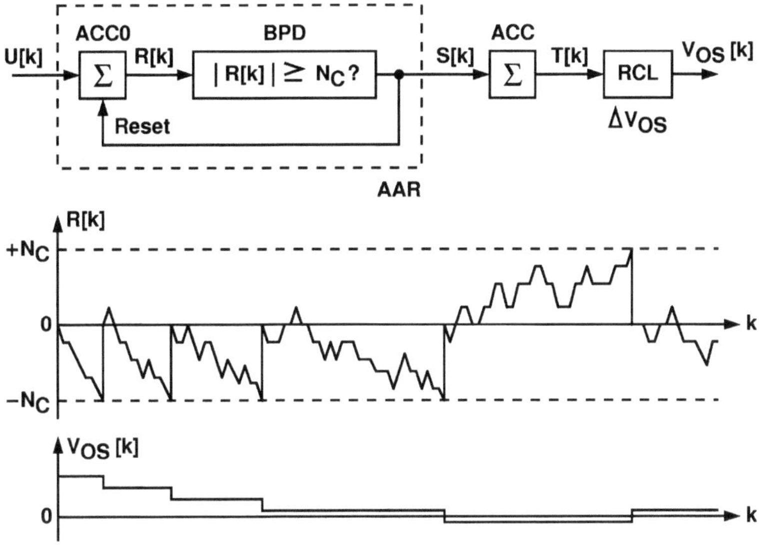

Fig. 5.9 CP Operation

The area ΔP is proportional to $V_{OS,j}$. Since the PDF of $V_i[k]$ is not known, the CP is designed to detects only the polarity of $V_{OS,j}$ by finding the polarity of ΔP.

Figure 5.9 shows the operation of the CP. It includes an accumulation-and-reset (AAR) block to detect the ΔP polarity. Its input is the product of comparator output and $q[k]$, i.e., $U[k] = D_c[k] \times q[k]$. The average of $U[k]$ is proportional to ΔP. The AAR uses an accumulator (ACC0) to integrate $U[k]$, yielding $R[k]$. A bilateral peak detector (BPD) constantly compares $R[k]$ against a positive threshold $+N_C$ and a negative threshold $-N_C$. If $-N_C < R[k] < +N_C$, its output is $S[k] = 0$. Whenever $R[k] \geq +N_C$, it issues an output $S[k] = +1$ for one clock cycle and then resets $R[k]$ to 0. Whenever $R[k] \leq -N_C$, it issues an output $S[k] = -1$ for one clock cycle and also resets $R[k]$ to 0. The CP output $T[k]$ is generated by accumulating $S[k]$. The CP contains only two simple accumulators. There is no switching in ACC0 and ACC if $D_c[k] = 0$. The output $T[k]$ changes only when the polarity of $V_{OS}[k]$ is detected with confidence.

If the input sample $V_i[k]$ appears uniformly across the entire ADC input range, V_{FS}, the offset calibration loop can be modeled as a single-pole feedback system with a time constant

$$\tau_{c,os} = N_C \times \frac{V_{FS}}{\Delta V_{OS}} \times T_c \tag{5.5}$$

This time constant indicates the calibration agility. For examples, consider the RCL of Fig. 5.6. Its initial offset standard variation is 28.5 mV, which is equal to $2.2\Delta V_R$. It will takes $\ln(2.2/0.25) \times \tau_{c,os} = 2.17\tau_{c,os}$ for the calibration to decrease the offset standard variation from $2.2\Delta V_R$ to $0.25\Delta V_R$.

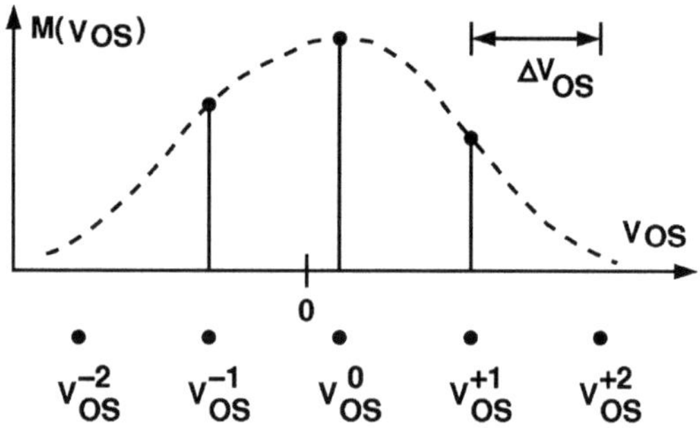

Fig. 5.10 Probability mass function of $V_{OS}[k]$

As shown in Fig. 5.9, after the calibration has converged, $V_{OS}[k]$ fluctuates around zero. This is due to the granular property of the digital offset control. Figure 5.10 shows a probability mass function (PMF) of $V_{OS}[k]$ around zero. Possible values for $V_{OS}[k]$ are V_{OS}^{0}, $V_{OS}^{\pm 1}$, $V_{OS}^{\pm 2}$, etc. They are spaced by the offset control step size ΔV_{OS}. The standard deviation of the V_{OS} fluctuation can be found from the PMF.

Note that the PMF varies with the position of 0. We define $\sigma(V_{OS})$ as the standard deviation of $V_{OS}[k]$ averaged over possible position of 0. The averaged standard deviation $\sigma(V_{OS})$ is a function of the offset control step size ΔV_{OS} and the AAR threshold N_C. For $N_C > 4(\Delta V_R/\Delta V_{OS})$, $\sigma(V_{OS}) \approx \Delta V_{OS}/\sqrt{6}$.

This $V_{OS}[k]$ fluctuation can be treated as a noise imposed on the ADC input. The offset standard deviation $\sigma(V_{OS})$ should be less than that of the ADC quantization noise, i.e., $\sigma(V_{OS}) < \Delta V_R/\sqrt{12}$. It can be shown that this $V_{OS}[k]$ fluctuation is related to the area P illustrated in Fig. 5.8. Large P leads to large $\sigma(V_{OS})$, for which the calibration needs large N_C to suppress. To reduce P, we use the TCED output $D_{e,j}[k]$ as the CP input, replacing the comparator output $D_{c,j}[k]$. As shown in Fig. 5.8, the ΔP region remains the same but the P region is drastically shrunk. The area P for $D_{e,j}[k] = 1$ is from $V_{R,j} + V_{OS,j}$ to $V_{R,j+1}$.

For this 6-bit ADC design, $V_{FS} = 2^6 \times \Delta V_R$, $N_C = 16$, and $\Delta V_{OS} = 0.25\Delta V_R$, yielding $\tau_{c,os} = 4096T_c = 2\mu\text{sec}$ and $\sigma(V_{OS}) = 0.13\Delta V_R$.

5.3.3 Multi-phase Clock Generator

Figure 5.11 shows the multi-phase clock generator that generates the 8 equal-phase-spaced sampling clocks, ϕ_1 to ϕ_8. It contains a delay-locked loop (DLL), comprising a phase detector (PD), a charge pump (CP), and 8 identical variable-delay delay cells, D1 to D8. The 8 sampling clocks from the DLL are delivered to the A/D channels through 8 variable-delay clock buffers, B_1 to B_8.

Figure 5.12 shows the schematic of a variable-delay clock buffer. It is a cascade of three inverters. Its delay is adjusted by changing the capacitance of the MOS varactors attached to the outputs of the first and second inverters. Both n-channel and p-channel MOSTs are used as varactors. The capacitances of the varactors are

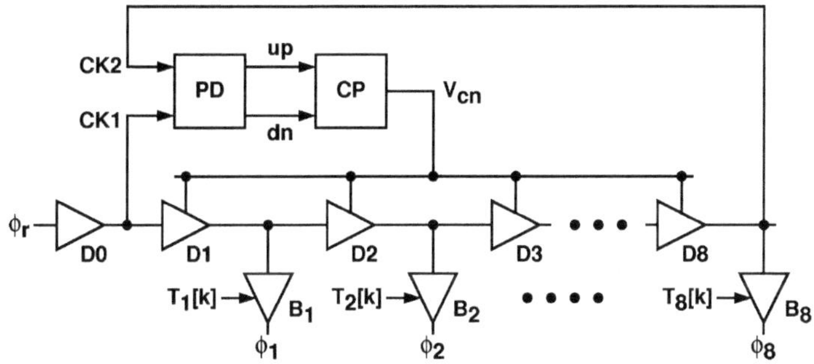

Fig. 5.11 DLL multi-phase clock generator

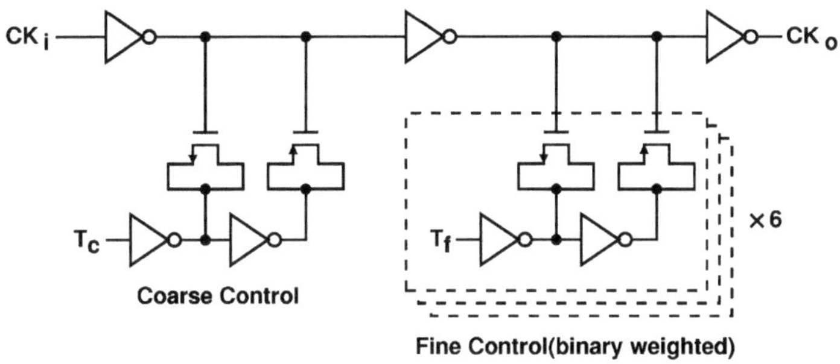

Fig. 5.12 Variable-delay clock buffer

binary weighted. They are controlled by digital control signals, T_c and T_f. The digital delay control has a step size of $\mu_t = 0.4$ ps.

The 8 sampling clocks from the clock generator are sent to the 8 A/D channels respectively. Each clock controls an input sampler. Timing skews occur due to the device variation in the DLL and in the 8 clock buffers, and also due to the mismatches among the clock distribution routes. We use calibration to minimize the timing skew. Figure 5.13 shows the configuration for timing-skew calibration [30]. A calibration signal x(t) is sampled by the 8 sampling clocks, yielding $x_1[k]$ to $x_8[k]$. The sampled data are used to detect the timing skew. The calibration generates digital control signals, $T_1[k]$ to $T_8[k]$. They control the delay of clock buffers, B_1 to B_8, to minimize the skew.

The principle of timing-skew detection is based on the zero-crossing (ZC) detection [30]. Figure 5.14 shows the concept of ZC in a four-phase sampling system. A ZC occurs when x(t) changes polarity. In Fig. 5.14, there is a ZC between $x_2[0]$ and $x_3[0]$, and another one between $x_2[1]$ and $x_3[1]$. If x(t) is a periodic signal and asynchronous to the sampling clocks, the probability of a ZC between two adjacent samples, $x_j[k]$ and $x_{j+1}[k]$, is $2f_x \times T_s$, where f_x is the frequency of x(t) and T_s is the sampling interval between $x_j[k]$ and $x_{j+1}[k]$. Thus, the number of ZCs between $x_j[k]$ and $x_{j+1}[k]$ over a fixed time period is proportional to their sampling interval.

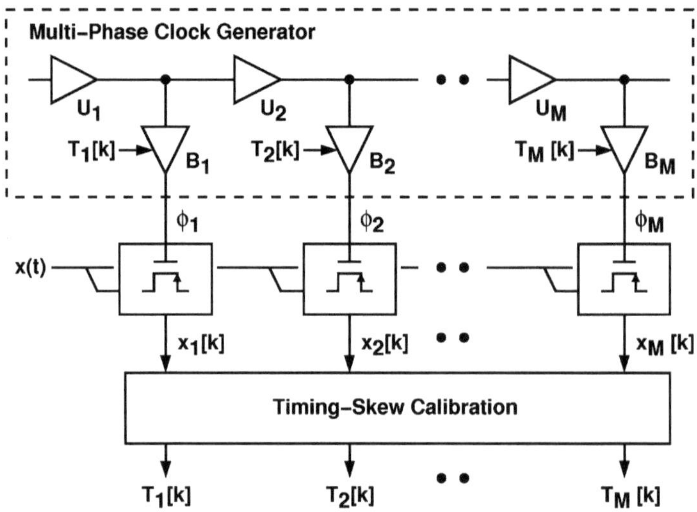

Fig. 5.13 Timing-skew calibration configuration

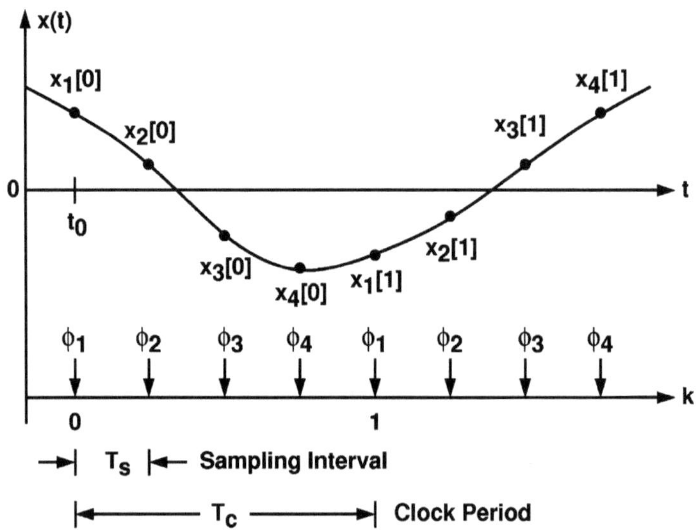

Fig. 5.14 Zero crossing (ZC)

Figure 5.15 shows a simple ZC detector, ZCD1. The polarities of $x_j[k]$ and $x_{j+1}[k]$ are determined by two comparators, yielding $c_j[k] \in \{0, 1\}$ and $c_{j+1}[k] \in \{0, 1\}$. The output $z_j[k] = 1$ when $c_j[k] \neq c_{j+1}[k]$, i.e., when a ZC occurs. ZCD1 is sensitive to comparator offsets. If the two internal comparators exhibit offsets, detection errors may occur. Figure 5.16 shows a ZC detector, ZCD2, which is less sensitive to the comparator offsets. Each of the comparator outputs first goes through an 1-bit

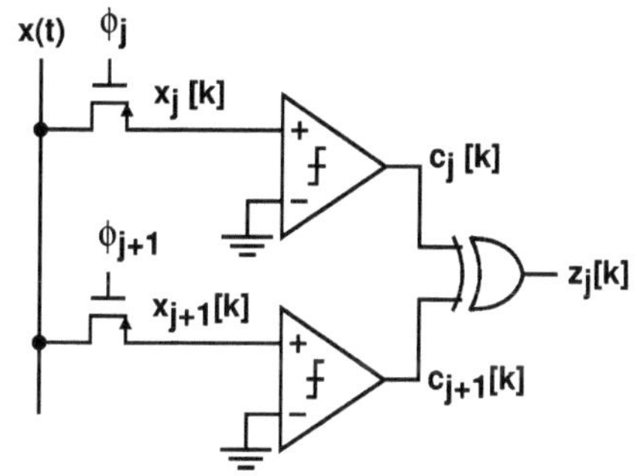

Fig. 5.15 ZC detector, ZCD1

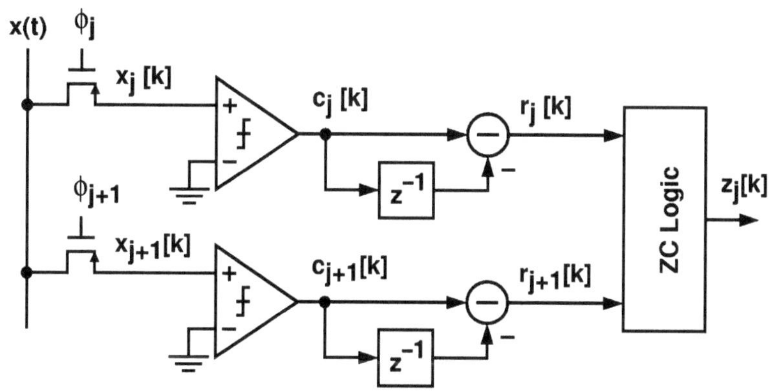

Fig. 5.16 ZC detector, ZCD2

high-pass filter, $1 - z^{-1}$, yielding $r_j[k] \in \{-1, 0, +1\}$ and $r_{j+1}[k] \in \{-1, 0, +1\}$. The output $z_j[k] = 1$ if $r_j[k] \times r_{j+1}[k] \leq 0$, otherwise $z_j[k] = 0$. In other words, $z_j[k] = 0$ if both $r_j[k]$ and $r_{j+1}[k]$ are +1 or both are -1. ZCD2 is no longer a simple detector for ZCs in x(t). It detects events that yield $z_j[k] = 1$. However, its behavior in the timing-skew calibration is similar to that of a ZCD1.

The TI ADC shown in Figure 5.4 uses the above ZC detection scheme in its timing-skew calibration. To facilitate background calibration, an on-chip clock generator x(t) is added as the calibration signal. In each A/D channel, x(t) is sampled by a replica sampler similar to the s(t) sampler in the flash ADC. A comparator is used to determine the polarity of the sampled signal $x_j[k]$, yielding $c_j[k]$. A timing-skew calibration processor (TSCP) collects the digital data $c_1[k]$ to $c_8[k]$ from all A/D channels. It detects the sampling intervals among the sampling clocks, ϕ_1 to ϕ_8. Its digital outputs, $T_1[k]$ to $T_8[k]$, adjust the delay of the clock buffers, B_1 to B_8, such that all sampling intervals can remain uniform. The calibration does not use the regular ADC input s(t) as the calibration signal. A separate calibration signal x(t) ensures robust calibration. Only the polarity of x(t) is used in

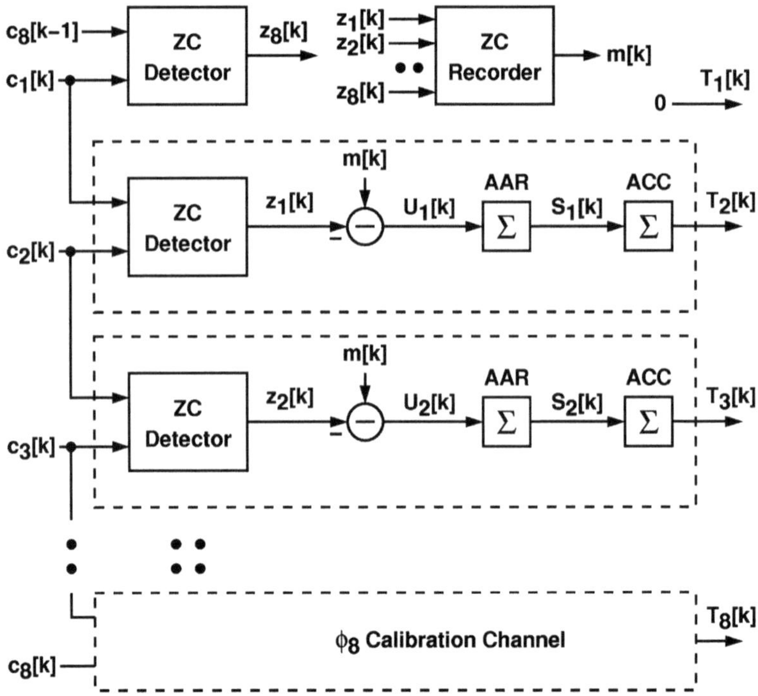

Fig. 5.17 Timing-skew calibration processor

the calibration. Its frequency and waveform shape are not crucial. In each A/D channel, the s(t) sampler and the x(t) sampler should be placed in close proximity to minimize mismatch. This mismatch will introduce uncorrected timing skew.

Figure 5.17 shows the TSCP block diagram. It chooses ϕ_1 as the reference phase by setting $T_1[k] = 0$. All other clocks are aligned to ϕ_1. There are seven calibration channels controlling the phase of clocks ϕ_2 to ϕ_8. Consider the ϕ_2 calibration channel which generates $T_2[k]$. Its ZC detector is identical to the ZCD2 shown in Fig. 5.16. The average of its output $z_1[k] \in \{0, 1\}$ represents the sampling interval between ϕ_1 and ϕ_2. The measured sampling interval is compared against the nominal sampling interval, which is the average of $m[k] \in \{0, 1\}$ generated from a ZC recorder. The polarity of the averaged $(m[k] - z_1[k])$ is extracted by an accumulation-and-reset (AAR) similar to the one used in offset calibration, but with a threshold of N_T. The AAR output $S_1[k]$ updates the following accumulator (ACC) whose content is $T_2[k]$. The signal $T_2[k]$ controls the delay of clock buffer B_2 with a delay-control step size of μ_t.

The $m[k]$ sequence represents the average of the ZC occurrences among all sampling intervals. The $m[k]$ is generated by the ZC recorder shown in Fig. 5.18. It counts every ZC in x(t), and issues an $m[k] = 1$ every 8 ZCs. The recorder accumulates all ZCs from all ZC detectors. The ZC detector in the top-left corner of Figure 5.17 is added to detect the ZCs in the missing interval between ϕ_1 and the

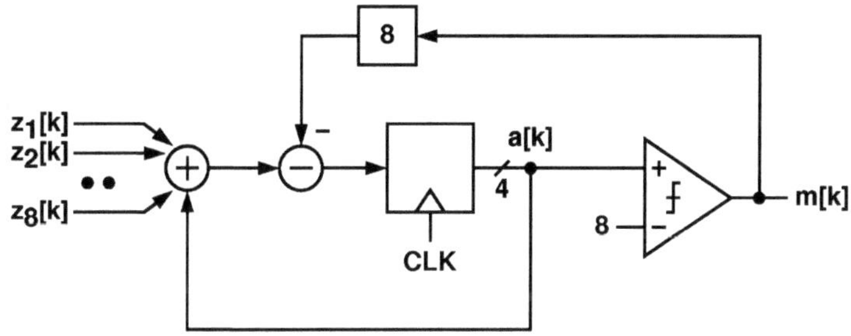

Fig. 5.18 Variable-delay clock buffer

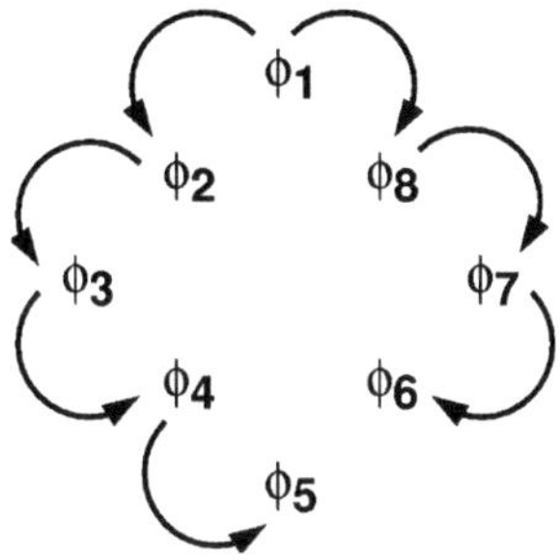

Fig. 5.19 Clock alignment scheme

ϕ_8 immediately before ϕ_1. A comparator compares the accumulation result $a[k]$ with integer 8, yielding a binary $m[k] \in \{0, 1\}$ every clock cycle. Whenever the comparator issues a $m[k] = 1$, an amount of 8 is subtracted from $a[k]$during the following clock cycle. The digital stream $m[k]$ is a sequence of 0 and 1. Its mean value represents the nominal sampling interval. The proposed ZC recorder is simple and its hardware cost is low.

The dynamic behavior of this timing-skew calibration is similar to that of the offset calibration. The calibration loop in each calibration channel can be modeled as a single-pole feedback system with a time constant of

$$\tau_{c,ts} = N_T \times \frac{1}{2f_x \times \mu_t} \times T_c \tag{5.6}$$

where N_T is the AAR threshold, μ_t is the delay-control step size, and $f_x$is the frequency of the calibration signal x(t).

After the calibration has converged, the timing skew fluctuates around zero similar to Fig. 5.10. We define $\sigma(\tau)$ as the averaged standard deviation of the skew fluctuation. This $\sigma(\tau)$ is a function of the skew control step size μ_t and the AAR threshold N_T. For $N_T > 2(T_s/\mu_t)$, $\sigma(\tau) \approx \mu_t/\sqrt{6}$.

Figure 5.19 shows the clock alignment scheme. For clock ϕ_j, its timing skew is denoted as $\tau_j[k]$. The average of $\tau_j[k]$ is 0 and the standard deviation of $\tau_j[k]$ is

$\sigma(\tau_j)$. Clock ϕ_1 is used as reference. Its phase is not adjusted by calibration, thus $\sigma(\tau_1) = 0$. Both clocks ϕ_2 and ϕ_8 are calibrated and aligned to ϕ_1, yielding $\sigma(\tau_2) = \sigma(\tau_8) = \sigma(\tau)$. Clock ϕ_3 is aligned to ϕ_2 and clock ϕ_7 is aligned to ϕ_8. We have $\sigma(\tau_3) = \sigma(\tau_7) = \sqrt{2} \times \sigma(\tau)$. It can be found that $\sigma(\tau_4) = \sigma(\tau_6) = \sqrt{3} \times \sigma(\tau)$ and $\sigma(\tau_5) = \sqrt{4} \times \sigma(\tau)$. The overall averaged timing skew is defined as

$$\sigma(\tau_T) = \sqrt{\frac{1}{8}[\sigma^2(\tau_1) + \sigma^2(\tau_2) + \cdots + \sigma^2(\tau_8)]} = 2\sigma(\tau) \tag{5.7}$$

The skew fluctuation causes noise-like conversion errors. As a rule of thumb, we want $\sigma(\tau_t) < \Delta T_R / \sqrt{12}$, where ΔT_R is the ADC timing resolution.

For this 6-bit 8-channel TI ADC, the clock frequency is $f_c = 2$ GHz, the clock period is $T_c = 500$ ps, and sampling interval is $T_s = 62.5$ ps. The timing resolution is $\Delta T_R = T_s/2^6 \approx 1$ ps. We choose x(t) frequency $f_x = 500\text{MHz}$, $N_T = 2^{10}$, and $\mu_t = 0.25\Delta T_R$, yielding a timing-skew calibration time constant $\tau_{c,ts} = 2^{22} T_c = 2.$ 1 ms and a skew fluctuation standard deviation $\sigma(\tau_T) = 0.22\,\Delta T_R$.

5.3.4 Experimental Results

This TI ADC was fabricated using a 65 nm CMOS technology. All ADC circuits are realized with standard MOSTs. The supply voltage is raised to 1.5 V to obtain a better SNDR performance out of the s(t) samplers and to increase the speed of the comparators. Figure 5.20 shows the chip micrograph. The ADC active area is 0.9 3 × 1.58 mm^2. To ensure that all A/D channels exhibit identical conversion gain, the external reference V_{RT} and V_{RB} shown in Fig. 5.5 must be the same when

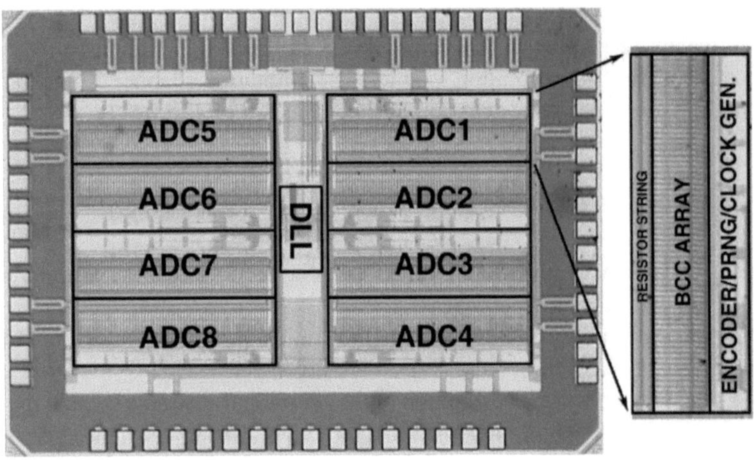

Fig. 5.20 TI ADC chip micrograph

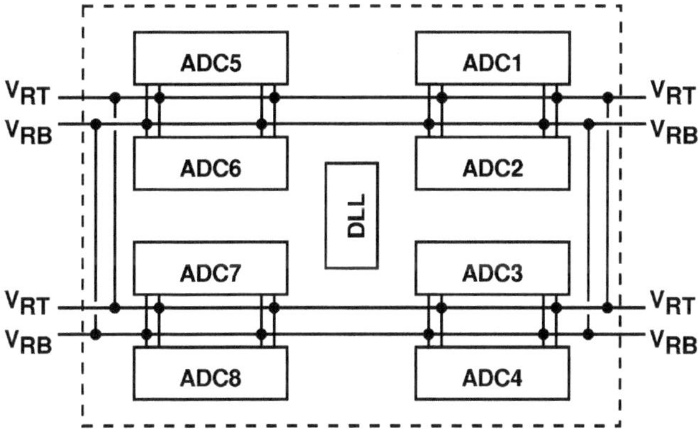

Fig. 5.21 Voltage reference floorplan

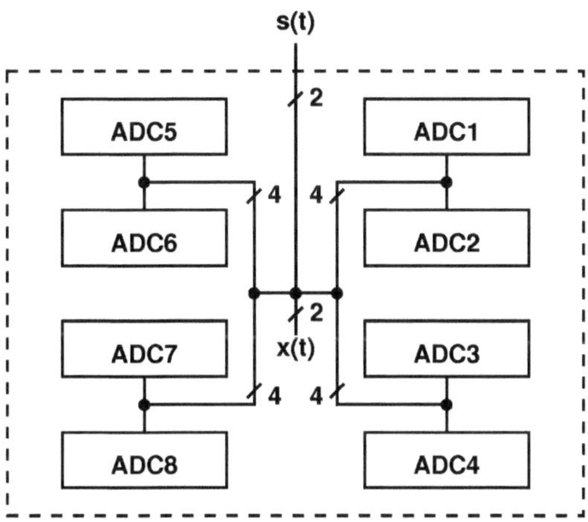

Fig. 5.22 Signal routes floorplan

received by each A/D channel. Figure 5.21 shows the floorplan for the V_{RT} and V_{RB} routes. These reference routes are realized with multi-layer metals to reduce resistances. The TI ADC also requires that all A/D channels receive the same analog input, which is the signal s(t) shown in Fig. 5.4. Figure 5.22 shows the floorplan for the s(t) signal routes. The differential s(t) is first directed to the center of the chip through two metal lines. It is then sent to each A/D channel through routes of identical length and shape.

The multi-phase clock generator, including DLL and clock buffers, is placed near the center of the chip. The on-chip timing-skew calibration processor can correct timing skews among the clocks caused by devices mismatches and clock routes mismatches. However, the calibration requires that the x(t) samplers in all A/D channels receive the same calibration signal x(t). As shown in Fig. 5.22, the x(t)

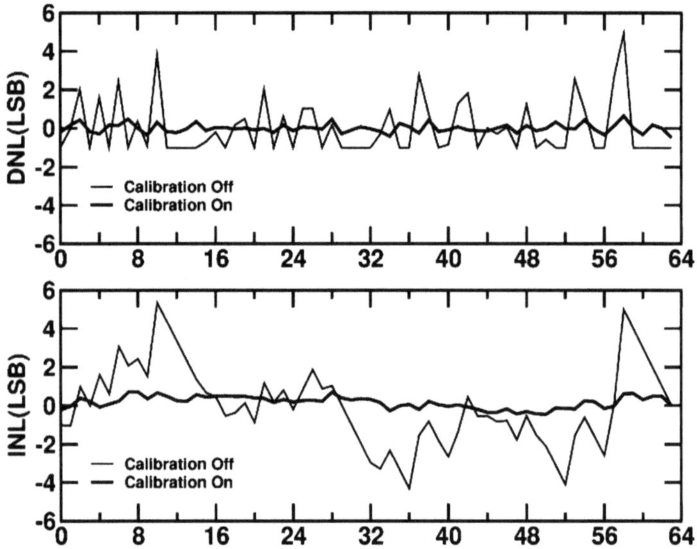

Fig. 5.23 Measured DNL and INL

generator is located near the center of the chip. It is a simple free-running ring oscillator. The differential x(t) is sent to each A/D channel through routes similar to those for s(t). The x(t) and s(t) routes are shielded separately to avoid coupling between the two signals. Finally, the timing-skew calibration dictates that the s(t) and x(t) samplers in the same A/D channel have the same turn-off instants. They are placed in close proximity. To avoid leaking x(t) to the s(t) sampler, both samplers are surrounded by separate guard rings. The two samplers are driven by the same clock driver, whose output impedance is made low to minimize x(t) leakage.

The reference clock ϕ_r of frequency $f_c = 2$ GHzis generated from an off-chip signal generator. It is converted into a differential signal using a power splitter. The ADC test input signal s(t) is generated from another signal generator synchronized with ϕ_r. The ADC chip is mounted directly on a printed circuit board. Digital outputs from all A/D channels, $s_1[k]$ to $s_8[k]$, are first downsampled by a ratio of 1/64 and then sent off-chip to a logic analyzer. The final TI ADC digital output stream s[l] is constructed by resampling the acquired data. The equivalent down-sampling ratio is 1/64.125.

Figure 5.23 shows the measured differential nonlinearity (DNL) and integral nonlinearity (INL) of a single A/D channel. Before activating the calibration, the DNL is -1.0/+4.9 LSB and the INL is -4.3/+5.4 LSB. There are missing codes. After activating the offset calibration, the DNL becomes -0.5/+0.6 LSB and the INL is reduced to -0.4/+0.7 LSB.

Figure 5.24 is the measured ADC output spectra with and without the timing-skew calibration. The sampling rate is 16 GS/s. The input signal is a full-swing 2.9 GHz sine wave. Without the calibration, there are many spurious tones caused by timing skews.

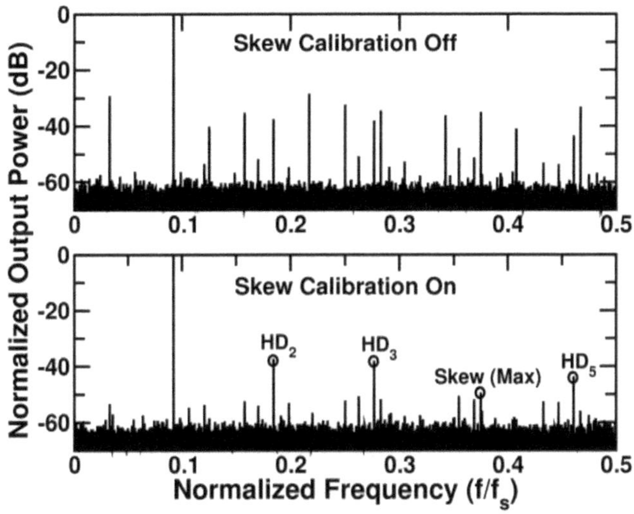

Fig. 5.24 Measured output spectra

After activating the timing-skew calibration, most of skew-related spurious tones are eliminated. Note that the locations of the skew-related spurious tones are shuffled because of downsampling and resampling of the output codes. The remaining harmonic tones in the spectrum is mainly due to the non-ideal input signal paths, including the distortion of the power splitter and the mismatches of the wire parasitics and the sampling switches. The harmonic distortion of the ADC can be improved by employing a chip layout of better symmetry and using better signal sources.

Figure 5.25 shows the measured TI ADC signal-to-distortion-plus-noise ratio (SNDR) versus input frequencies. The sampling rate is 16 GS/s. The effective resolution bandwidth (ERBW) is 3 GHz, which is limited by the bandwidth of the ADC input sampling switches. At frequencies near ERBW, the SNDR is improved from 19.8 to 28.0 dB by the timing-skew calibration.

Table 5.1 summarizes the measured specifications of this TI ADC chip. The input capacitance is 1.8 pF for each input pin. The power consumption is 435 mW, excluding I/O. Each A/D channel consumes 54 mW. Most of the dissipated power is dynamic power.

Table 5.2 compares this work with other recently published TI ADCs that achieve a sampling rate exceeding 10 GS/s. In this table, the ADC figure-of-merit (FOM) is defined as

$$\mathrm{FOM} = \frac{\mathrm{Power}}{2^{\mathrm{ENOB}} \times 2\mathrm{ERBW}} \tag{5.8}$$

where ENOB is the effective number of bits at low input frequencies, and ERBW is the effective resolution bandwidth at which ENOB drops by 0.5 bit. FOM for this work is 2.6 pJ/conversion-step. The competitive FOM of this chip is obtained by

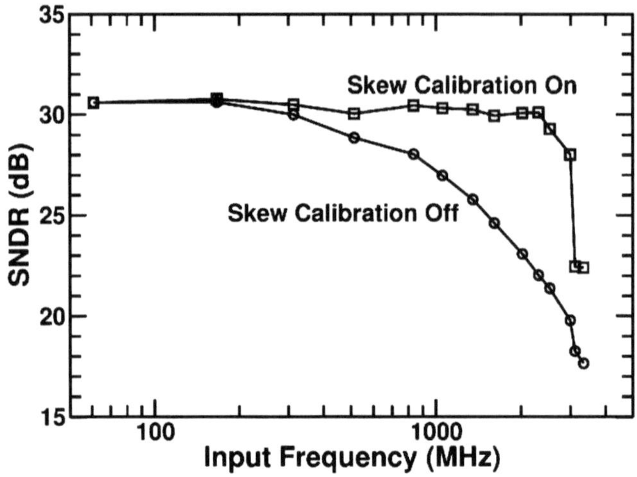

Fig. 5.25 Measured DNL and INL

Table 5.1 Performance summary

Technology	65 nm CMOS
Resolution	6 bit
Input loading	1.8 pF
Supply voltage	1.5 V
Sampling rate	16 GS/s
Differential input range	0.813 V_{pp}
SNDR (f_{in} = 170 MHz)	30.8 dB
SNDR (f_{in} = 3 GHz)	28.0 dB
SNDR (f_{in} = 170 MHz)	37.4 dB
SNDR (f_{in} = 3 GHz)	40.4 dB
Power consumption	435 mW
Active area	1.47 mm^2

Table 5.2 Comparison of high-speed TI ADCs

Publication	This work [6]	[5]	[4]	[3]	[2]	[1]
Technology (nm)	65	65	65	90	90	180
Resolution (Bit)	6	5	6	6	6	8
TI channels	8	8	16	8	16	80
Speed (GB/s)	16	12	40	10.3	24	20
Supply (V)	1.5	1.1	1.0/2.5	N.A.	1.0/2.5	N.A.
Power (mW)	435	81	1,500	1,600	1,200	9,000
ERBW (GHz)	3	6	7	4	6	2
ENOB (bit)	4.9	4.3	5.5	5.8	5.5	6.5
FOM (pJ/step)	2.6	0.35	2.4	3.6	2.0	24.8

using the latch-type comparators with automatic offset calibration. Better FOM can be achieved if the input samplers are realized with bootstrapped switches to improve ERBW. In addition, this chip includes a timing-skew calibration that can continuously operate in the background.

5.4 Conclusions

An 8-channel 6-bit 16-GS/s time-interleaved ADC was fabricated using a 65 nm CMOS technology. The chip demonstrates our proposed digital background calibration techniques, including comparator offset calibration and timing-skew calibration. The calibrations relax the matching requirements for devices and layout, and also provide robustness against process-voltage-temperature (PVT) variations.

Advanced CMOS technologies can produce ultra-high-speed TI ADCs. The sampling rates of TI ADCs are limited only by the sampling switches and the timing resolution of the technologies. The design challenges are high-frequency input sampling network, low-jitter multi-phase clock generator, and inter-channel mismatch calibration. CMOS ADCs with sampling rates over 100 GS/s should appear soon.

Acknowledgements The authors thank Taiwan Semiconductor Manufacturing Company (TSMC), Hsin-Chu, Taiwan, for chip fabrication. This research was supported by the National Science Council of Taiwan, R.O.C., and by the MediaTek Research Center at National Chiao-Tung University.

References

1. Poulton K, Neff R, Setterberg B et al (2003) A 20 GS/s 8b ADC with a 1 MB memory in 0.18 μm CMOS. In: International solid-state circuits conference, pp 318–319
2. Schvan P et al (2008) A 24 GS/s 6b ADC in 90 nm CMOS. In: International solid-state circuits conference, pp 544–545
3. Nazemi A et al (2008) A 10.3 GS/s 6 bit (5.1 ENOB at Nyquist) time-interleaved pipelined ADC using open-loop amplifiers and digital calibration in 90 nm CMOS. In: Symposium on VLSI circuits, Digest of technical papers, pp 18–19
4. Greshishchev Y et al (2010) A 40 GS/s 6b ADC in 65 nm CMOS. In: International solid-state circuits conference, pp 390–391, Feb 2010
5. El-Chammas M, Murmann B (2011) A 12-GS/s 81-mW 5-bit time-interleaved flash ADC with background timing skew calibration. IEEE J Solid-St Circ 46:838–847
6. Huang C-C, Wang C-Y, Wu J-T (2011) A CMOS 6-bit 16-GS/s time-interleaved ADC using digital background calibration techniques. IEEE J Solid-St Circ 46:848–858
7. Yu W et al (1999) Distortion analysis of MOS track-and-hold sampling mixers using time-varying Volterra series. IEEE Trans Circuits Syst-II 46:101–113
8. Dessouky M et al (1999) Input switch configuration suitable for rail-to-rail operation. IEE Electron Lett 35:8–9

9. Abo A et al (1999) A 1.5-V, 10-bit, 14.3-MS/s CMOS pipeline ADC. IEEE J Solid-St Circ 34:599–606
10. Kurosawa N et al (2001) Explicit analysis of channel mismatch effect in time-interleaved ADC systems. IEEE Trans Circuits Syst-1 48:261–271
11. Vogel C (2005) The impact of combined channel mismatch effects in time-interleaved ADCs. IEEE Trans Instrum Meas 54:415–427
12. Sin S-W et al (2008) Statistical spectra and distortion analysis of time-interleaved sampling bandwidth mismatch. IEEE Trans Circuits Syst-2 55:648–652
13. El-Chammas M, Murmann B (2009) General analysis on the impact of phase-skew in time-interleaved ADCs. IEEE Trans Circuits Syst-1 56:902–910
14. Elbornsson J, Gustafsson F, Eklund J (2004) Blind adaptive equalization of mismatch errors in a time-interleaved A/D converter system. IEEE Trans Circuits Syst-1 51:151–158
15. Prendergast R, Levy B, Hurst P (2004) Reconstruction of band-limited periodic nonuniformly sampled signals through multirate filter banks. IEEE Trans Circuits Syst-1 51:1612–1622
16. Seo M, Rodwell M, Madhow U (2005) Comprehensive digital correction of mismatch errors for a 400-Msamples/s 80-dB SFDR time-interleaved analog-to-digital converter. IEEE Trans Microw Theory Tech 53:1072–1082
17. Tsai T, Hurst P, Lewis S (2005) Bandwidth mismatch and its correction in time-interleaved analog-to-digital converters. IEEE Trans Circuits Syst-2 53:1133–1137
18. Huang S, Levy B (2007) Blind calibration of timing offsets for four-channel time-interleaved ADCs. IEEE Trans Circuits Syst-I 54:863–876
19. Divi V, Wornell G (2009) Blind calibration of timing skew in time-interleaved analog-to-digital converters. IEEE J Sel Topics Signal Process 3:509–522
20. Marelli D, Mahata K, Fu M (2009) Linear LMS compensation for timing mismatch in time-interleaved ADCs. IEEE Trans Circuits Syst-1 56:2476–2486
21. Jamal S, Fu D, Singh M, Hurst P, Lewis S (2004) Calibration of sample-time error in a two-channel time-interleaved analog-to-digital converter. IEEE Trans Circuits Syst-1 51:130–139
22. Haftbaradaran A, Martin K (2008) A background sample-time error calibration technique using random data for wide-band high-resolution time-interleaved ADCs. IEEE Trans Circuits Syst-2 55:234–238
23. Camarero D, Kalaia K, Naviner J, Loumeau P (2008) Mixed-signal clock-skew calibration technique for time-interleaved ADCs. IEEE Trans Circuits Syst-I 55:3676–3687
24. Saleem S, Vogel C (2011) Adaptive blind background calibration of polynomial-represented frequency response mismatches in a two-channel time-interleaved ADC. IEEE Trans Circuits Syst-1 58:1300–1310
25. McNeill JA, David C, Coln M, Croughwell R (2009) Split ADC calibration for all-digital correction of time-interleaved ADC errors. IEEE Trans Circuits Syst-II 56(5):344–348
26. Sandner C, Clara M, Santner A, Hartig T, Kutter F (2005) A 6-bit 1.2-GS/s low-power flash-ADC in 0.13-um digital CMOS. IEEE J Solid-St Circ 40:1499–1505
27. Ismail A, Elmasry M (2008) A 6-bit 1.6-GS/s low-power wideband flash ADC converter in 0.13-um CMOS technology. IEEE J Solid-St Circ 43:1982–1990
28. Van der Plas G, Decoutere S, Donnay S (2006) A 0.16 pJ/conversion-step 2.5 mW 1.25 GS/s 4b ADC in a 90 nm digital CMOS process. In: International solid-state circuits conference, pp 2310–2312
29. Huang C-C, Wu J-T (2005) A background comparator calibration technique for flash analog-to-digital converters. IEEE Trans Circuits Syst-I 52:1732–1740
30. Wang C-Y, Wu J-T (2009) A multiphase timing-skew calibration technique using zero-crossing detection. IEEE Trans Circuits Syst-I 56:1102–1114

Chapter 6
CMOS ADCs for Optical Communications

Yuriy M. Greshishchev

Abstract This paper provides a systematic view of ADCs embedded in DSP receivers of coherent optical communications systems. The functionality, performance and CMOS implementation trade-offs are discussed with the focus on techniques achieving high sampling rate and bandwidth. High conversion rate is efficiently addressed by massive interleaving of lower speed SAR ADCs, while the bandwidth limitation is dealt with on both architectural and circuit design levels. In conclusion, results of a 40 Gs/s 6b-ADC implemented in 65 nm CMOS are demonstrated.

6.1 Introduction

An optical communication receiver for a long time was essentially a single bit data converter with the main amplifier and decision flip-flop in a clock and data recovery circuitry performing the conversion. Signal processing options were limited to analog or analog-digital techniques for example: DFE. Gaining the full DSP functionality was absolutely critical for long haul systems in order to overcome the roadblocks of exponentially growing system cost and the impact of fiber impairments. A 40 Gb/s Dual Polarization (DP) QPSK coherent optical system was the first entry point in the industry, and later became the modulation standard for 40 Gb/s and 100 Gb/s data rates [1]. The 90 nm CMOS DSP ASIC receiver (Fig. 6.1) had four embedded 24 Gs/s 6-b CMOS ADCs with total power of 21 W, of which 25% was consumed by the ADCs [2].

The ADC showed, at the time of announcement, a breakthrough in CMOS speed-power performance at 24 Gs/s, as well as, in power of 6-b SAR converter

Y.M. Greshishchev (✉)
Ciena Corporation, 3500 Carling Avenue, Ottawa, ON K2H 8E9, Canada
e-mail: ygreshis@ciena.com

A.H.M. van Roermund et al. (eds.), *Nyquist AD Converters, Sensor Interfaces, and Robustness: Advances in Analog Circuit Design, 2012*,
DOI 10.1007/978-1-4614-4587-6_6,

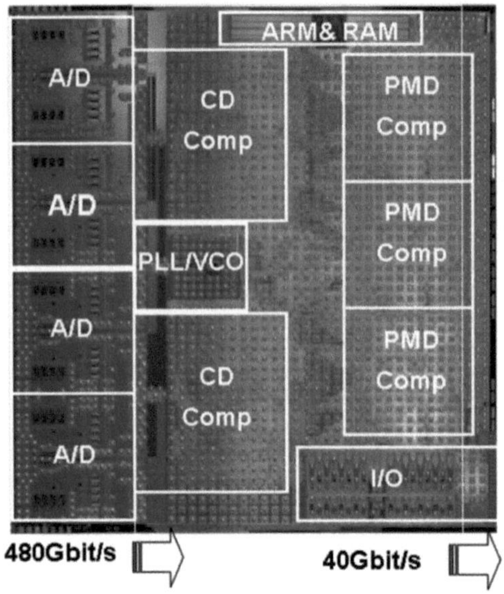

ADC performance

Resolution	6 bits
Conversion rate	0.1 - 24GS/s
Input range	$1.2V_{p\text{-}p}$ diff.
ENOB average cal / cal each freq	4.2/4.8, F_{in}= 8GHz 3.5/4.1, F_{in}= 12GHz
SFDR	40dB @ 8GHz 35dB @ 12GHz
Power	1.2W @ 1V and 2.5V
ADC core	4 x 4 mm^2
Process	90nm CMOS

Fig. 6.1 A 40 Gb/s optical DP QPSK Rx ASIC in 90 nm CMOS and embedded ADC performance

used with aggressive interleaved factor of 160 [3]. SAR architecture, traditionally viewed as the slowest, paved the way to the fastest ADC. A similar approach was later used in 65 nm CMOS for sampling rate up to 40 Gs/s [4], and up to 56 Gs/s with 320 SAR converters, but with new sampling technique [5]. The ADC also surpassed the performance of SiGe BiCMOS technology that was viewed as the most suitable option for an optical communication receiver [6].

6.2 DP QPSK Functionality and ADC Requirements

A simplified diagram of a 40 Gb/s DP QPSK receiver is shown in Fig. 6.2 [2]. The optical polarization beam splitter (PBS) segregates the light into two polarization components X,Y, and then each polarization is mixed with the local oscillator (LO) laser. It is converted to quadrature electrical signals I, Q using pin-detectors. A traditional TIA and VGA functionality is followed by the DSP. Each I, Q channel carries approximately a quarter of the bit rate; the symbol rate is 10 Gs/s in the case of a 40 Gb/s system. In order to avoid spectral aliasing and improve SNR, the transmitter and receiver sampling rates are selected at twice the Nyquist rate (2× oversampled). The ADC basic requirements are captured in Table 6.1.

The linear equalizer in the DSP compensates chromatic dispersion (CD) and polarization mode dispersion (PMD) in the fiber. It also equalizes the overall Rx frequency response, including the ADC. To some degree, the equalizer relaxes flatness requirements for the ADC frequency response. The DSP digitally peaks the

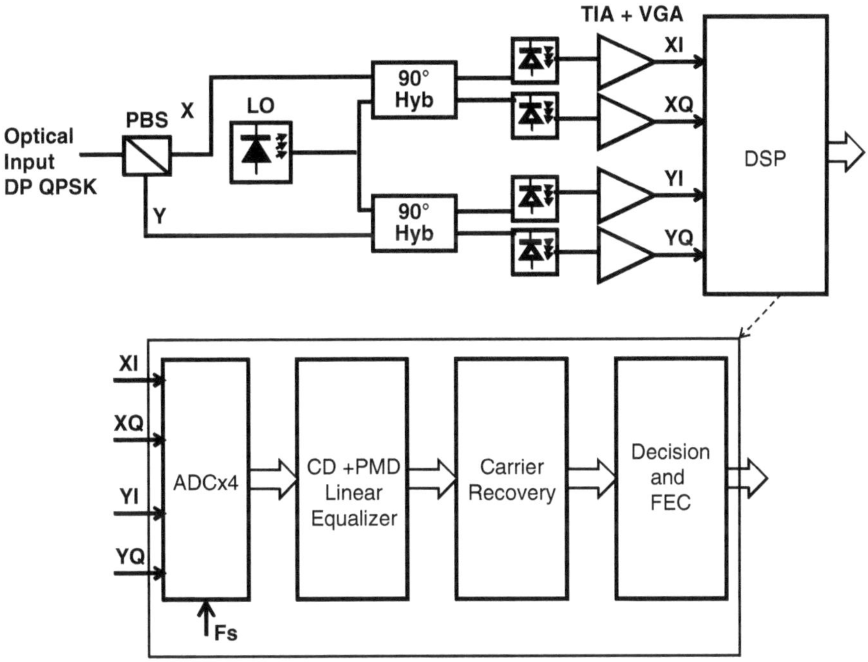

Fig. 6.2 Coherent optical DP QPSK receiver functionality

Table 6.1 ADC requirements in a *DP QPSK* receiver

Modulation	Resolution (bit)	Sampling rate (Gs/s)	Electrical bandwidth (GHz)
40G DP QPSK	6–8	23	5–10
100G DP QPSK dual carrier		29	5–10
100G DP QPSK single carrier		56–65	15–20

receiver bandwidth, if necessary, which may degrade overall SNR. Designers must ensure that the whole receiver analog bandwidth, including the ADC, is within the target range.

Forward-error correction (FEC) improves the BER approximately from 10^{-3} to 10^{-15}. Because of FEC, the ADC intrinsic BER is not critical. In an interleaved architecture, due to reduced sampling frequency in the comparator, the metastability rate is well below the required minimum.

Because of FEC, ADC resolution above 5-b gives a diminishing return in optical SNR improvement [2]. Additional resolution, however, is useful for correcting ADC imperfections, VGA functionality in the DSP, or improving production margin. Future system modulation formats with higher-complexity constellations like 16QAM will also benefit from 6- to 8-b resolution ADCs.

Table 6.2 Crest factor impact on ENOB

ADC input	CF	ENOBarb
Sine-wave	$2\surd 2$	6.02
Random uniform	$2\surd 3$	5.73
Random Gaussian	6 (arbitrary)	4.94

6.2.1 ADC's SNDR and ENOB

The signal to noise and distortion ratio (SNDR) is defined for a sinusoidal input signal with peak-to-peak amplitude equal to the full scale of the ADC [7]. SNDR accounts for all distortions in the ADC: quantization, harmonic distortions, thermal noise and jitter.

The effective number of bits is given:

$$ENOB = \frac{(SNDR - 1.76)}{6.02}\ bit \tag{6.1}$$

While this definition is widely used as a standard for ADC characterization, for optical ADCs with an arbitrary input signal, a more general formula is required:

$$ENOBarb\,(f) = \left[SNDR\,(f) - 20\log\left(\frac{CF}{2\sqrt{2}}\right) - 1.76)\right]/6.02\,bit, \tag{6.2}$$

where CF is the ADC input signal crest factor.

We define CF as the peak-to-peak signal to RMS ratio normalized to peak-peak magnitude equal to an ADC full scale of 1. Table 6.2 shows CF values for three types of signals and corresponding ENOBarb number for an ADC with sinusoidal SNDR = 38 dB.

For the unbounded Gaussian signal, CF is set by the ADC quantizer clipping level. CF is arbitrarily chosen to be equal to 6 in order to represent the case of a dispersive optical channel. Note that signal properties impact the ADC SNDR, and for Gaussian input the ideal 6-bit ADC is equivalent to 5-bit (SNDR = 32 dB) of sine-wave input performance. This has been known since the early years of digital communication. To overcome this degradation, a non-uniform quantizer with step size inversely proportional to the signal probability density was recommended [8].

6.2.2 ADC's SNR and Jitter

There are two dominant sources of SNDR degradation at high frequencies: bandwidth limitations and sampling jitter. The signal amplitude at the ADC quantizer follows the frequency response and ENOB may degrade accordingly because of ADC under-fill or clipping. For example, ENOB degrades by 0.5 bit at the −3 dB bandwidth point [7]. If the input amplitude is kept at 100% of ADC scale over the input frequency range, then the remaining source of degradation is sampling jitter.

The jitter typically has two components: bounded (random and deterministic) and random unbounded. In an interleaved ADC, deterministic jitter is due to timing misalignment of individual channels. With a large number of interleaved channels, it appears as bounded random jitter with a peak-to-peak value defined by maximum calibration inaccuracy. For instance, +/−0.5 ps calibration error will create a 0.29 ps_rms bounded random jitter (assuming that the error distribution is uniform). With a small number of interleaved channels, misalignment creates visible signatures in the ADC spectrum as mixing products at multiples of ADC sampling frequency divided by the interleaved factor.

Thermal noise in the clock generation and distribution circuitry is the source for Gaussian jitter. SNR is a parameter that helps to evaluate jitter impact. It is derived from SNDR by subtracting harmonic distortions and deterministic jitter spurs. For an input signal uncorrelated with the jitter, SNR degrades proportionally to the RMS value of the signal slew rate according to:

$$SNR = 20\log(S_{IN\_RMS}/SR_{RMS} \cdot \sigma_a), \tag{6.3}$$

where S_{IN_RMS}, SR_{RMS}, are the RMS values of the ADC input signal and its derivative; σ_a – RMS sampling jitter.

For a sine-wave input [9] we have:

$$SR_{RMS} = A\sqrt{2}\pi f,\ S_{IN\_RMS} = A\big/\sqrt{2}, \tag{6.4}$$

and the signal to noise ratio is given by:

$$SNR = -20\log\ (2\pi f \sigma_a) \tag{6.5}$$

The ENOB degradation over frequency according to Eq. 6.5 is shown in Fig. 6.3 for a 6b and 8b ADC. One can observe that relatively large jitter makes ENOB at high frequencies almost independent of ADC initial resolution. This does not mean, however, that there is no advantage from a higher ENOB at lower frequencies. The ENOB performance over the whole input signal bandwidth is important. System level simulations for a given modulation format and optical channel model must be performed to find an accurate answer.

For a Gaussian input, relationship Eq. 6.3 is also valid. The SNR can be represented similar to Eq. 6.5:

$$SNRGauss = -20\log\ (2\pi F_{BW}\sigma_a/K), \tag{6.6}$$

where F_{BW} is the ADC input signal bandwidth, and K *is* a coefficient that depends on signal bandwidth roll-off.

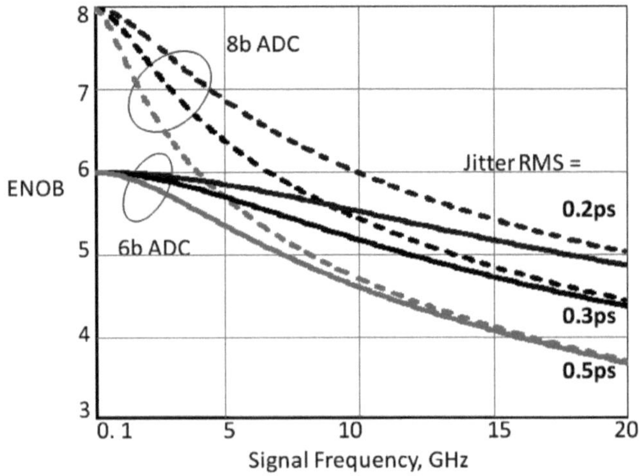

Fig. 6.3 ENOB of 8-b and 6-b ADCs versus sin wave input frequency and sampling jitter

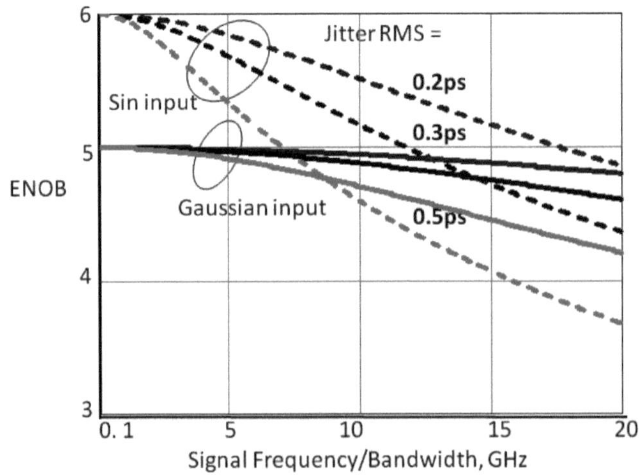

Fig. 6.4 ENOB of a 6-b ADC for sine-wave and Gaussian input versus frequency and bandwidth correspondingly

An approximation of K = 1.7 was found in behavioral simulation for the raised-cosine spectrum roll off. Comparison for 6-b ADCs ENOB with sine-wave and Gaussian input shows (Fig. 6.4.) that in the latter case the ADC efficiency starts from 5-b, and is less sensitive to the jitter amplitude.

6.3 Time Interleaved ADC

Time interleaving is the holy grail of CMOS technology. It allows increasing sampling rate by aggregating lower rate ADCs. The maximum conversion rate achieved to date with this technique is 56 Gs/s (Table 6.3). For the interleaved concept to work, trade-offs for the other parameters must be considered: bandwidth, jitter, power dissipation, and the die area. One-stage interleaving, shown in Fig. 6.5, results in the highest impact of the number of interleaved channels, M, on the ADC dynamic performance. Capacitive loading at the ADC input and jitter generation in clock generation and distribution circuitry are both strong functions of the ratio M.

A two stage interleaved ADC (Fig. 6.6) relaxes the dependence on M. It is possible when front-end T&H circuits, TH-1. . .TH-M1, provide buffering (current gain) to reduce capacitance at the ADC input. The combined interleaving ratio M equals M1·M2. The critical circuitry that determines the jitter generation is reduced to M1 channels.

The interface blocks, shown in Figs. 6.5 and 6.6 are the least critical parts of the design. Power dissipation in these blocks is a small fraction of the total power. Functionality depends on converter speed and DSP core clock frequency: both multiplexor or demultiplexor functions may be required.

The first 20 Gs/s 8-b CMOS ADC (see Table 6.3), built for digital oscilloscopes, is a one-stage design and completely relies on the dynamic performance of 80 ADC

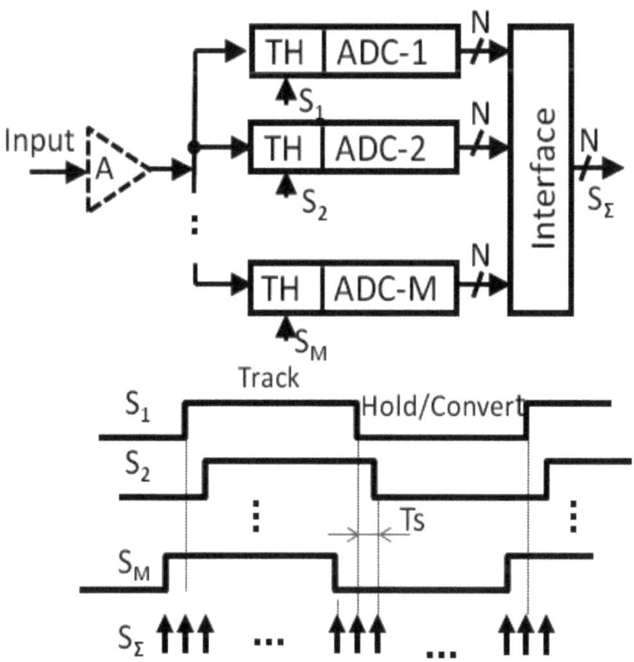

Fig. 6.5 ADC architecture with one-stage time interleaving

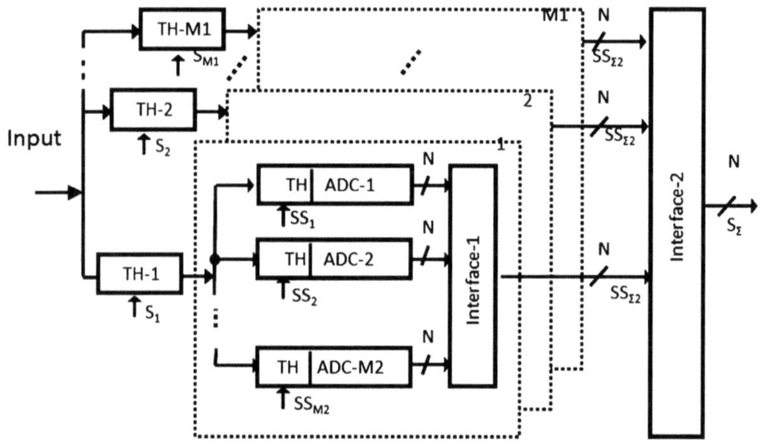

Fig. 6.6 ADC architecture with two-stage time interleaving

Table 6.3 CMOS interleaved ADCs with sampling rate > 20 Gs/s

ADC	Interleaving ratio: Front-end T&H	Interleaving ratio: Elementary T&H – ADC	Total number of interleaved ADCs and speed
8-b 20 Gs/s [10]	–	80 pipeline, radix = 1.6	80 @ 250 MHz
8-b 56 Gs/s [5]	Four Samplers-demultiplexors	80 SAR with demultiplexed inputs	320 @ 175 MHz
6-b 24 Gs/s [3] 6-b 40 Gs/s [4]	16	10 SAR with combined inputs	160 @ 250 MHz
6-b 25 Gs/s [17]	8	Flash ADC	8 @ 3.125 GHz

slices. An external 25–Ohm wideband buffer is used to overcome high capacitive load at the input.

The 24 Gs/s and 40 Gs/s ADCs have an advantage of a front-end T&H interleaving ratio of 16. The 56 Gs/s ADC, as described in [5], has a front-end with four charge samplers (instead of T&H) followed by a 4:80 demultiplexor driving the input of one of the 320 ADC slices. This approach addresses capacitive loading and makes it a weak function of the interleave ratios M1, M2. In addition, the reduced number of front-end samplers (M1 = 4) simplifies clock generation and improves the overall ADC jitter.

6.4 ADC Interleaved Slice

The first CMOS interleaved ADC was built with SAR ADC slices [11]. The fastest ADCs reported today are also based on SAR converters. The feedback capacitive DAC with charge redistribution technique makes it ideally suitable for sub-nanometer

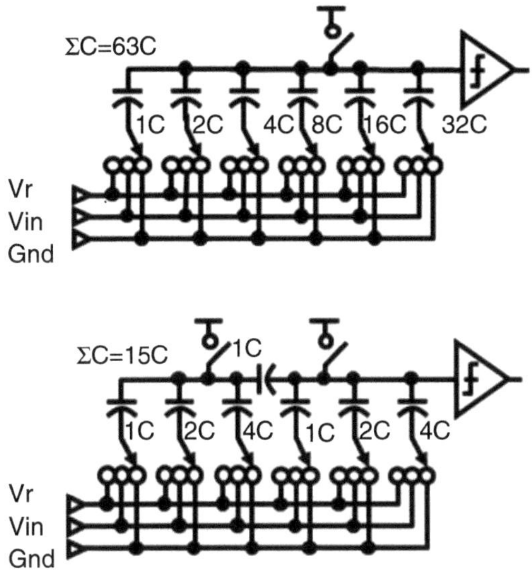

Fig. 6.7 Switched-capacitor DAC structures: binary weighted (*top*) and segmented (*bottom*)

CMOS technologies. Apart from the DAC, the ADC has only one analog component: a comparator that typically consumes a dominant portion of the power. From the total capacitance point of view, two DAC structures are of interest (Fig. 6.7): binary weighted (top [3, 12]) and segmented (bottom [13]). The C-2C structure [14] would help to further reduce input capacitance, however the capacitance then becomes comparable to the parasitics and a gain error would be introduced. For 6-b ADCs the kT/C noise is not a factor even at $\Sigma C = 10$fF. (kT/C $= 0.63$ mV rms).

There are reported architectures with better power efficiency than SAR, but they are algorithmically more complex and demand higher accuracy analog components, for instance the folding type [7, 15]. The pipeline ADC architecture is the other type that may serve in the interleaved slice (and was used in 20 Gs/s ADCs). Designs where a traditional analog gain function is implemented with switched capacitor circuits are of special interest. This may be more suitable because, similar to SAR, they rely on charge redistribution. An example of a 10-b pipeline design with a charge pump technique in residue formation circuitry was demonstrated at 50 Mb/s in 0.18 μm CMOS [16]. Most recently, flash sub-ADC architecture was introduced in 40 nm CMOS 25 Gs/s ADC showing comparable to 65 nm CMOS 40 Gs/s ADC (with SAR sub-ADCs) energy per conversion -step performance [17].

A very important factor is the ADC slice conversion accuracy: offset, gain, and timing mismatch. Timing error is a lesser problem in two-stage architectures. In SAR ADCs, comparator device mismatch is the single source of ADC error (offset only) that is easily addressed with auto- calibration. One approach is to use an additional conversion cycle and create an auto-zeroing loop with comparator inputs temporarily shorted together [3]. The other popular solution is ADC background calibration [5].

6.5 Track and Hold Circuit

6.5.1 T&H with Source Followers

Differential T&H with MOS switches, followed by source followers (Fig. 6.8), is the simplest topology that delivers tens of GHz of bandwidth. This circuit can drive a capacitive load, C_L, with one order of magnitude higher value than the hold capacitance, which is the input capacitance of the source follower itself. The M1, M2 switches are the main switches, and M3, M4 are for charge compensation during a transition to a hold mode. In the ADC, the input signal bandwidth easily exceeds 10 GHz. At such high frequencies, a source follower preceded with a MOS switch exhibits gain peaking in its fundamental frequency response with a mechanism that cannot be explained with continuous (switch is permanently ON) operation or AC response simulation. The next section analytically addresses this behavior.

6.5.2 T&H Gain Peaking in Time Domain

If one performs AC analyses of the circuit in Fig. 6.8 with switch condition ON driven from a 50-Ohm-terminated voltage source, no peaking in frequency response will be found due to low impedance driving conditions. In contrast, T&H transient simulation with switch condition alternating from ON (tracking mode) to OFF (hold mode) clearly shows peaking in the fundamental frequency response. This is a very

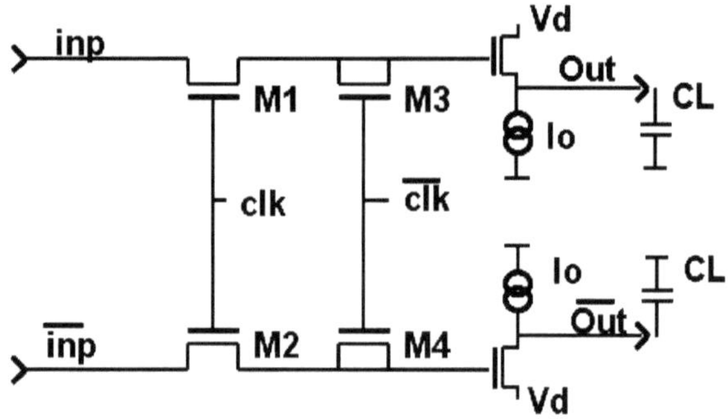

Fig. 6.8 Differential T&H circuit with source followers buffers

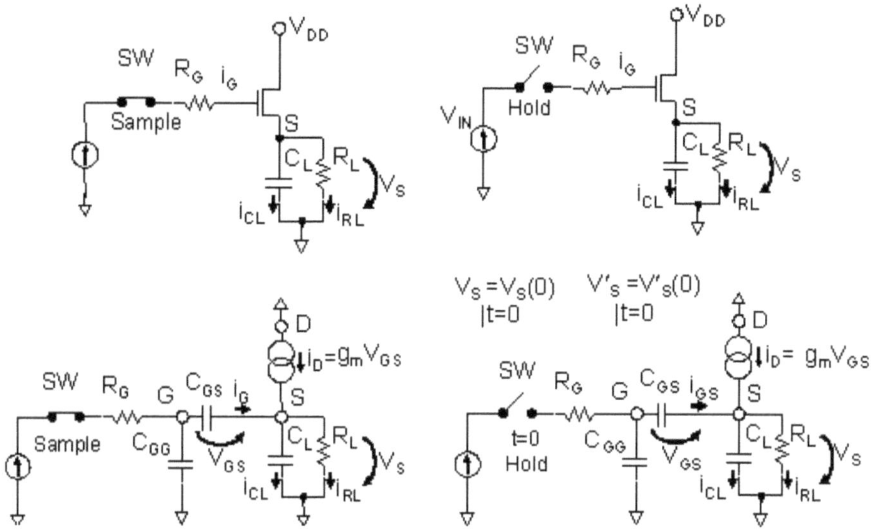

Fig. 6.9 T&H modes (*top*) and corresponding equivalent circuits (*bottom*)

useful behavior that extends the interleaved ADC bandwidth. Because it can be explained with time domain analyses only, we call it time domain gain peaking.

First, let's find differential equation that describes T&H operation. Figure 6.9 depicts equivalent circuits for sampling and hold modes of a T&H circuit based on a simple MOS transistor model. We start with equations for the currents in the hold mode (see Fig. 6.9 for definitions):

$$i_D = i_{CL} + i_{RL} + i_{GS}, \tag{6.7}$$

where the branch currents are:

$$i_D = gm \cdot V_{GS}, \, i_{GS} = \frac{C_{GS} \cdot C_{GG}}{C_{GS} + C_{GG}} \cdot \left(\frac{dV_S}{dt}\right), \, i_{RL} = \frac{V_S}{R_L} \tag{6.8}$$

After substitution, (6.8) can be represented in ODE form:

$$\tau \cdot \frac{dV_S}{dt} + V_S = 0 \tag{6.9}$$

Where:

$$\tau = \frac{a}{b}, \, a = \frac{C_{GS} \cdot C_{GG}}{C_{GS} + C_{GG}} + C_L, \, b = gm \cdot \frac{C_{GG}}{C_{GS} + C_{GG}} + \frac{1}{R_L} \tag{6.10}$$

Combining Eq. 6.10 with initial conditions for the voltage Vs (t) and its derivative at t = 0 (start of transitioning to a hold mode), one can find solution for (6.9)

$$V_S\,(t) = V_S(0) + \tau \cdot V'_S(0) \cdot \left(1 - exp\left(\frac{-t}{\tau}\right)\right) \tag{6.11}$$

or

$$\Delta V_S\,(t) = \tau \cdot V'_S(0) \cdot \left(1 - exp\left(\frac{-t}{\tau}\right)\right), \tag{6.12}$$

where $V_S\ (0),\ V'_S(0)$ are the source follower output voltage and its derivative at the start of switching to a hold mode.

Equations 6.11 and 6.12 reveal a remarkable property of a source follower with a voltage switch at the gate: the sampled output voltage is the sum of the input signal sampled value plus a component proportional to its derivative. The derivative component settles with time constant τ (Eq. 6.12). If the time constant τ is small compared to the T&H clock period (typical case for interleaved ADCs), then the circuit frequency response exhibits gain peaking at frequencies close to ~ $1/2\pi\tau$.

The peaking component Eq. 6.12 can be made very large, as long as the ADC interleaving ratio is also large enough for the output transient process to settle.

To verify the validity of model Eq. 6.12, a Spice circuit simulation was performed with the waveforms and circuit parameters shown in Fig. 6.10. The analytical model and simulation yield practically identical results (Fig. 6.10 bottom).

In general, in this type of ADC without proceeding DSP, an additional peaking component (6.12) would probably be undesirable and treated as a dynamic error. In embedded ADC applications, the impact of component Eq. 6.12 being a linear transformation is easily corrected in a follow-up DSP.

6.6 40 Gs/s 6b ADC Example

The ADC architecture (Fig. 6.11) is a two stage time-interleaved design with 160 SAR converters aggregated in 16 subADCs. Each subADC is self-contained and only requires a 2.5 GHz clock to operate. The output data from subADCs are collected by an interface block (see Fig. 6.6). The 6-dB signal attenuation at the input is a result of physical implementation with subADCs split in even and odd banks fed via a 50- Ohm star- type power splitter.

To characterize the ADC, it is combined with a memory and PLL synthesizer as shown in Fig. 6.12 along with the measured ADC performance.

Measurements are performed with the ADC mounted on a printed circuit board with sine-wave input generated from a low-jitter signal source (Fig. 6.13).

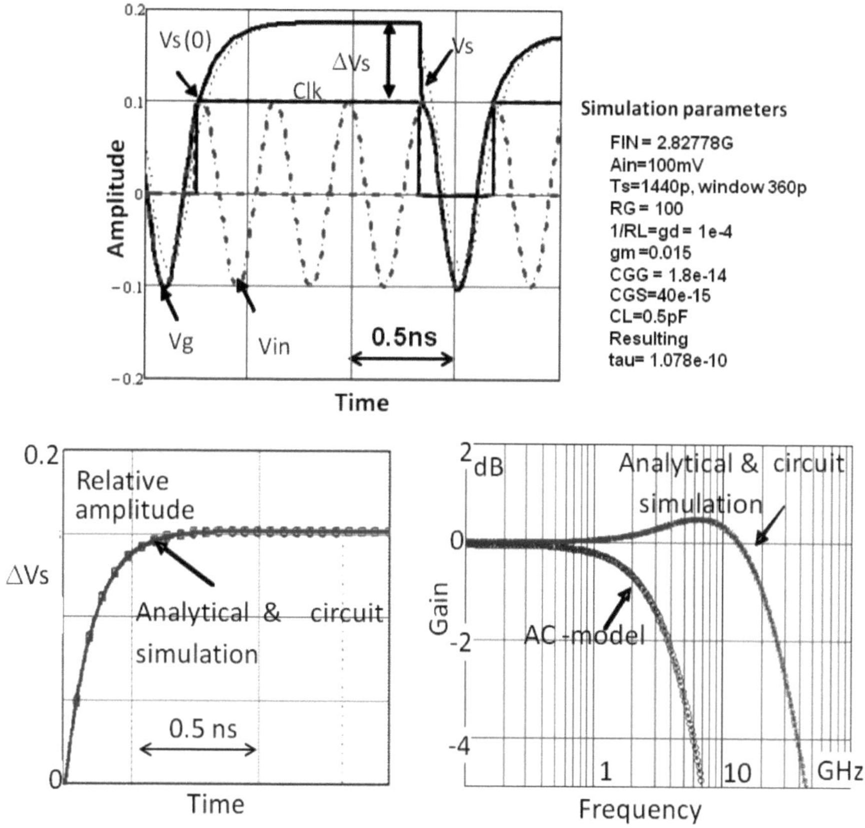

Fig. 6.10 Model (6.12) verification versus circuit Spice simulation: the waveforms (*top*); the exponential portion fit after transition to hold mode (*bottom left*); and frequency response (*bottom right*)

The ENOB frequency response shows (Fig. 6.14) performance above 3.9-b up to 18 GHz. Note that this result does not include the impact of bandwidth degradation. Removing harmonics and deterministic spurs allows evaluation of ENOB based on SNR and random Gaussian jitter as demonstrated in Fig. 6.15. The SNR model (6.5) fit yields 0.25 ps rms jitter. To verify this number, time domain analyses of the measured results is performed. A three-parameter sine-fit algorithm for 16,384 points (instead of FFT) was used to reconstruct the input sine-wave and calculate ADC residue (Fig. 6.16).

An ADC Monte-Carlo model was created to fit the residue statistics and find ADC jitter and other parameters. In order to separate the ADC input voltage noise and jitter component, two 1 GHz and 10 GHz inputs were captured and analyzed.

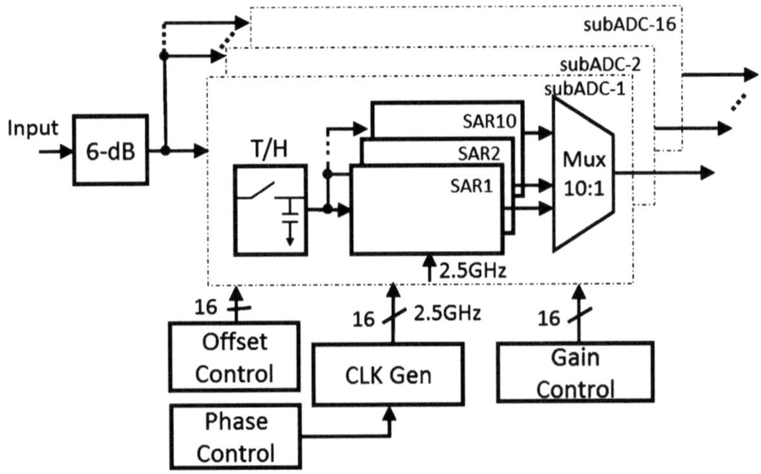

Fig. 6.11 40 Gs/s 6b-ADC architecture

Resolution	6 bits
Conversion rate	0.1 – 40 GS/s
Input range	1.2 V_{p-p} diff.
ENOB 1-point cal Cal each freq	 4.5 b, F_{in}= 10 GHz 3.9 b, F_{in}= 18 GHz
SFDR	40 dBc @ 10 GHz 35 dBc @ 18 GHz
Power	≤ 1.5 W
ADC core	4 x 4 mm^2
Process	65nm CMOS

Fig. 6.12 40 Gs/s 6b-ADC die microphotograph and packaged performance summary

One can notice timing jitter impact by the high slew-rate and amplitude-noise in low slew-rate regions. The error probability distribution, as shown in Fig. 6.17, fitted with amplitude noise and the jitter Monte-Carlo model, yields a result similar to the frequency domain result: $\sigma_a = 0.25$ ps rms. The model is accurate over three orders of magnitude (limited by the number of samples).

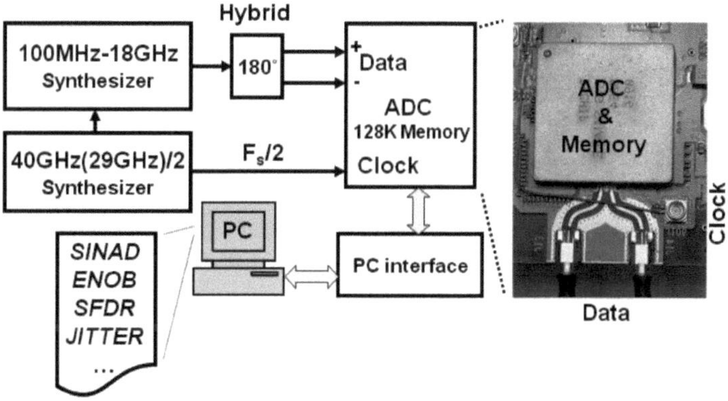

Fig. 6.13 ADC test setup. Parameter extraction according to IEEE Standard 1241-2000 and Draft P1057/D7.6

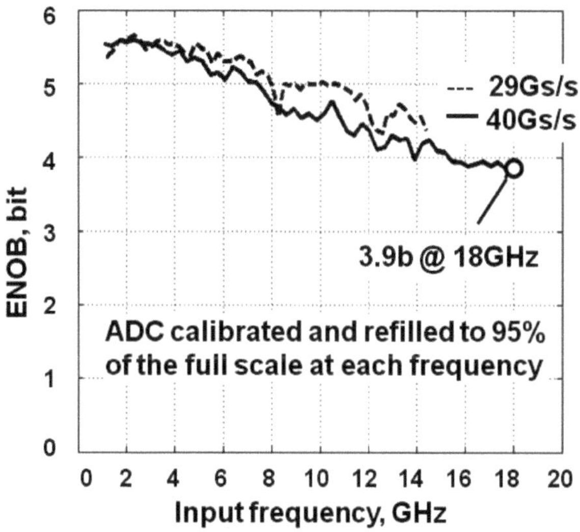

Fig. 6.14 40 Gs/s 6b-ADC measured ENOB over input frequency

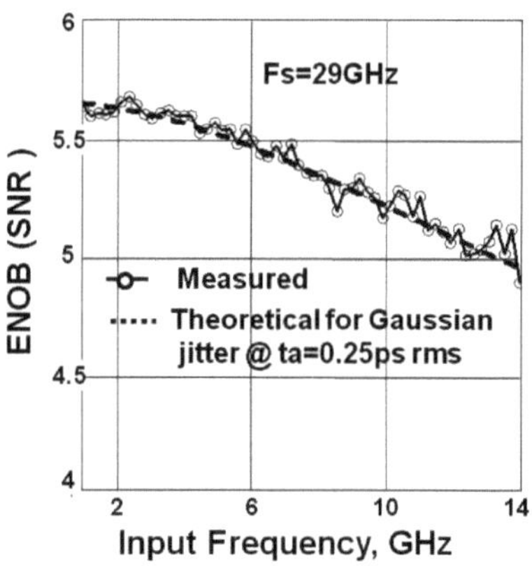

Fig. 6.15 40 Gs/s 6b-ADC measured SNR and theoretical fit with jitter 0.25 ps rms

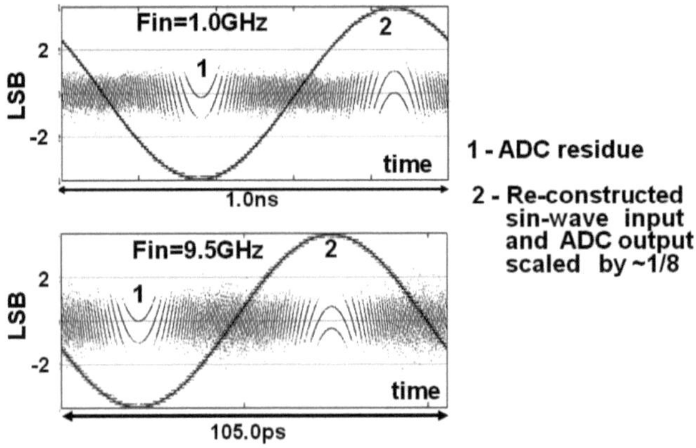

Fig. 6.16 Measured ADC residue (no averaging)

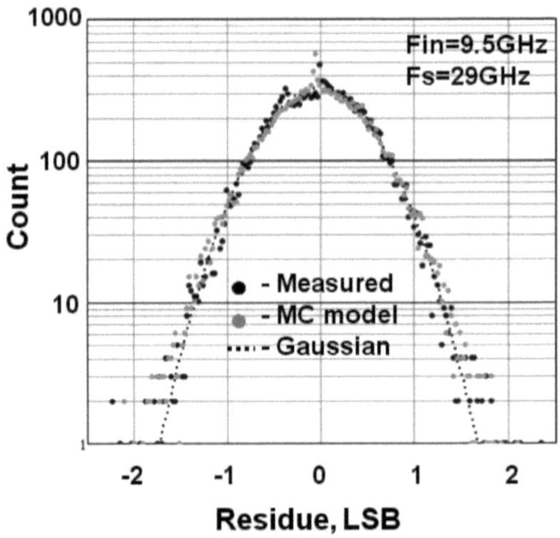

Fig. 6.17 Measured ADC residue histogram @ Fin = 9.5 GHz with Monte Carlo and Gaussian models fit @ 0.25 ps rms jitter

6.7 ADC Bandwidth Horizons

The bandwidth of an ADC embedded into a DSP ASIC for optical communications is limited by two main factors: package interconnects and circuit performance. While circuit speed improves with every new CMOS technology node, there is less progress in packaging interconnect performance. With an improved substrate material, a bandwidth of 30 GHz is probably achievable [5]. The existing trend for sub-nanometer CMOS ASICs to use organic substrate materials for better manufacturability and shorter bump pitch complicates bandwidth expansion; an organic substrate has inferior frequency-dependent loss compared to ceramic.

The high-end real time digital oscilloscope industry provides good guidance for feasible ADC sampling-rate and bandwidth. A 160 Gs/s sampling rate with 60 GHz bandwidth is achieved with digital bandwidth interleaving technology by combining two 35 GHz channels [18, 19].

6.8 Conclusion

Optical communications is a driving application for ultra high speed medium-resolution ADCs embedded in CMOS DSP ASICs. The interleave technique is a critical factor in achieving high speed and bandwidth that, in conjunction with innovative circuit design, can satisfy 40–100 Gs/s system needs.

Acknowledgement The author is grateful to colleagues P. Schvan, J. Aguirre, M. Besson, R. Gibbins, C. Falt, P. Flemke, N. Ben-Hamida, D. Pollex, S.-C. Wang, and J. Wolczanski for contribution to the ADC design and characterization. Special thanks to K. Roberts, B. Beggs, and J. Sitch for system insight and support.

References

1. Roberts K, Beckett D, Boertjes D, Berthold J, Laperle C (2010) 100 G and beyond with digital coherent signal processing. IEEE Commun Mag 48(7):62–69
2. Ben-Hamida N, Greshishchev Y, Beggs B (2010) Advances in transceiver circuits to suit new optical modulation formats – a historical perspective. In: Forum F3-Transciever circuits for optical communications, Digest of technical papers international solid- state conference (ISSCC). IEEE, pp 514–515, Feb 2010 San-Francisco, USA
3. Schvan P, Bach J, Falt C, Flemke P, Gibbins R, Greshishchev Y, Ben-Hamida N, Pollex D, Sitch J, Wang S-C, Wolczanski J (2008) A 24 GS/s 6b ADC in 90 nm CMOS. In: International solid-state conference, Digest technical papers, pp 544–545, Feb 2008 San-Francisco, USA
4. Greshishchev YM, Aguirre J, Besson M, Gibbins R, Falt C, Flemke P, Ben-Hamida N, Pollex D, Schvan P, Wang S-C (2010) A 40 GS/s 6b ADC in 65 nm CMOS. In: International solid-state conference, Digest technical papers, pp 390–391, Feb 2010 San-Francisco, USA
5. Bower P, Dedic I (2011) High speed converters and DSP for 100 G and beyond. Opt Fiber Technol 17(5):464–471
6. Schvan P, Pollex D, Wang S-C, Falt C, Ben-Hamida N (2006) A 22 GS/s 5b ADC in 0.13 um SiGe BiCMOS. In: International solid-state conference, Digest technical papers, pp 572–573, Feb 2006 San-Francisco, USA
7. Van de Plassche R (2003) CMOS integrated analog-to-digital and digital-to-analog converters. Kluwer, Boston
8. Cattermole KW (1973) Principles of pulse-code modulation. Iliffe Books, London
9. Shinagawa M, Akazawa Y, Wakimoto T (1990) Jitter analysis of high-speed sampling systems. IEEE J Solid-St Circ 25(1):220–224
10. Poulton K, Neff R, Setterberg B, Wuppermann B, Kopley T, Jewett R, Pernillo J, Tan C, Montijo A (2003) A 20 GS/s 8b ADC with a 1 MB Memory in 0.18um CMOS. In: International solid-state conference, Digest technical papers, pp 318–319, Feb 2003 San-Francisco, USA

11. Black WC, Hodges DA (1980) Time interleaved converter array. In: International solid-state conference, Digest technical papers, pp 14–15, Feb 1980 San-Francisco, USA
12. Draxelmayr D (2004) A 6b 600 MHz 10 mW ADC array in digital 90 nm CMOS. In: International solid-state conference, Digest technical papers, pp 264–265, Feb 2004 San-Francisco, USA
13. Agnes A, Bonizzoni E, Malcovati P, Maloberti F (2008) A9.4-ENOB 1 V 3.8 μW 100 kS/s SAR ADC with time–domain comparator. In: International solid-state conference, Digest technical papers, pp 246–247, Feb 2008 San-Francisco, USA
14. Alpman E, Lakdawal H, Carley LR, Soumyanath K (2009) A 1.1 V 50 mW 2.5 GS/s 7b time interleaved C-2C SAR ADC in 45 nm LP digital CMOS. In: International solid-state conference, Digest technical papers, pp 76–78, Feb 2009 San-Francisco, USA
15. Verbruggen B, Craninckx J, Kuijk M, Wambacq P, Van der Plas G (2010) A 2.6 mW 6b 2.2 GS/s 4-times interleaved fully dynamic pipelined ADC in 40 nm digital CMOS. In: International solid-state conference, Digest technical papers, pp 296–297, Feb 2010 San-Francisco, USA
16. Ahmed I, Mulder J, Johns DA (2009) A 50MS/s 9.9 mW pipelined ADC with 58 dB SNDR in 0.18 μm CMOS using capacitive charge-pumps. In: International solid-state conference, Digest technical papers, pp 164–165, Feb 2009 San-Francisco, USA
17. Crivelli D, Hueda M, Carrer H, Zachan J, Gutnik V, Del Barco M, Lopez R, Hatcher G, Finochietto J, Yeo M, Chartrand A, Swenson N, Voois P, Agazzi O (2012) A 40 nm CMOS single-chip 50 Gb/s DP-QPSK/BPSK transceiver with electronic dispersion compensation for coherent optical channels. In: International solid-state conference, Digest technical papers, pp 328–329, Feb 2012 San-Francisco, USA
18. LeCroy product datasheet (2012) LabMasters 10 Zi high bandwidth oscilloscopes, LeCroy San-Francisco, USA
19. Pupalaikis PJ (2007) An 18 GHz bandwidth, 60 GS/s sample rate real-time waveform digitizing system. In: Microwave symposium, IEEE/MTT-S international, pp 195–198, June 2007 Honolulu, USA

Part II
Sensor Interfaces

The second part of this book is dedicated to the topic "Sensor Interfaces". The first Chapter is from Tommaso Ungaretti. He presents the main MEMS sensor features and their relative market importance and market trends. After this, he focuses his discussion on possible application of MEMS sensors like accelerometers, gyroscope, magnetic field sensor, and microphone showing their past, present and future applications.

In the second Chapter Michiel Pertijs discusses new solutions for capacitive sensors, addressing the power minimization. He identifies in the capacitance-to-digital converter (CDC) the key block for the power minimization and deeply analyzes its performance and its power reduction limitation. Moreover he proposes two case-studies: a period-modulator-based CDC and a delta-sigma CDC, demonstrating the validity of the proposed power minimization approach.

In the third Chapter, Piero Malcovati introduces electronics front end design and application for a MEMS microphone. He describes the basic capacitive front-end building block for it. Then he passes to two design examples. The first is based on a constant-charge approach. He analyzes the design of the input buffer and of the following SD modulator. The second example describes the possibility of using a force-feedback approach, and the design of the preamplifier, of the SD modulator and of the force-feedback logic is focused.

Finally the fourth Chapter from Jan Kaplon presents circuit design techniques currently employed for the development of analog front-end electronics dedicated to the readout of radiation semiconductor sensors used in tracking detectors for High Energy Physics experiments. Due to the very large numbers of channels, power consumption becomes a critical issue in the analog front-end design, without compromising SNR and speed requirements. A selection of amplifier circuits were discussed in the context of the evolution of the CMOS technologies requiring the adaptation of design techniques to the new deep scaled MOS transistors properties.

Andrea Baschirotto

Chapter 7
Motion MEMS and Sensors, Today and Tomorrow

Benedetto Vigna, E. Lasalandra, and T. Ungaretti

Abstract Only 6 years passed since the dawn of the Motion Sensors Era in the Consumer market and now the world of MEMS Motion Sensors is completely different and it's going to change even more in the future. If 2006 was the year of the Accelerometer adoption by Nintendo^{TM} in the Wii^{TM} Controller, 2010 was the year of the Gyroscope adoption by Smartphone manufacturers. And in both cases STMicroelectronics triggered the big volume production of those two micro-machined devices, previously known only by automotive customers and used only for active and passive safety applications. Moreover nowadays combinations of these two inertial products, also known as Six Degree-Of-Freedom Motion Sensors 6XDOF, are starting to appear in the market and most likely they will coexist with standalone accelerometers and gyroscopes, depending on customer needs in terms of compactness and performances. This paper reports the details of an innovative tri-axis silicon MEMS Coriolis' gyroscope that fulfills the pressing market requirements for low power consumption, small size, slim form factor, high performances and low cost, but it addresses also the recent trends of the Six Degree-Of-Freedom 6XDOF and Nine-Degree-of-Freedom 9XDOF systems, realized by integrating the 6XDOF with a compass.

7.1 Introduction

More intuitive user interface has been the key application driving the adoption of the Motion Sensors in the Mobile Phones and Remote Controllers, Personal Computers and Tablets, Games and Portable Multimedia Players.

Image Stabilization, Context Awareness, Location Based Services and Remote Monitoring are the new emerging applications that will continue to drive the adoption

B. Vigna (✉) • E. Lasalandra • T. Ungaretti
STMicroelectronics s.r.l., Via Tolomeo 1, 20010 Cornaredo, MI, Italy
e-mail: benedetto.vigna@st.com

A.H.M. van Roermund et al. (eds.), *Nyquist AD Converters, Sensor Interfaces, and Robustness: Advances in Analog Circuit Design, 2012*,
DOI 10.1007/978-1-4614-4587-6_7, © Springer Science+Business Media New York 2013

of Motion Sensors in the market segments mentioned above as well in new ones, still unknown to most of the people.

Linear motion detection is requested to be combined with angular rate sensing along pitch, yaw and roll axis in the space to address a wide range of applications where three to nine degree of freedom sensing feature is required. One example is related to Inertial Measurement Unit (IMU) or Inertial Navigation Unit (INU) where accelerometer and gyroscopes are each other complementary to detect linear and angular motion in several applications including image stabilization, enhanced user interface for mobile phones, games and pointers/remote controllers, pedestrian and car navigation for location based services and fitness/wellness people monitoring. As seen for the accelerometer standalone, once again size, performances and price are the driving factors for gyroscope and 6XDOF/9XDOF success in Consumer & Industrial market.

In Optical Image Stabilization (OIS) applications, IMU can be used to track an object where in the images the subject slurs due to vibrations applied to the devices. OIS system can be implemented using pixel tracking by intensity movement or frame tracking. The IMU is usually embedded in video or into system with Pan-Tilt and Zoom capabilities or mounted on moving platforms in surveillance [1]. Motion activation and image stabilization are features aimed to improve picture or video quality.

Inertial Navigation Unit (INU) in pedestrian/car navigation is used for Dead Reckoning (DR) to estimate user's current location based upon a previously determined position, when no GPS signal is available or its quality is very poor. The information coming from the sensor cluster is processed through complex Kalman filters to estimate the current position of the user. This powerful feature open the doors to a wide variety of applications based on the user's actual location (the so called LBS, Location Based Services), making easier our everyday life. Motion Sensors are not enough, but they can work together with short range radios like Wifi to enable an outstanding indoor navigation experience.

7.2 THELMA™ and SMERALDO™: The Fabrication Processes of Motion MEMS

All gyroscopes and all the accelerometers manufactured by STMicroelectronics are composed of two chips stack assembled in one plastic Land Grid Array Package (Fig. 7.1). The first chip contains the micro-machined sensing element, that converts the angular rate and/or the acceleration in a capacitance variation, and it's manufactured in a proprietary fabrication process with 1.0 μm minimum lithography (THELMA™ or SMERALDO™); the second chip is using an advanced CMOS process (0.13 μm node) to amplify the small signal of the mechanical transducer and to enable an easy integration of the product in the final customer board. The companion pure CMOS chip of the motion sensor comes in two versions, analog

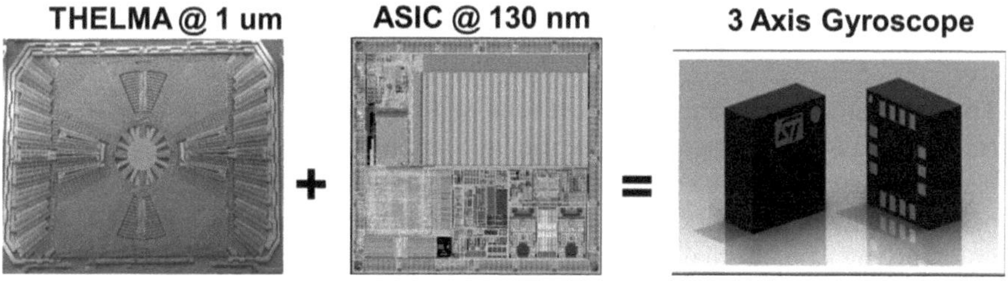

Fig. 7.1 The micro-machined element on the left is assembled with the high performance ASIC in a single pkg $4 \times 4 \times 1\ \mathrm{mm}^3$ on the right

Fig. 7.2 In THELMA™ process (*left picture*) the connection pads for wire bonding to ASIC are on the same side of the micro-machined structure and the cap has some clearance window during the bonding process. In SMERALDO™ instead the connections are on the back side of the micro-machined structure and the wafer sandwich is flat (*right picture*) during wafer to wafer bonding and at the end

and digital, depending on the integration of the Analog Digital Converter circuit block and on the need to have a dedicated microcontroller for signal processing.

THELMA™ process begins with a standard silicon wafer onto which a layer of first oxide (~2.5 μm) is grown for electrical isolation. A thin poly-silicon layer used for interconnections and a second sacrificial oxide (~2.5 μm) are then deposited. Into this layer, holes are etched at the points corresponding to the supports for fixed elements and anchors for moving elements. A thicker poly-silicon epitaxial layer (~25 μm) is grown on top of this, and into this third layer the structures for the moving and fixed elements of the device are etched with a single mask. Finally the sacrificial oxide layer beneath the structures is removed by an isotropic etching operation to free the moving parts. The open space around the structures is filled with a gas, usually dry nitrogen or argon to tune the damping factor and the resonance frequency. A second wafer with patterned getter layer and glass frit is then bonded to the first wafer to protect the tiny structures during an injection molding process during which high pressures are applied. *SMERALDO™* process is very much similar to *THELMA™*, but the first wafer has embedded through silicon vias (Fig. 7.2) enabling further die shrinkage.

7.3 The Tri-axial Coriolis' Gyroscope

The innovative compact micro-mechanical transducer is named "*The Beating Heart*" because it combines a triple tuning-fork structure within a single vibrating mass and because of its working mode. The Beating Heart achieves excellent performances in terms of thermal stability, cross-axis error and acoustic noise immunity by using a small die size. Furthermore, the presence of a single primary vibration mode for the excitation of the three tuning-forks simultaneously, together with the possibility of sensing the pickoff modes in a multiplexing fashion, allow to design a small area, low power companion chip. The advantage of having a single driving mass is the main reason of the big commercial success of STMicroelectronics gyroscopes. Infact, the presence of a single vibrating mass instead of multiple vibrating masses avoids the problems of mechanical frequency mixing and interference, extremely deleterious in the gyroscopes of other suppliers. The mechanical structure design is explained with the aid of Fig. 7.3.

The structure comprises four suspended plates coupled each other by means of four folded springs connected to their outer corners, and elastically connected to a central cross-shaped hinge by an additional set of central coupling springs. The primary mode of vibration (driving mode) consists of an in-plane inward/outward radial motion of the plates: on the whole, the structure cyclically expands and contracts, similarly to a "*Beating Heart*" (hence, the name). Primary actuation is provided with a set of comb-finger electrodes placed only on a pair of opposite plates; the mechanical motion is then propagated to the second pair by means of the coupling folded springs at the corners. The secondary modes of vibration (sensing modes) consist of an in-plane, opposite-phase motion of the second pair of plates (yaw mode), and two out-of-plane, opposite-phase motions of both pairs (pitch & roll modes). The yaw mode is sensed by a set of parallel-plate electrodes located on the second pair of plates, whereas the pitch and roll motions are detected by sensing the capacitive variations between each plate and an electrode placed underneath; additional comb-fingers electrodes are reserved for sensing the vibrating motion of the driving mode. The mechanical coupling between the two proof masses of each sensing pair allows to read the secondary vibrating motions in differential mode,

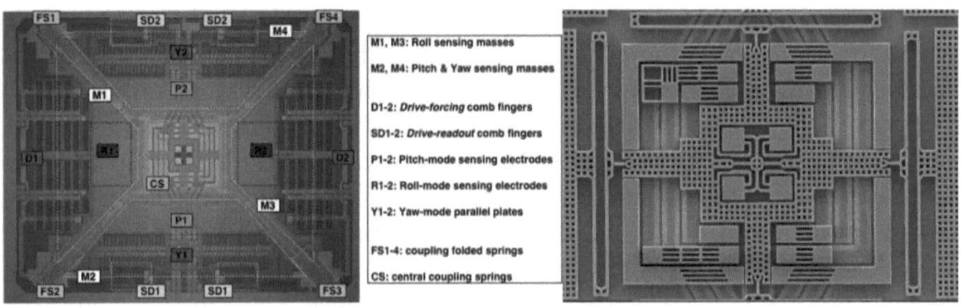

Fig. 7.3 The poly-silicon micro-machined structure on the left is expanding and contracting at a frequency of 20 kHz around the suspended center of mass highlighted in the *right picture*

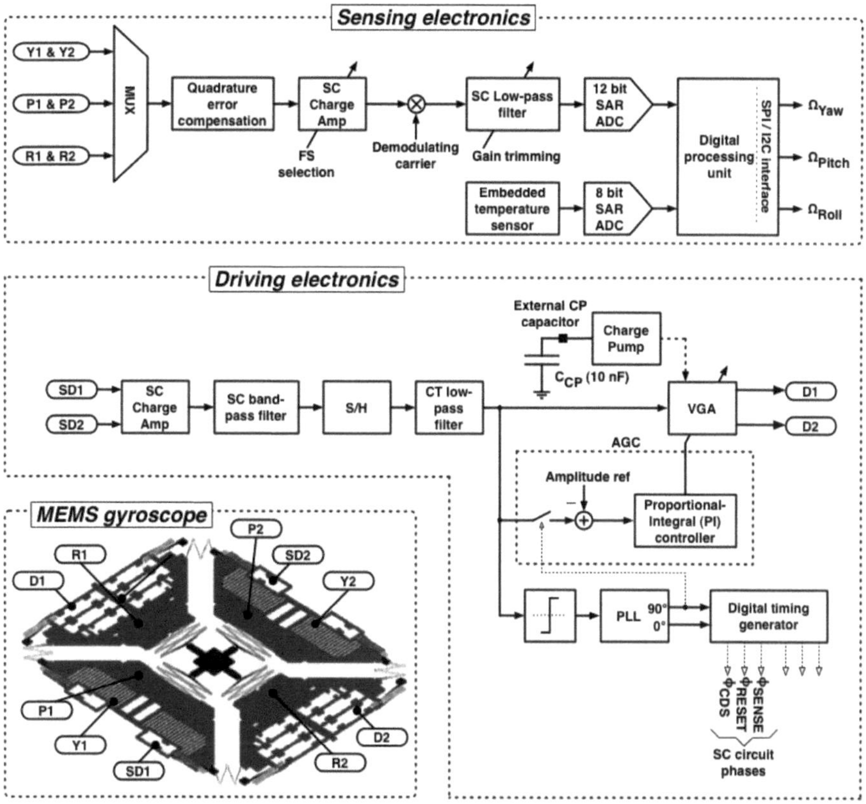

Fig. 7.4 System architecture block diagram

thus improving rejection of external linear accelerations and vibrations; moreover, each secondary mode has a single resonant sensing frequency, instead of two independent frequencies that require accurate matching to avoid performance degradations, such as in designs with uncoupled proof masses. The overall mechanical structure has frequency-unmatched primary and secondary modes, with a nominal primary resonant frequency of 20 kHz and the peak-to-peak movement of the driving mass is about 20 μm. This design choice, combined with a high and time stable quality factor (Q-Factor) guaranteed by the presence of a gettering material in the wafer-level cavity, enables an excellent level of acoustic noise isolation.

The Coriolis force exciting a secondary mode is proportional to the velocity of the driving mode and the input angular rate, and directed orthogonally to both the driving axis and sensor rotation axis. The angular rate measurement is obtained from the sensed Coriolis acceleration by demodulation, once the driving mode is oscillated at constant amplitude [2] (Fig. 7.4).

The primary mode is excited to oscillate at resonance by closing a feedback loop around the micro-resonator made up of the resonating masses and the drive-readout/drive-forcing comb-finger electrodes. In the feedback path, the capacitive

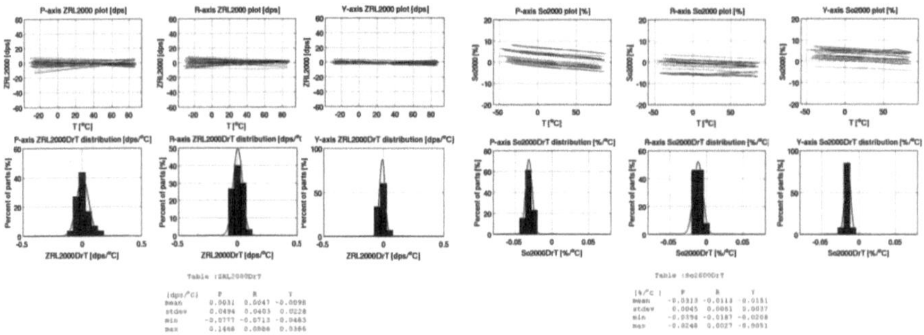

Fig. 7.5 ZRO stability over temperature and statistical distribution of the ZRO over temperature (First 6 graphs). Sensitivity (So) stability over temperature and statistical distribution of So over temperature

unbalancing generated by the oscillating motion of the primary mode is transduced into a voltage signal by a differential charge amplifier; then, a band-pass (BP) switching-capacitor (SC) amplifier removes the residual offset and provides the necessary phase adjustment to have a total loop phase shift of 360° at the resonant frequency, which is required for enforcing a sustained oscillation in the electromechanical loop. The BP amplifier output is interpolated by a second order continuous-time (CT) low-pass (LP) Chebyshev filter and it is amplified by a variable-gain amplifier (VGA). The VGA gain is automatically tuned by an outer automatic gain control (AGC) loop to regulate and verify the amplitude of the sustained oscillation at the CT-LP filter output to a constant set-point value. Finally, the VGA output, boosted by a charge-pump, is fedback to the comb-drive actuating electrodes. All internal timings are generated by a PLL synchronized with the CT-LP filter output.

A single, time-division multiplexed open-loop readout interface is used to retrieve the angular rate measurements out of the Coriolis accelerations along the three sensing axes. A differential charge-amplifier front-end converts the capacitive unbalancing induced by the Coriolis movement into a voltage signal, which is then synchronously AM-demodulated using a carrier in-phase with the velocity of the primary mode motion. A 12 bit SAR ADC performs internal A/D conversion at a rate of 6.06 kHz/axis; a 100/200/400/800 Hz output data rate (ODR) is selected by changing the decimation factor of the output sync-decimators. The final output is equal to 16 bit. The compensation of the quadrature error, the biggest enemy of the gyroscope due to a non-uniform deep silicon etching profile, is performed passively thanks to an extremely good control of the spring cross section.

The characterization results of 33 different samples are reported in Fig. 7.5 and tests are all done with a Full Scale of 2,000 degree per second.

The overall performances are excellent: the average noise density level is less than 0.02 dps/$\sqrt{}$Hz (with BW $=$ 40 Hz and ODR $=$ 200 Hz – note: dps $=$ degree per second), and the zero-rate output (ZRO) and sensor scale factor (So) are very stable over temperature – with FS $=$ 2,000 dps, the ZRO temperature sensitivity is less than $\pm$0.04 dps/°C, while the scale factor change over the temperature range

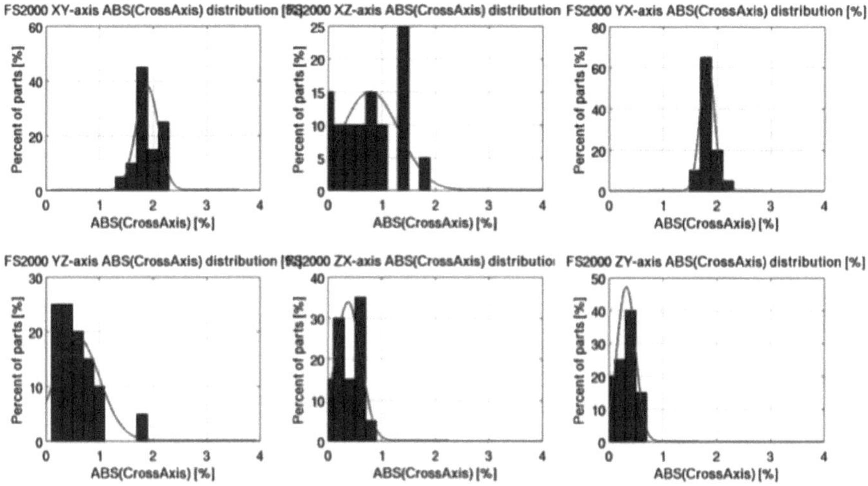

Table :ABS(CrossAxis)

[%]	XY	XZ	YX
mean	1.8949	0.7815	1.8230
stdev	0.2115	0.5324	0.1264
min	1.4640	0.0644	1.6441
max	2.2857	1.8636	2.1897

Table :ABS(CrossAxis)

[%]	YZ	ZX	ZY
mean	0.5896	0.3742	0.3215
stdev	0.4212	0.2361	0.1691
min	0.1248	0.0630	0.0287
max	1.8369	0.8828	0.6507

Fig. 7.6 Statistical distributions of the cross-axis sensitivities over a set of 33 samples. Tests have been performed with FS = 2,000 dps

−40 ÷ 85 °C is within ±2 % of the factory trimmed value (So = 70 mdps/LSB with FS = 2,000 dps). The design robustness is certified by the tight statistical distributions of the ZRO and So temperature sensitivities. Thanks to the highly symmetric mechanical design, the cross-axis sensitivities, measured as a percentage of the nominal selected full scale, are always below ±2 % and mainly due to mounting tolerances during packaging (Fig. 7.6).

High immunity to acoustic noise is evident from data reported in Fig. 7.7 (Test conditions FS = 2,000 dps, ODR = 200 Hz, output BW = 50 Hz)

Regarding power consumptions, with a supply voltage in the range 2.16 ÷ 3.6 V, the current absorbtion is 6 mA during normal operation, 1.5 mA in sleep mode (sensing electronics switched off, but with the driving microresonator still operative to reduce the power-on time), and 5 μA in power-down mode.

7.4 Six and Nine Degree of Freedom Motion Sensors

Depending on the application need of the customers, on the layout constraints for the final product and on customer's purchasing strategy there are cases in which Six Axes and Nine Axes Modules are preferred to a discrete solution. Being able to design and produce in the same technology platform both tri-axis accelerometers

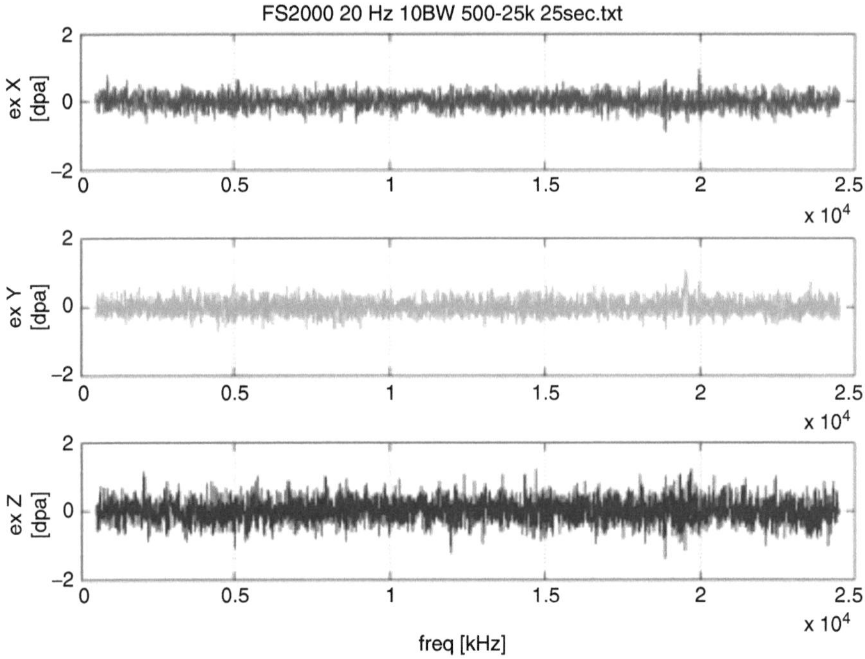

Fig. 7.7 Gyroscope output response to an acoustical stimulus (sine noise at 90 dBSPL with frequency sweeping in the range 500 Hz ÷ 25 kHz with steps of 5 Hz). The plots show the average values of the pitch/roll/yaw angular rate outputs for each frequency of the sinusoidal acoustical stimulus

and tri-axis gyroscopes, the Six Degree-Of-Freedom Combos are easily available for the market. STMicroelectronics is currently in production with an High Performance Six Axis Motion Sensors in $3.5 \times 5.0 \times 1.0\ \text{mm}^3$ package and it's aiming to futher miniaturize it.

The Six Axis Motion Sensor results from the combination of volume proven micro-mechanical structures realized on two different silicon dice or on the same die, but in any case assembled in the same package by means of System in Package techniques (Figs. 7.8 and 7.9).

At this point it's clear that by adding a 3X compass to the 6XDOF, a 9XDOF sensor can be easily realized. There are many different technologies available on the market to realize the compass, tecniques that range from Hall effect devices to MagnetoImpedance based sensors. STMicroelectronics decided to use the Anistotropic MagnetoResistance based compass thanks to its very high resolution, high temperature stability and low current consumption. Thanks to STMicroelectronics proven assembly techniques customers nowadays have the option to select a discrete approach as in Fig. 7.10 or an integrated solution as in Fig. 7.11.

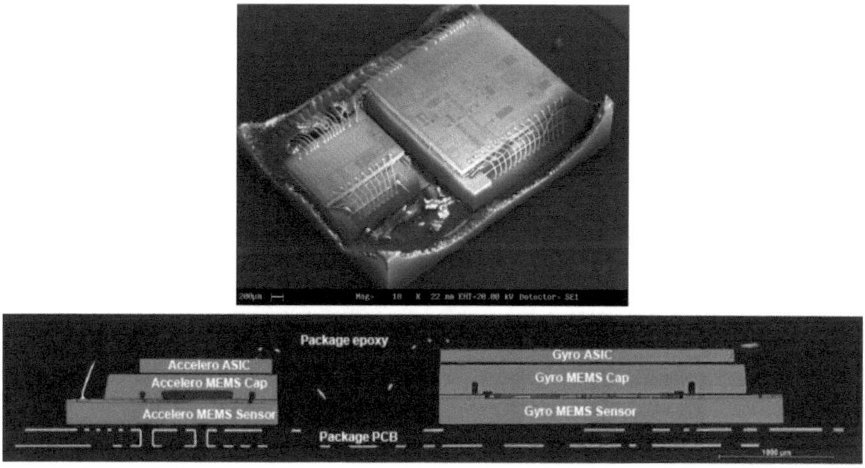

Fig. 7.8 Six axis motion sensor in $3 \times 5.5 \times 1.0\ \text{mm}^3$ LGA Package seen in perspective and in cross section. The accelerometer and gyroscope mechanical structures are located on two different silicon dice

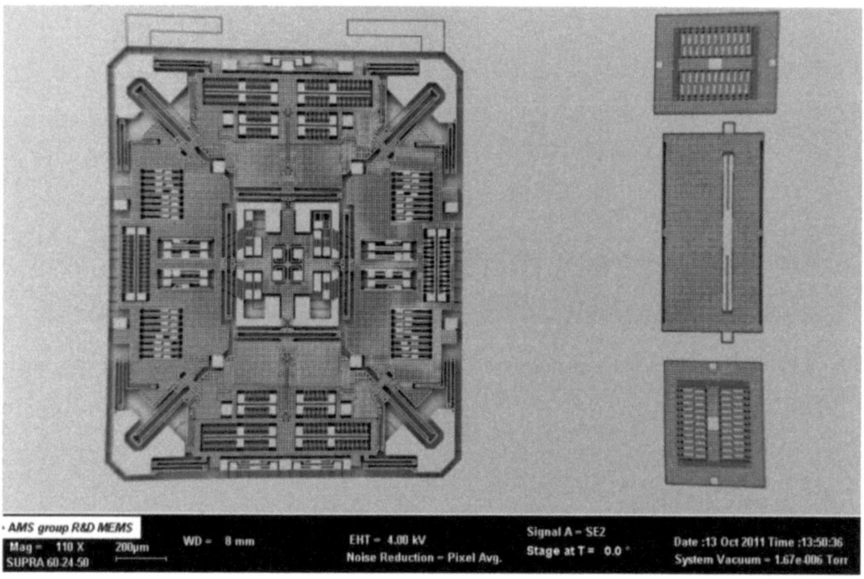

Fig. 7.9 3X gyroscope and 3X acceleroemeter realized on the same silicon die with THELMA™ micromachining process

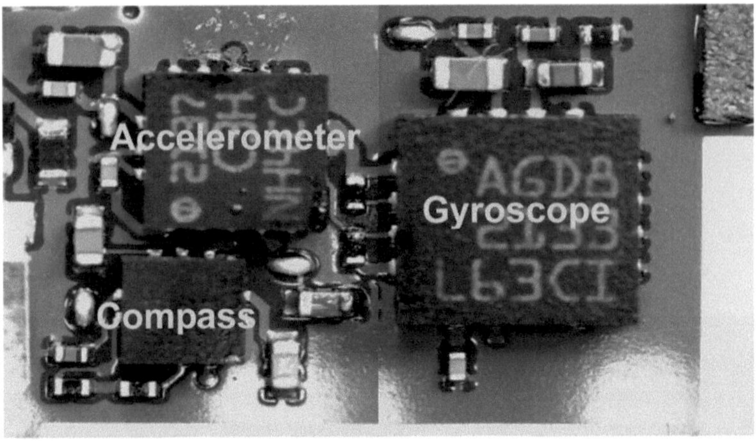

Fig. 7.10 The three basic components of the 9XDOF are each housed in three different packages. Accelerometer package footprint is $3 \times 3\ mm^2$, gyroscope one measures $4 \times 4\ mm^2$ and compass is $2 \times 2\ mm^2$, totalling a footprint of $29\ mm^2$ + the passive overhead

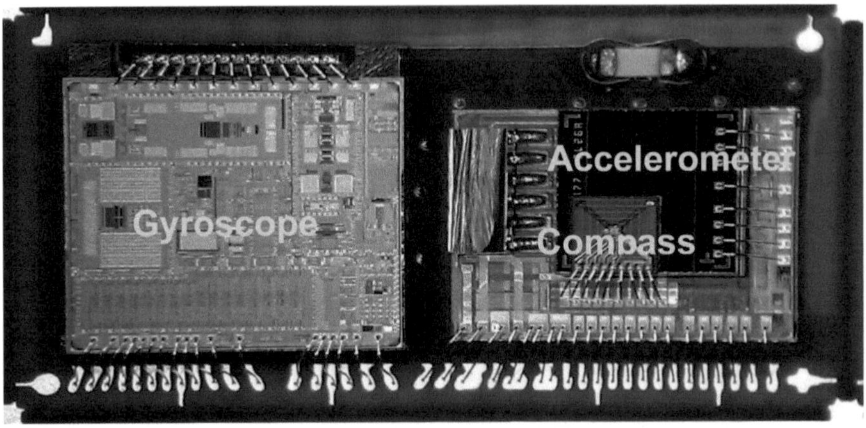

Fig. 7.11 The three basic components of the 9XDOF are packaged alltogether in one single package measuring only $3.0 \times 6.5 \times 1.0\ mm^3$

7.5 Conclusions

Motion Sensors, whether discrete or fully integrated in one single package, will continue to be an important block of all present and future consumer products. Enhanced User Interface, Optical Image Stabilization, Remote Monitoring and Location Based Services are the identified applications that will benefit of these technologies...but many more will come. Motion MEMS applications are only limited by imagination!

References

1. Vigna B (2011) STMicroelectronics masters the art and the science of MEMS. Semicon, Taiwan
2. Geen J et al (2002) Single-chip surface micromachined integrated gyroscope with 50°/h Allan deviation. IEEE J Solid-St Circ 37:1860–1866
3. Antonello R, Oboe R, Prandi L, Biganzoli F, Caminada C (2011) Open a low power 3-axis digital output MEMS gyroscope with single drive and multiplexed angular rate readout. In: International solid-state circuits conference, San Francisco, Feb 2011
4. Vigna B (2012) Consumer device go SMART with MEMS. Semicon, China
5. Johnson C (2012) Ten technologies that will change consumer devices. EETimes, 23 Jan 2012

Chapter 8
Energy-Efficient Capacitive Sensor Interfaces

Michiel A.P. Pertijs and Zhichao Tan

Abstract Capacitive sensor systems are potentially highly energy efficient. In practice, however, their energy consumption is typically dominated by that of the interface circuit that digitizes the sensor capacitance. Energy-efficient capacitive sensor interfaces are therefore a prerequisite for the successful application of capacitive sensors in energy-constrained applications, such as battery-powered devices and wireless sensor nodes. This chapter derives lower bounds on the energy consumption of capacitive sensor interfaces. A comparison of these bounds with the state-of-the-art suggests that there is significant room for improvement. Several approaches to improving energy efficiency are discussed and illustrated by two design examples.

8.1 Introduction

In low-power and energy-constrained applications, such as battery-powered systems and wireless sensor networks, capacitive sensors are an attractive choice since they do not consume static power, and the energy required to read them out can be very low [1, 2]. Energy-efficient capacitive sensor systems for the measurement of e.g. pressure [3], acceleration [4] and humidity [5] have been reported. In most cases, however, the energy consumption of such systems will be dominated by that of the interface circuit that digitizes the sensor capacitance. Hence, energy-efficient capacitance-to-digital converters (CDCs) are essential to make the most of the low-energy potential of capacitive sensors.

This chapter explores the limits on the energy consumption of CDCs. It will be shown in Sect. 8.2 that there is a significant gap between these limits and the energy

M.A.P. Pertijs (✉) • Z. Tan
Electronic Instrumentation Laboratory, DIMES, Delft University of Technology, Mekelweg 4, 2628 CD Delft, The Netherlands
e-mail: pertijs@ieee.org

A.H.M. van Roermund et al. (eds.), *Nyquist AD Converters, Sensor Interfaces, and Robustness: Advances in Analog Circuit Design, 2012*,
DOI 10.1007/978-1-4614-4587-6_8,

consumption of most designs reported in literature. Various reasons for this gap will be explored, some of which are linked to the fundamental limitations of practical capacitive sensors, others to the limitations of practical interface circuits. In Sect. 8.3, various ways to realize more energy-efficient interface circuits will be discussed. In Sect. 8.4, these approaches will be illustrated using two design examples, one based on period modulation, the other based on delta-sigma modulation. Finally, conclusions are presented in Sect. 8.5.

8.2 Limits on the Energy Consumption of CDCs

8.2.1 The Energy Efficiency Potential of Capacitive Sensors

To better appreciate the energy efficiency of practical CDCs, it is helpful to first establish a fundamental lower bound on their energy consumption. For this, we will consider the class of CDCs based on switched-capacitor techniques, a simplified representation of which is shown in Fig. 8.1a. It consists of a voltage source V_{ref} that charges a sensor capacitor C_x during a first phase ϕ_1. The resulting signal charge $Q_{sig} = V_{ref} \cdot C_x$ is measured by the interface circuit during a second phase ϕ_2.

Irrespective of the implementation of the interface circuit, the energy E_1 associated with charging the capacitor during phase ϕ_1 will be a lower bound on the overall energy consumption. This energy equals

$$E_1 = C_x V_{ref}^2 \tag{8.1}$$

The noise charge on the capacitor after phase ϕ_1 will equal $\overline{q_{n1}^2} = kTC_x$, giving a signal-to-noise ratio (SNR) of

$$SNR = \frac{Q_{sig}^2}{\overline{q_{n1}^2}} = \frac{C_x V_{ref}^2}{kT} = \frac{E_1}{kT} \tag{8.2}$$

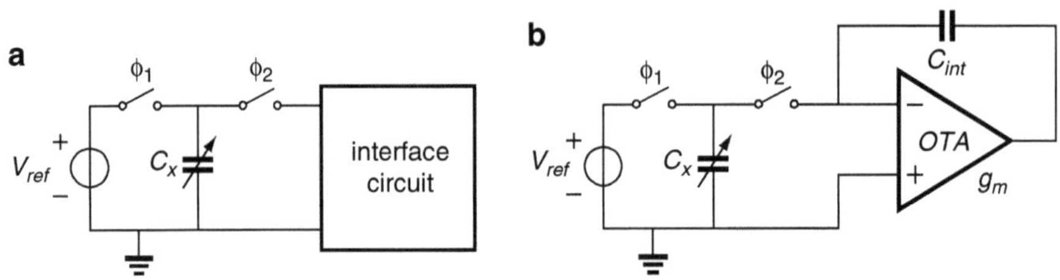

Fig. 8.1 (**a**) Simplified capacitive-sensor interface; (**b**) simplified interface based on an active integrator

This equation shows that, in order to obtain a given SNR, a minimum energy consumption of

$$E_1 = kT \cdot SNR \tag{8.3}$$

is required. For a given sensor capacitance C_x, V_{ref} can, in principle, be appropriately scaled to reach this minimum. If the required V_{ref} exceeds the available supply voltage, multiple samples of Q_{sig} could, again in principle, be averaged to obtain the desired SNR. For a given C_x and V_{ref}, this will increase the both the energy consumption and the SNR by a factor N, so that (8.3) remains valid. Equivalently, this lower bound can be expressed in terms of the effective number of bits (ENOB) of the interface by using the following equation [6]:

$$SNDR = 6.02\ \text{dB/bit} \cdot ENOB + 1.76\ \text{dB}, \tag{8.4}$$

which links $ENOB$ to the signal-to-noise-and-distortion (SNDR) ratio (in dB). For simplicity, we ignore the effect of distortion or non-linearity and equate SNDR to SNR, which leads to

$$E_1 = \frac{3}{2} kT \cdot 2^{2ENOB} \tag{8.5}$$

8.2.2 Energy Efficiency of State-of-the-Art CDCs

Figure 8.2 shows a plot of the energy consumption bound given by (8.5) as a function of $ENOB$, as well as the energy consumption of a variety of capacitive sensor interfaces reported in literature. These include interfaces based on successive approximation [7–9], delta-sigma modulation [4, 5, 10–15], pulse-width modulation [16–18] and period modulation [19–22]. Figure 8.2 shows that there is a substantial gap between the theoretical lower bound and the energy consumption of actual designs. In itself, this should not be a surprise, since it takes much more than just charging a capacitor to digitize its value. Nevertheless, the gap of more than four orders of magnitude is somewhat surprising.

Since many CDCs are structurally very similar to voltage-input ADCs, it is instructive to compare their energy efficiencies. In order to do this, the following well-known figure-of-merit (FOM) can be used, which relates energy consumption to the converter's effective number of bits (ENOB):

$$FOM = \frac{P}{BW \cdot 2^{ENOB}} = \frac{E_{meas}}{2^{ENOB}}, \tag{8.6}$$

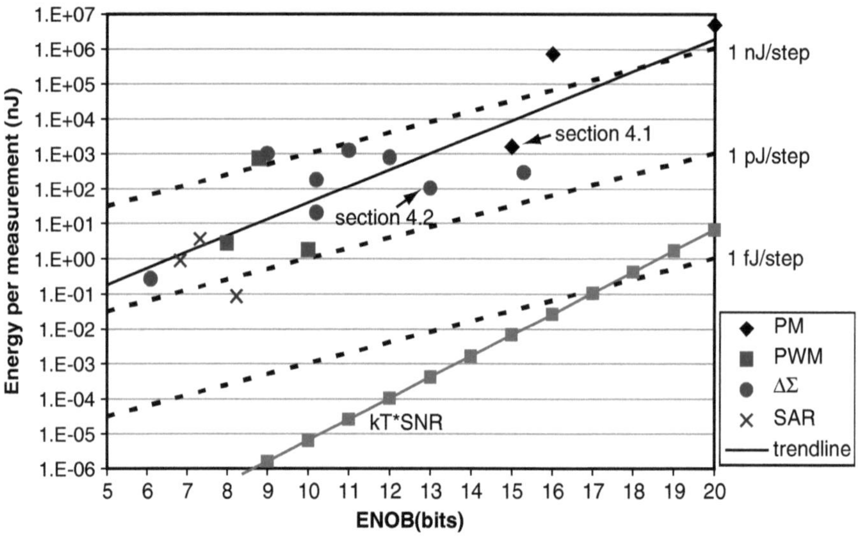

Fig. 8.2 Energy consumption of published CDCs based on period modulation (*PM*), pulse-width modulation (*PWM*), ΔΣ modulation (*ΔΣ*) and successive approximation (*SAR*), along with the lower bound given by Eq. 8.5. The example designs discussed in Sect. 8.4 are indicated

where P and BW is the ADC's power consumption and bandwidth, and E_{meas} is the energy consumed per measurement [23]. Figure 8.2 shows three dotted lines representing FOM values of 1 fJ/step, 1 pJ/step and 1 nJ/step. State-of-the-art voltage-input ADCs achieve FOMs as low as 10 fJ/step [24, 25], with average designs sitting around 1 pJ/step [23]. In contrast, most reported CDCs fail to achieve 1 pJ/step. This shows that CDCs, in spite of their similarity to ADCs, are at least an order of magnitude less energy efficient. In the following, we will discuss various explanations for this gap, and explore potential remedies.

8.2.3 *The Effect of Baseline and Parasitic Capacitance*

In the discussion so far, we have assumed that the information of interest is the total sensor capacitance C_x, and, moreover, that the only capacitor that needs to be charged to obtain this information is C_x. In practice, however, capacitive sensors are associated with a certain parasitic capacitance C_p, which also needs to be charged, but contributes no signal. Moreover, for many practical capacitive sensors, only a relatively small variation ΔC_x on a much larger baseline capacitance $C_{x,base}$ carries information about the measurand. Hence, the relevant signal charge Q_{sig} is then $V_{ref} \cdot \Delta C_x$ rather than $V_{ref} \cdot C_x$, while the noise charge increases to $\overline{q_n^2} = kT\left(C_x + C_p\right)$. Moreover, the energy consumption increases to $E_{1,p} = (C_x + C_p)V_{ref}^2$. Combining these equations, a new lower bound for the energy consumption can be found:

$$E_{1,p} = kT \cdot SNR \cdot \left(\frac{C_x + C_p}{\Delta C_x}\right)^2 \tag{8.7}$$

This shows that in the presence of baseline and parasitic capacitances, the minimum energy consumption required to achieve a given SNR increases substantially. For example, for a sensor whose capacitance only changes by 10% in response to a measurand and that is moreover associated with a parasitic capacitance of 10× the nominal sensor capacitance, the minimum energy consumption is 10,000× higher than that given by (8.3).

For a fair comparison of the energy efficiency of CDCs, their ability to handle parasitic capacitors should therefore be taken into account. To do so, one could define a modified FOM, taking the last term of (8.7) into account. Unfortunately, many papers in literature do not report the information required to calculate such a FOM. An example of a paper that does present this information is [20], which describes a high-resolution period-modulator-based interface; it is represented by the data point in the upper-right corner of Fig. 8.2. In terms of the ADC FOM given by (8.6), this design initially seems to be quite poor, achieving only 5 nJ/step. However, it can handle parasitic capacitance of more than 30 times the sensor capacitance while maintaining its resolution. Taking this into account via the last term in (8.7) results in a modified FOM of 5 pJ/step, which is much more in line with the state-of-the-art.

8.2.4 Energy Efficiency of Charge-Balancing CDCs

We will now attempt to estimate the energy consumption of practical CDCs, starting with the charge-balancing approach. In a charge-balancing CDC, a modulator converts the sensor capacitance into an intermediate signal, by balancing the signal charge Q_{sig} associated with the sensor capacitance C_x against a reference charge. A digital counter or decimation filter is then used to convert the intermediate signal into a final digital output. The intermediate signal can take the form of a quasi-digital signal, such as a pulse-width or period modulated signal, see e.g. [16, 19–22]. In this case, Q_{sig} is compared to the charge accumulated over time by integrating a reference current, so that effectively the sensor capacitance is converted into a time interval. In CDCs based on $\Delta\Sigma$ modulation, see e.g. [5, 10, 12, 15], Q_{sig} is balanced by the charge associated with a reference capacitor C_{ref}, so that an intermediate signal in the form of a bitstream is obtained, whose bit-density is proportional to the ratio of C_x and C_{ref}.

Here, we assume that the energy consumption of charge-balancing CDCs is dominated by the modulator. Moreover, we assume that the charge-balancing in the modulator is implemented using an active integrator built around an operational transconductance amplifier (OTA), as shown in Fig. 8.1b. After the sensor capacitance C_x has been charged to a reference voltage V_{ref} in a first phase ϕ_1, the resulting charge Q_{sig} is transferred to an integration capacitor C_{int} in a second phase ϕ_2, from which it is then removed by a current or by the charge of a reference capacitor to generate the intermediate signal. Since charge-balancing CDCs differ in how this is implemented, further details about this step are omitted here, and the energy consumption associated with charging C_x and transferring the charge from C_x to C_{int} is taken as a lower bound on the overall energy-consumption of charge-balancing CDCs using an active integrator.

To find this lower bound, the relation between the resolution and energy consumption for the circuit in Fig. 8.1b has to be found. For phase ϕ_1, the same equations discussed in Sect. 8.2.1 apply: the noise charge and energy consumption associated with this phase are $\overline{q_{n1}^2} = kTC_x$ and $E_1 = C_x V_{ref}^2$, respectively.

Assuming the bandwidth in phase ϕ_2 is determined by the transconductance g_m of the OTA and the sensor capacitance C_x (i.e. assuming that the load capacitance of the OTA and the on-resistance of the switches have a negligible effect on the bandwidth), the noise charge associated with this phase is $\overline{q_{n2}^2} = \gamma kTC_x$, where the factor γ accounts for the noise contribution of the OTA, and equals, for example, 4/3 for an OTA whose noise is dominated by an input differential pair in strong inversion [26]. By adding the noise charges of the two phases, and assuming the CDC is thermal-noise limited, we obtain the following expression for the SNR:

$$SNR = \frac{Q_{sig}^2}{\overline{q_{n1}^2} + \overline{q_{n2}^2}} = \frac{C_x V_{ref}^2}{kT} \cdot \frac{1}{1+\gamma} \tag{8.8}$$

The energy consumption during phase ϕ_2 can be found from the minimum supply current required by the OTA to ensure accurate charge transfer within the duration T_2 of this phase. Assuming exponential settling with a time constant $\tau = C_x/g_m$, it can be shown that the minimum g_m required equals

$$g_m = \frac{C_x}{\tau} = \frac{C_x \cdot K}{T_2}, \tag{8.9}$$

where $K = \ln(2^{ENOB+1})$ is the number of time constants required for a settling error less than 0.5LSB for a given ENOB. Assuming further that the supply current I_{OTA} of the OTA is proportional to its g_m, i.e. that I_{OTA}/g_m is a circuit-topology-dependent constant, the energy consumption during phase ϕ_2 can be written as

$$E_2 = T_2 \cdot V_{sup} \cdot I_{OTA} = T_2 \cdot V_{sup} \cdot \frac{I_{OTA}}{g_m} \cdot g_m = C_x \cdot V_{ref} \cdot \frac{I_{OTA}}{g_m} \cdot K \tag{8.10}$$

where, for simplicity, it is assumed that $V_{ref} = V_{sup}$. The total energy consumption E_{AI} associated with the use of an active integrator can then be written, using (8.8), as

$$E_{AI} = E_1 + E_2 = C_x \cdot V_{sup}^2 \cdot \left(1 + \frac{I_{OTA}}{V_{sup} \cdot g_m} \cdot K\right) \quad (8.11)$$

$$= kT \cdot SNR \cdot (1 + \gamma) \cdot \left(1 + \frac{I_{OTA}}{V_{sup} \cdot g_m} \cdot K\right) \quad (8.12)$$

Compared to (8.3), two extra terms appear in (8.12): a factor representing the noise contribution of the OTA, and a factor representing the OTA's energy consumption. Note that the factor K depends (logarithmically) on the SNR.

The front-end of CDCs based on a capacitance-to-voltage conversion followed by a voltage-input ADC typically employ essentially the same structure as the active integrator shown in Fig. 8.1b (see e.g. [10, 20]). Hence, the same energy-consumption lower bound applies to these CDCs.

Thus far, we have only considered the energy consumption associated with a single charge transfer from C_x to C_{int}. Most charge-balancing CDCs, in contrast, perform multiple charge transfers within a conversion, i.e. they oversample C_x. Nevertheless, in principle, (8.12) remains valid, because N-fold oversampling increases both the energy consumption and the SNR by a factor N.

The energy consumption represented by (8.12) is plotted in Fig. 8.3 for the example values of $\gamma = 4/3$, $g_m/I_{OTA} = 5\ V^{-1}$ (corresponding to a differential pair in strong inversion) and $V_{sup} = V_{ref} = 1$ V. Even though this energy consumption is about an order-of-magnitude higher than the lower bound given by (8.3), it is still well below that of practical charge-balancing CDCs. In part, this can be explained by the fact that practical charge-balancing CDCs, particularly those at the low-end of the resolution scale, are often not thermal-noise limited but quantization-noise limited. This is because practical sensor capacitances tend to be much larger than those required to arrive at a thermal-noise-limited design. For example, according to (8.8), a 10-bit CDC with $V_{ref} = 1$ V and $\gamma = 4/3$ would require an impractically small C_x of about 15 fF. An even smaller value would be required if the sensor capacitance is oversampled!

To get a feeling for the energy consumption of quantization-noise limited designs, Fig. 8.3 also shows curves representing a first-order and a second-order $\Delta\Sigma$ CDC for $C_x = 1$ pF. It is assumed that these CDCs require an oversampling ratio N of 2^{ENOB} and $2^{ENOB/2}$, respectively. The energy consumption is then calculated by multiplying (8.11) by N, using the same parameters as before. The first-order $\Delta\Sigma$ CDC remains quantization-noise limited across the full resolution range, while the first-order $\Delta\Sigma$ CDC becomes thermal-noise limited above roughly 17 bits, after which C_x needs to be increased. The values thus obtained come much closer to those of published designs.

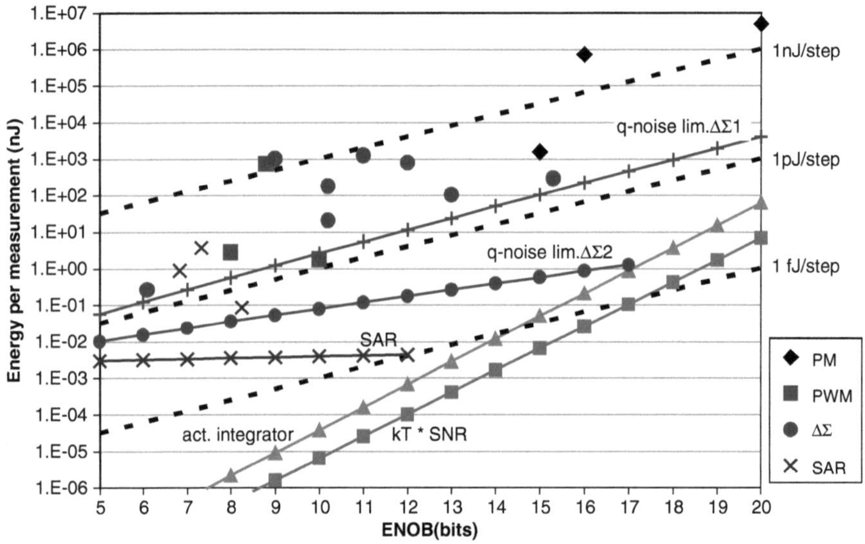

Fig. 8.3 Extension of Fig. 8.2 (energy consumption of published CDCs) with the lower bounds for an active integrator, given by Eq. 8.12, a quantization-noise limited first- and second-order ΔΣ modulator, and a SAR CDC, given by Eq. 8.13

8.2.5 *Energy Efficiency of Charge-Redistribution CDCs*

Rather than converting the sensor capacitance C_x to an intermediate signal by means of a charge-balancing approach, a direct comparison to the capacitance of a capacitive DAC C_{DAC} by means of charge redistribution is also possible. If a successive approximation algorithm is used, this approach leads to a SAR CDC [7–9]. The operating principle of SAR CDCs is very similar to that of general-purpose voltage-input SAR ADCs. While many SAR ADCs have been published recently [27], only a few SAR CDCs have been reported in literature [7–9]. Like most SAR ADCs, their ENOB is limited by component matching to less than 10 bits. They are thus not very suited for high-resolution applications, and care must be taken to use their limited resolution effectively by preventing the sensor's baseline capacitance from consuming too much of their input range [8].

Like SAR ADCs, SAR CDCs are potentially very energy efficient, since the only active elements required are a comparator and a SAR register. The most energy-efficient SAR CDC, however, has a FOM of 290 fJ/step [9], which is quite respectable in comparison with other CDCs (see Fig. 8.2), but still rather high compared to state-of-the-art SAR ADCs, which achieve FOMs around 10 fJ/step [24, 25]. One reason for this is that the sensor capacitance used in this design (10 pF) is significantly larger than that used in typical SAR ADCs (sub 1 pF).

In a typical implementation, SAR ADCs and CDCs will both employ a dynamic comparator, so that they do not consume any static current. The energy consumed per bit decision by the comparator and the logic is then associated with charging and discharging of gate capacitances, can be hence be expressed as $C_{gates}V_{sup}^2$. SAR ADCs have been reported with dynamic comparators that consume less than 100 fJ per bit decision, and with logic consuming similar amounts of energy [24, 25]. On a typical supply voltage of 1 V, this corresponds to an effective gate capacitance around 200 fF.

An approximate lower bound on the energy consumption of SAR CDCs can then be found by assuming that, in addition to the sensor capacitance, also the capacitive DAC has to be charged to the reference voltage throughout the conversion (ignoring, for simplicity, the benefits of more sophisticated charging schemes, e.g. [24]). Assuming further that the total capacitance of the DAC equals the sensor capacitance C_x, the energy per conversion is then

$$E_{SAR} = \left(2C_x + nC_{gates}\right)V_{ref}^2 \tag{8.13}$$

where n is the number of steps in the SAR algorithm, which we assume (optimistically) to be equal to the ENOB, and it is assumed that $V_{ref} = V_{sup}$.

The energy consumption represented by (8.13) is plotted in Fig. 8.3 for the example values of $C_x = 1$ pF, $C_{gates} = 200$fF, and $V_{ref} = 1$ V. This shows that the energy consumption significantly exceeds the lower bound given by (8.3), in particular for lower resolutions. This is because, for most practical sensor capacitances, SAR CDCs are quantization-noise limited, since capacitor values well below 1 pF would be required to obtain a thermal-noise limited design [25], while practical sensor capacitances tend to exceed this level. The energy consumption depends only weakly on the resolution, leading to considerably better FOMs for higher resolutions. Note that to achieve higher resolutions in practice, the energy consumption may increase if a larger DAC capacitance is needed to meet the matching requirements. Also note that any parasitic capacitance, if significant compared to C_x, should be included in (8.13).

8.3 Approaches for Improving Energy Efficiency

The discussion in the previous section shows that energy efficiency can be improved on the side of the sensor by reducing the sensor capacitance as much as possible to obtain a thermal-noise-limited design, by minimizing parasitic capacitances, by minimizing the baseline capacitance, and by maximizing the sensitivity, i.e. the relevant change ΔC_x with respect to the total sensor capacitance C_x. In some cases, however, such measures are not possible because of practical constraints associated with the sensor design. Equally important are the measures that can be taken on the side of the interface circuit to maximize energy efficiency. This section will discuss several examples of such measures.

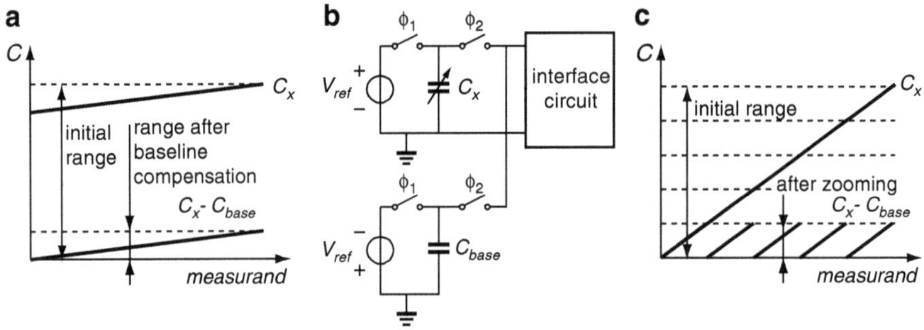

Fig. 8.4 Compensation for baseline capacitance: (**a**) principle; (**b**) implementation using a baseline-compensation capacitor; (**c**) extension to zooming

8.3.1 Baseline Compensation and Zooming

In quantization-noise-limited CDCs that oversample the sensor capacitance, methods that reduce the oversampling ratio can help to improve energy efficiency. For CDCs based on $\Delta\Sigma$ modulation, for example, this can be done by using higher-order loop filters (as already illustrated in Fig. 8.3), multi-bit feedback or cascaded architectures [28].

Here, we discuss an alternative approach that is based on preventing baseline or offset capacitance from consuming part of the CDC's input-capacitance range. Thus, the CDC's resolution can be reduced without increasing the smallest capacitance change that can be measured. This leads to a reduction in the required degree of oversampling, and hence, at least in principle, a reduced energy consumption. This concept is illustrated in Fig. 8.4a.

A straightforward implementation of this approach is shown in Fig. 8.4b: a compensation capacitor C_{base} is driven with a voltage of opposite polarity as that driving the sensor capacitor, so that the effective charge delivered to the interface is proportional to $(C_x - C_{base})$. This approach can be applied both to interfaces based on charge balancing, e.g. $\Delta\Sigma$ CDCs [5, 12, 15] as well as to interfaces based on charge redistribution, e.g. SAR CDCs [8].

While this approach relaxes the resolution requirements of the CDC, it does increase thermal noise, because the noise charge sampled on C_{base} adds to that on C_x. Moreover, charging and discharging C_{base} consumes extra energy. In this sense, C_{base} plays a similar role to that of the parasitic capacitance C_p in (8.7). Nevertheless, it still makes sense to add such a capacitor, as long as these drawbacks do not outweigh the energy reduction associated with the lower oversampling ratio.

If the baseline value is an invariable part of the sensor capacitance, a fixed compensation capacitor can be used, or a capacitor that is adjusted once after fabrication [5]. The baseline-compensation concept can also be extended by adjusting the compensation capacitor dynamically to track variations in the sensor capacitance [8, 12, 15]. This leads to a two-step conversion process that is sometimes referred to as zooming (Fig. 8.4c) [8]: in a first step, C_{base} is adjusted to approximate C_x; in a second step, $(C_x - C_{base})$ is digitized. Depending on how fast C_x changes, it may not be necessary to repeat the first step for every measurement. Typically, a successive approximation algorithm is used in the first step, while the second step may consist of a SAR conversion [8] or a conversion based on $\Delta\Sigma$ modulation [12, 15]. The resolution requirement of the second step is relaxed by the number of sub-ranges that can be distinguished in the first step.

In a CDC based on zooming, the limited linearity of the adjustable compensation capacitor may cause discontinuities in the digital output as the sensor capacitance moves from one sub-range to the next. This problem can be addressed by introducing some overlap between the sub-ranges so that C_x can be measured in two adjacent sub-ranges when it moves from one sub-range to the next, allowing the discontinuity to be digitally corrected. An alternative is to employ an auto-calibration approach, in which the CDC itself is used to accurately measure C_{base} [15]. While such an auto-calibration step may be relatively energy-hungry due to the high resolution required, it need not be performed for every measurement, and thus need not dominate the system's energy consumption.

8.3.2 Auto-calibration

The discussion so far has focused primarily on the relation between energy consumption and resolution, but in practice, energy is also spent to achieve a given level of accuracy, e.g. to reduce offset errors, gain errors, non-linearity, long-term drift. If the energy consumption of a CDC is limited by accuracy constraints rather than resolution constraints, it could be desirable to apply auto-calibration techniques to obtain the desired level of accuracy at the system level while using inaccurate, energy-efficient building blocks. An example of this is the period-modulator-based interface that will be discussed in Sect. 8.4.1.

Figure 8.5 shows an auto-calibration concept that can be applied in principle to any capacitive-sensor interface [3, 29]. At the input of the CDC, a multiplexer is added that selects between the sensor capacitance C_x, a reference capacitance C_{ref}, and an offset capacitance C_{off}. While an explicit offset capacitor C_{off} can be used, often the associated terminals of the multiplexer are left floating, so that only the parasitic capacitances and the offset of the CDC are measured if C_{off} is selected. The CDC is assumed to provide a digital output D_{out} that is a linear function of the capacitor C_i applied to its input:

$$D_{out} = a \cdot C_i + b, \tag{8.14}$$

where the coefficients a and b can be poorly defined, e.g. subject to device-to-device variation, long-term drift, power-supply variations etc. If the multiplexer is

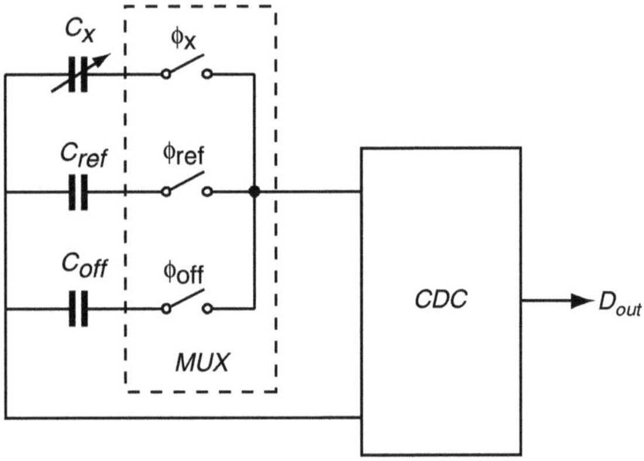

Fig. 8.5 Auto-calibrated capacitive-sensor interface

employed to successively digitize $(C_x + C_{off})$, $(C_{ref} + C_{off})$ and C_{off}, the uncertainty due to a and b can be eliminated by means of the following digital post-processing:

$$M = \frac{D_{out,Cx+Coff} - D_{out,Coff}}{D_{out,Cref+Coff} - D_{out,Coff}} = \frac{C_x}{C_{ref}} \tag{8.15}$$

where it is assumed that the coefficients a and b, while poorly defined, do not change between the three successive conversions. Thus, C_x is measured ratiometrically with respect to C_{ref}, and a measurement result M is obtained that is independent of a, b and C_{off}.

This auto-calibration approach comes at the cost of the extra energy consumed by the conversions of C_{off} and C_{ref}. In return, it strongly relaxes the offset- and gain-accuracy requirements of the CDC. In some cases this can be translated into a significant reduction in energy consumption. In interfaces based on period modulation, for instance, the propagation delay of the comparator introduces a poorly-defined offset error. Without auto-calibration, this error needs to be reduced to a level that agrees with the overall accuracy requirements of the interface. By means of auto-calibration, the error is canceled, so that a slow, but energy-efficient, comparator can be used [21].

8.3.3 Energy-Efficient Building Blocks

The implementation of the building blocks of a CDC also has a significant impact on its energy efficiency, since different circuit topologies may require substantially different amounts of energy to implement essentially the same functionality. As an example, consider a charge-balancing CDC that employs an active integrator built around an operational transconductance amplifier (OTA), as in Fig. 8.1b. To obtain the same transconductance, which is the key design parameter, different topologies require quite different supply-current levels. This relates to the g_m/I_{OTA} term in (8.12).

A telescopic OTAs, for instance, consumes roughly half the supply current of a current-mirror OTA or a folded-cascode OTA with the same transconductance. This saving obviously comes at the cost of a significantly reduced output swing. If it is possible, however, to deal with this reduced swing at the architectural level, it may be possible to save substantial energy by applying a telescopic structure.

An example of this will be discussed in the context of the period modulator presented in Sect. 8.4.1. A further example is the use of a feed-forward loop-filter topology in $\Delta\Sigma$ modulators to reduce the output swing and gain requirements of the OTAs, thus opening the way to more energy-efficient circuit implementations. A very promising direction in energy-efficient OTAs is the use of complementary transconductance topologies, such as inverters [30]. In these topologies, both NMOS and PMOS transistors contribute transconductance while sharing the same supply current, making them twice as energy-efficient as an equivalent topology in which only the NMOS or the PMOS transistors contribute transconductance.

8.4 Examples of Energy-Efficient Capacitive Sensor Interfaces

In this section we will discuss two examples of energy-efficient capacitive-sensor interfaces, one based on period modulation [21, 22], and one based on delta-sigma modulation [8].

8.4.1 An Energy-Efficient Interface Based on Period Modulation

The first design example illustrates the use of auto-calibration [29] as well as the use of a negative feedback architecture [19, 31] to enable the use of inaccurate, but energy-efficient, building blocks in a period-modulator-based CDC. Figure 8.6 shows its operating principle. During phase ϕ_1 of a two-phase non-overlapping clock, the sensor capacitor C_x is connected between the supply voltage V_{dd} and a mid-supply common-mode reference V_{cm}. During the subsequent phase ϕ_2, it is connected between V_{ss} and the virtual ground of an active integrator, which is also biased at V_{cm}. As a result, a charge $V_{dd} \cdot C_x$ is transferred to the integration capacitor C_{int}, causing the output voltage V_{int} of the integrator to step up. A constant integration current I_{int} then removes the charge from C_{int}, bringing V_{int} back to its original level. A comparator at the output of the integrator detects the moment when this happens. The time interval T_{msm} between the start of phase ϕ_2 and this moment is then proportional to C_x:

$$T_{msm} = \frac{V_{dd}}{I_{\text{int}}} C_x \tag{8.16}$$

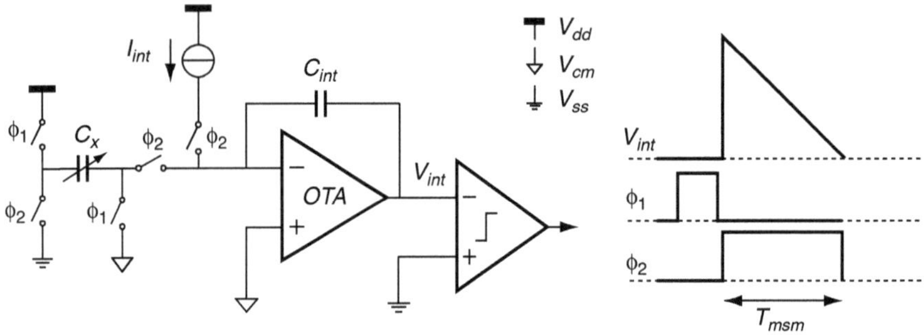

Fig. 8.6 Operating principle of the period-modulator-based capacitive-sensor interface [21]

This time interval can thus be used as a measure of C_x and can be digitized by counting its duration in terms of clock cycles of a faster reference clock. This digitization can be easily performed, for instance, by a counter in a microcontroller, leading to a digital output proportional to C_x, as in (8.14).

Due to process variability and various circuit non-idealities, such as comparator delay, supply-voltage variations, component tolerances and their temperature dependencies, the relation between T_{msm} and C_x is affected by offset and gain errors. While some of these non-idealities, such as comparator delay, could be mitigated at the expense of increased energy consumption, this design employs simple, energy-efficient building blocks and removes the errors by means of the auto-calibration approach discussed in Sect. 8.3.2.

A further step in reducing energy consumption is taken by employing a simple cascoded telescopic OTA in the integrator, as opposed to the more current-hungry Miller-compensated opamp used in a previous period-modulator-based design [19]. This can only be done if the output swing is kept within the limited range of this telescopic topology. As illustrated in Fig. 8.6, the output swing of the OTA is directly proportional to the ratio of C_x and C_{int}, which implies that the swing could be reduced by increasing C_{int}. This, however, would come at the cost of increased die size. Instead, we employ a negative feedback loop, shown in Fig. 8.7, that controls the charge transfer between C_x and C_{int} in such a way as to limit the swing at the OTA's output [31].

This negative feedback loop works as follows. Rather than hard-switching C_x to V_{ss} in phase ϕ_2, it is driven by the output of an amplifier OTA_F operated at a current well below that of the main OTA. As soon as OTA_F detects that V_{int} exceeds a maximum level V_b, it limits the current to I_{int}, keeping V_{int} constant. This continues until the drive-side of C_x has almost reached V_{ss}, at which point V_{int} ramps down to the comparator's threshold level. Since no charge is lost during this entire operation, the total amount of charge transferred from C_x to the integrator is not affected by the feedback loop. Thus, the capacitance-to-time conversion remains the same, i.e. (8.16) still holds.

This approach enables the use of a telescopic OTA without using a large C_{int}. It comes at the cost, however, of a reduced ability to handle parasitic capacitors: in the presence of large parasitic capacitors around the sensor capacitor, the

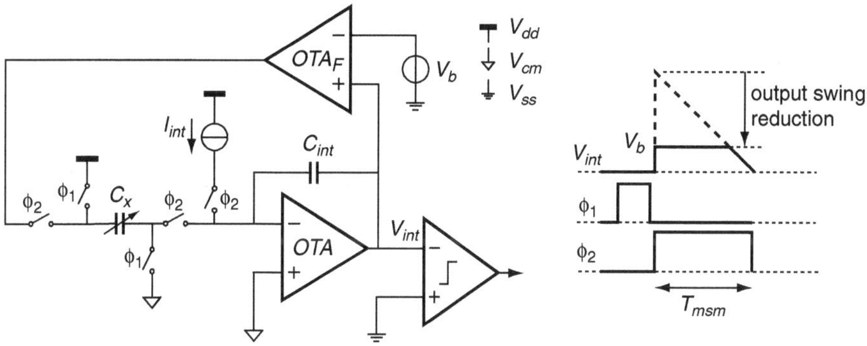

Fig. 8.7 Use of a negative feedback loop to limit the OTA's output swing

period modulator may fail to oscillate, or the negative feedback loop can become unstable. The associated trade-offs are discussed in detail in [22]. The prototype discussed here was designed to be able to handle parasitic capacitances of at most five times C_x.

Experimental results show that the interface achieves 15-bit resolution and 12-bit linearity within a measurement time of 7.6 ms for sensor capacitances up to 6.8 pF, while consuming 64 μA from a 3.3 V power supply. This corresponds to an energy per measurement of 1.6 μJ and a FOM of 49 pJ/step. As can be seen in Fig. 8.2, the design is considerably more energy efficient than previous period-modulation-based interfaces. For the same ENOB, however, it is outperformed by ΔΣ-based interfaces.

8.4.2 An Energy-Efficient Interface Based on Delta-Sigma Modulation for a Capacitive Humidity Sensor

The second design example illustrates the use of baseline capacitance compensation in a capacitive-sensor interface based on ΔΣ modulation [5]. The interface was designed to readout a co-integrated capacitive humidity sensor. The 100 μm × 100 μm sensor consists of an array of meander-fork electrodes implemented in the top metal layer of a CMOS process, and coated with a humidity-sensitive dielectric layer (Fig. 8.8). The sensor has a sensitivity of about 1 fF/%RH on a nominal capacitance of 0.8 pF. Hence, the capacitance variation associated with changes in relative humidity is on the order of 0.1 pF, which is substantially smaller than the sensor's baseline capacitance. Therefore, the CDC employed in this design can benefit from the baseline-capacitance compensation techniques discussed in Sect. 8.3.1.

Figure 8.8 illustrates the operating principle of this design. It is based on a second-order switched-capacitor ΔΣ modulator in which the sensor capacitor C_x is embedded: it serves as the sampling capacitor of the first stage [32]. In Fig. 8.8, the

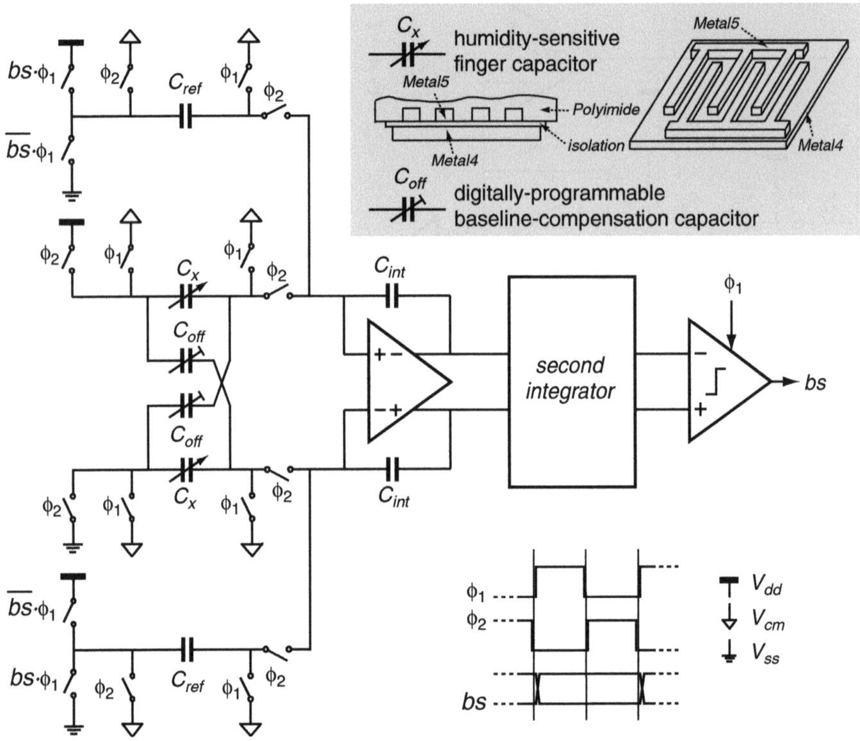

Fig. 8.8 Operating principle of the capacitive humidity sensor with ΔΣ CDC

details of the second integrator have been omitted for simplicity. A fully-differential design is employed, in which the sensor capacitance has been split into two equal parts, which are then driven by voltages of opposite polarity. To remove the baseline capacitance and thus relax the resolution requirements for the modulator, programmable cross-coupled compensation capacitors C_{off} are employed. These offset capacitors can be programmed digitally from 0.1 pF to 1.5 pF in steps of 0.1 pF.

As a result of the cross-coupling, a charge proportional to $C_x - C_{off}$ is integrated every clock cycle. In addition, a charge proportional to a reference capacitor C_{ref} is integrated with a polarity that depends on the bitstream output *bs*. The negative feedback in the modulator ensures that the former charge is balanced by the latter, resulting in a net zero average charge flowing into the first integrator:

$$\left(C_x - C_{off}\right) - \mu \cdot C_{ref} + (1 - \mu) \cdot C_{ref} = 0 \tag{8.17}$$

where μ is the bit-density of the bitstream ($0 \leq \mu \leq 1$). Solving for μ gives:

$$\mu = \frac{1}{2} + \frac{C_x - C_{off}}{2C_{ref}} \tag{8.18}$$

Thus, the modulator produces a bitstream *bs* whose bit-density is proportional to the ratio of $(C_x - C_{off})$ and C_{ref}. A simple digital sinc^2 decimation filter converts this bitstream into a digital number representing relative humidity.

A reference capacitor C_{ref} of 0.4 pF is used, which is larger than what is strictly necessary to digitize the expected variation in C_x. This value was nevertheless chosen to ensure that the interface can be used with more sensitive versions of the sensor obtained by expected improvements in the post-processing. With the present sensor, the interface needs to perform a 13-bit capacitance-to-digital conversion in order to achieve a humidity-sensing resolution of 0.1% RH. A second-order modulator is used to achieve this resolution at a modest number of clock cycles of 500. This is less than what would have been needed without compensation for the baseline capacitance, but a further reduction could be achieved by optimizing C_{ref}.

To achieve sufficient signal swing and DC gain, the two integrators of the modulator have been implemented using fully-differential folded-cascode OTAs with switched-capacitor common-mode feedback. At a 1.8 V supply and a bias current of 1.6 μA, the OTAs provide 85 dB DC gain. The 1-bit quantizer is implemented by a low-power dynamic comparator consuming 0.42 μA. Including a bias circuit, the complete modulator only consumes 5.85 μA from a 1.8 V power supply. The measurement results show that in a 10.2 ms measurement time, the interface achieves a resolution of 13 bits with respect to the full capacitance range. This corresponds to an energy consumption of 107 nJ per measurement, and a FOM of 13 pJ/step. As can be seen in Fig. 8.2, this level of energy efficiency is quite respectable for this ENOB. Only the design presented in [15] provides better energy efficiency, thanks to the fact that it employs the zoom approach discussed in Sect. 8.3.1.

8.5 Conclusions

This chapter has explored the limits on the energy consumption of capacitive sensor interfaces, starting from the fundamental lower bound given by the energy needed to generate a charge proportional to the sensor capacitance with sufficient SNR, while also taking its parasitic capacitance into account. Lower bounds on the energy consumption of practical interfaces based on charge balancing and charge redistribution have also been provided. It has been shown that the energy consumption of practical interfaces reported in literature is quite a bit higher than these lower bounds. Several approaches to reducing the energy consumption have been presented: compensation for baseline capacitance, either statically or dynamically by means of a zoom approach; auto-calibration at the system level to enable the use of energy-efficient yet inaccurate building blocks; and improved circuit topologies for these building blocks. Finally, two design examples have been presented to illustrate these approaches: an energy-efficient interface based on period modulaton, and an interface based on ΔΣ modulation.

Given the gap between the lower bounds on energy-consumption derived in this chapter and the energy consumption of practical designs, substantial improvements in energy efficiency should be possible. This will benefit applications such as autonomous data-logging applications and RFID-based wireless sensors, in which the energy consumption of the sensor and its interface dominates the overall energy consumption.

References

1. Li X, Meijer GCM (2008) Capacitive sensors. In: Meijer GCM (ed) Smart sensor systems. Wiley, Chichester
2. Baxter LK (1997) Capacitive sensor, design and applications. IEEE Press, New York
3. Smith MJS, Bowman L, Meindl JD (1986) Analysis, design, and performance of micropower circuits for a capacitive pressure sensor IC. IEEE J Solid-St Circ SC-21(6):1045–1056
4. Paavola M et al (2009) A micropower $\Delta\Sigma$-based interface ASIC for a capacitive 3-axis micro-accelerometer. IEEE J Solid-St Circ 44(11):3193–3210
5. Tan Z et al (2011) A 1.8 V 11μW CMOS smart humidity sensor for RFID sensing applications. In: Proceedings of IEEE asian solid-state circuits conference (A-SSCC), Jeju, korea, pp 105–108
6. van de Plassche R (2003) CMOS integrated analog-to-digital converters, 2nd edn. Kluwer, Boston
7. Bechen B, Weiler D, Boom T, Hosticka BJ (2006) A 10 bit very low-power CMOS SAR-ADC for capacitive micro-mechanical pressure measurement in implants. Adv Radio Sci 4:243–246
8. Tanaka K et al (2007) A 0.026 mm^2 capacitance-to-digital converter for biotelemetry applications using a charge redistribution technique. In: Proceedings of IEEE Asian solid-state circuits conference (A-SSCC), Jeju, korea, pp 244–247
9. Vo TM et al (2009) A 10-bit, 290 fJ/conv. steps, 0.13 mm^2, zero-static power, self-timed capacitance to digital converter. In: Proceedings of international conference on Solid State Devices and Materials (SSDM) Sendai, Japan
10. Bracke W, Puers R, Van Hoof C (2007) Ultra low power capacitive sensor interfaces. Springer, Dordrecht
11. Jawed SA, et al (2008) A 828 μW 1.8 V 80 dB dynamic range readout interface for a MEMS capacitive microphone. In: Proceedings of European solid-state circuits conference (ESSCIRC), Edinbourgh, UK, pp 442–445
12. Shin D-Y, Lee H, Kim S (2011) A delta–sigma interface circuit for capacitive sensors with an automatically calibrated zero point. IEEE Trans Circ Syst II 58(2):90–94
13. AD7156 datasheet, Analog devices (Online). http://www.analog.com
14. Danneels H, Coddens K, Gielen G (2011) A fully-digital, 0.3 V, 270 nW capacitive sensor interface without external references. In: Proceedings of European solid-state circuits conference (ESSCIRC), Helsiulei, Fiuland, pp 287–290
15. Xia S, Makinwa K, Nihtianov S (2012) A capacitance-to-digital converter for displacement sensing with 17 b resolution and 20 μs conversion time. In: International solid-state circuits conference (ISSCC), Digest of technical papers San Francisco, USA (in press)
16. Bruschi P, Nizza N, Piotto M (2007) A current-mode, dual slope, integrated capacitance-to-pulse duration converter. IEEE J Solid-St Circ 42(9):1884–1891
17. Bruschi P, Nizza N, Dei M (2008) A low-power capacitance to pulse width converter for MEMS interfacing. In: Proceedings of European solid-state circuits conference (ESSCIRC), Edinburgh, UK, pp 446–449
18. Lu JH-L et al (2011) A low-power, wide-dynamic-range semi-digital universal sensor readout circuit using pulsewidth modulation. IEEE Sens J 11(5):1134–1144

19. Heidary A, Meijer GCM (2008) Features and design constraints for an optimized SC front-end circuit for capacitive sensors with a wide dynamic range. IEEE J Solid-St Circ 43 (7):1609–1616
20. Heidary A, Heidary Shalmany S, Meijer G (2010) A flexible low-power high-resolution integrated interface for capacitive sensors. In: Proceedings of international symposium on industrial electronics (ISIE), Bari, Haly, pp 3347–3350
21. Tan Z, Pertijs MAP, Meijer GCM (2011) An energy-efficient 15-bit capacitive sensor interface. In: Proceedings of European solid-state circuits conference (ESSCIRC), Helsinki, Finland, pp 283–286
22. Tan Z, Heidary Shalmany S, Meijer GCM, Pertijs MAP (2012). An energy-efficient 15-bit capacitive-sensor interface based on period modulation. IEEE J Solid-St Circ 47(7): pp 17011–1711
23. Murmann B. ADC performance survey 1997–2011 (Online). http://www.stanford.edu/~murmann/adcsurvey.html
24. van Elzakker M et al (2010) A 10-bit charge-redistribution ADC consuming 1.9 μW at 1 MS/s. IEEE J Solid-St Circ 45(5):1007–1015
25. Harpe P et al (2011) A 26 μW 8 bit 10 MS/s asynchronous SAR ADC for low energy radios. IEEE J Solid-St Circ 46(7):1585–1595
26. Schreier R, Silva J, Steensgaard J, Temes GC (2005) Design-oriented estimation of thermal noise in switched-capacitor circuits. IEEE Trans Circ Syst-I 52(11):2358–2368
27. Cho S-H, Lee C-K, Kwon J-K, Ryu S-T (2011) A 550-μW 10-b 40-MS/s SAR ADC with multistep addition-only digital error correction. IEEE J Solid-St Circ 46(8):1881–1892
28. Norsworthy SR, Schreier R, Temes GC (eds) (1997) Delta-sigma data converters: theory, design and simulation. IEEE Press, Piscataway/New York
29. Meijer GCM (2008) Interface electronics and measurement techniques for smart sensor systems. In: Meijer GCM (ed) Smart sensor systems. Wiley, Chichester
30. Chae Y, Han G (2009) Low voltage, low power, inverter-based switched-capacitor delta-sigma modulator. IEEE J Solid-St Circ 44(2):458–472
31. Meijer GCM, Iordanov VP (2001) SC front-end with wide dynamic range. Electron Lett 37 (23):1377–1378
32. Malcovati P et al (1995) Combined air humidity and flow CMOS microsensor with on-chip 15 bit sigma-delta A/D interface. In: Symposium on VLSI circuits, Digest of technical papers, Kyoto, Japan, pp 45–46

Chapter 9
Interface Circuits for MEMS Microphones

Piero Malcovati, Marco Grassi, and Andrea Baschirotto

Abstract This paper presents an overview of interface circuits for capacitive MEMS microphones. The interface circuits and the building blocks are analyzed in detail, highlighting the most important design issues and trade-offs. Moreover, two design examples are reported, including circuit details and experimental results. The first example is based on a conventional constant-charge approach, while the second introduces the force-feedback concept. Both examples are implemented in a 0.35-μm CMOS technology and achieve a signal-to-noise and distortion ratio larger than 60 dB with a power consumption of about 1 mW from a 3.3-V power supply.

9.1 Introduction

The first microphone has been invented in 1876. The carbon microphones developed in 1878 were the core of early telephone systems. Ribbon microphones were invented in 1942 for radio-broadcasting. The introduction of a self-biased condenser microphone in 1962, namely the electret condenser microphone, combined high-sensitivity and broad frequency-range features with low-cost. Since then,

P. Malcovati (✉)
Department of Industrial and Information Engineering, Department of Electrical Engineering, University of Pavia, Via Ferrata 1, Pavia 27100, Italy
e-mail: piero.malcovati@unipv.it

M. Grassi
Department of Electrical Engineering, University of Pavia, Pavia, Italy
e-mail: marco.grassi@unipv.it

A. Baschirotto
Department of Physics, University of Milano Bicocca, Milano, Italy
e-mail: andrea.baschirotto@unimib.it

A.H.M. van Roermund et al. (eds.), *Nyquist AD Converters, Sensor Interfaces, and Robustness: Advances in Analog Circuit Design, 2012*,
DOI 10.1007/978-1-4614-4587-6_9, © Springer Science+Business Media New York 2013

electret microphones have dominated the market for high-volume applications, with a production of almost one billion parts per year (almost 90% of all microphones produced).

The first microphone based on silicon micro-machining (MEMS microphone) was introduced in 1983. Following the trend toward miniaturization in electronic devices, MEMS microphones are gaining market share in consumer applications [1]. The two major areas that are driving the penetration of MEMS microphones are hearing aids and mobile phones. Indeed, MEMS microphones offer several advantages with respect to electret devices: they are smaller in size, compatible with high-temperature automated printed circuit board (PCB) mounting processes, and less susceptible to mechanical shocks. Moreover, MEMS sensors can be integrated together with the CMOS electronics on the same chip or within the same package (micromodule approach [2]), thus reducing area, complexity and costs, while increasing efficiency and reliability [3–14]. MEMS microphones are based on different transduction principles, such as piezoelectric, piezoresistive, and optical detection. However, 80% of the produced MEMS microphones are based on capacitive transduction [15–18], since it achieves higher sensitivity, consumes less power, and is more inline with batch production [19].

9.2 MEMS Microphones

A microphone is a transducer, which translates a perturbation of the atmospheric pressure, hereinafter sound, into an electrical quantity. In a condenser microphone, the pressure variation leads to the vibration of a mechanical mass, which is transformed into a capacitance variation.

Sound pressure is typically expressed in dB_{SPL} (sound-pressure-level). A sound pressure of 20 μPa, corresponding to 0 dB_{SPL}, is the auditory threshold (the lowest amplitude of a 1-kHz signal that a human ear can detect). The sound pressure levels of a face-to-face conversation ranges between 60 dB_{SPL} and 70 dB_{SPL}. The sound pressure rises to 94 dB_{SPL} if the speaker is at a distance of 1 in. from the listener (or the microphone), which is the case, for example, in mobile phones. Therefore, a sound pressure level of 94 dB_{SPL}, which corresponds to 1 Pa, is used as a reference for acoustic applications. The performance parameters for acoustic systems, such as the signal-to-noise ratio (*SNR*), are typically specified at 1-Pa and 1-kHz.

A MEMS condenser microphone, whose simplified structure is shown in Fig. 9.1, consists of two conductive plates at a distance x. The top plate, in this case, is fixed and cannot move, while the bottom plate is able to vibrate with the sound pressure, producing a variation of x (Δx) with respect to its steady-state value (x_0), proportional to the instantaneous pressure level (P_S). Different arrangements of the electrodes and fabrication solutions are possible [16, 17, 20–24], but the basic principle does not change.

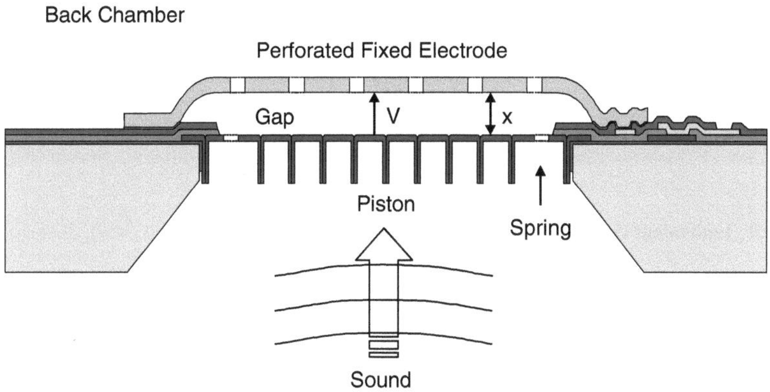

Fig. 9.1 Basic structure and working principle of a MEMS condenser microphone

The capacitance of a MEMS microphone can then be written as

$$C(P_S) = \frac{\epsilon_0 A}{x(P_S)} = \frac{\epsilon_0 A}{x_0 + \Delta x(P_S)}, \tag{9.1}$$

where A is the area of the smallest capacitor plate and ε_0 is the vacuum dielectric permittivity. Considering that $C = Q/V$, as for any capacitive sensor, there are two possible ways to transform thesound related capacitance variation into an electrical signal (a voltage$V$ or acharge Q):

- The constant-charge approach, for which the output signal is a voltage ($V = Q/C$);
- The constant-voltage approach, for which the output signal is a charge ($Q = CV$).

Denoting with C_0 the MEMS capacitance in the absence of sound, i.e. when $x = x_0$, and assuming linear the relationship between the sound pressure P_S and thedeformation $x(\Delta x = -\kappa \Delta P_S)$, which is actually truefor $\Delta x \ll x_0$, we can calculate the output signals ($\Delta V$and ΔQ) as afunction of ΔP_S in the two cases.

With the constant-charge approach, the MEMS capacitor is initially charged to a fixed voltage V_B. If the capacitor is well insulated, the charge stored on it ($Q = C_0 V_B$) remains constant, as in electret microphones, in which the charge is actually "trapped" within the device itself. As a consequence, the capacitance variation due to a sound pressure variation ΔP_S leads to a voltage signal (ΔV) given by

$$\Delta V = \frac{Q}{C(P_S)} - \frac{Q}{C_0} = \frac{Q \Delta x}{\epsilon_0 A} = -\frac{\kappa C_0 V_B \Delta P_S}{\epsilon_0 A} = -\kappa_V \Delta P_S, \tag{9.2}$$

where κ_V denotes the voltage sensitivity of the microphone.

With the constant-voltage approach, a fixed voltage V_B is applied and maintained across the MEMS capacitor. The capacitance variation due to a sound pressure variation ΔP_S, assuming $\Delta x \ll x_0$, leads to a charge signal (ΔQ) given by

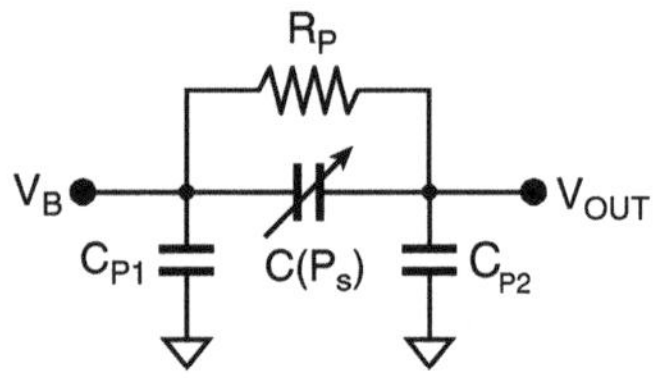

Fig. 9.2 Equivalent circuit of a MEMS condenser microphone

$$\Delta Q = V_B[C(P_S) - C_0] \approx -\frac{V_B C_0^2 \Delta x}{\epsilon_0 A} = \frac{\kappa C_0^2 V_B \Delta P_S}{\epsilon_0 A} = \kappa_Q \Delta P_S, \tag{9.3}$$

where $\kappa_Q = \kappa_V C_0$ denotes the charge sensitivity of the microphone.

According to (9.2) and (9.3), both κ_V and κ_Q depend on the bias voltage V_B. Therefore, in order to increase the microphone sensitivity and, hence, the *SNR*, the value of V_B has to be pretty high, typically ranging from 5 V to about 15 V. As a consequence, a charge pump is usually required to generate the desired value of V_B, starting from the standard CMOS power supply voltage (1.8, 2.5, or 3.3 V).

In practical implementations, a MEMS microphone is not just a capacitor, but some additional parasitic components have to be taken into account. The equivalent circuit of an actual MEMS microphone is shown in Fig. 9.2. Besides the variable capacitance $C(P_S)$, the equivalent circuit includes two parasitic capacitances C_{P1} and C_{P2}, connected between each plate of the MEMS microphone and the substrate, as well as a parasitic resistance R_P, connected in parallel to $C(P_S)$. The value of these parasitic components depends on the specific implementation of the microphone, but typically C_{P1} and C_{P2} are of the order of few pF, while R_P is in the GΩ range.

9.3 MEMS Microphone Interface Circuits

The interface circuit for a MEMS microphone has to read-out the electrical signal ΔV or ΔQ and convert it in the digital domain. Digital output is, indeed, a must for MEMS microphones, in order to gain a competitive advantage over electret devices, in terms of area and cost at system level. Therefore, the interface circuit for a MEMS microphone, whose block diagram is shown in Fig. 9.3, typically consists of some sort of preamplifier followed by an analog-to-digital converter (ADC). Moreover, a charge pump is usually required for generating the microphone bias voltage V_B.

9.3.1 Preamplifier

The topology and the functionality of the preamplifier in a MEMS microphone interface circuit is different depending on the approach used to read-out the microphone capacitance variation.

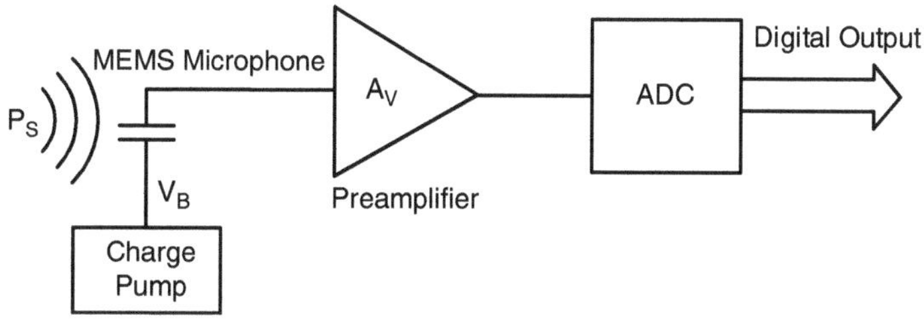

Fig. 9.3 Typical block diagram of the interface circuit for a MEMS condenser microphone

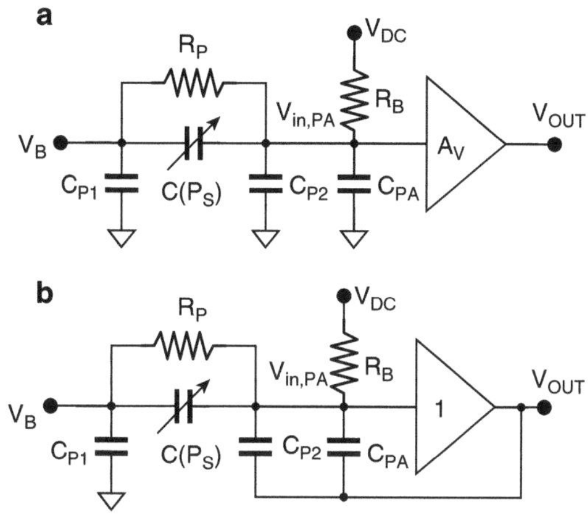

Fig. 9.4 Block diagram of the preamplifier for the constant-charge approach without (**a**) and with (**b**) parasitic capacitance bootstrapping

9.3.1.1 Constant-Charge Approach

With the constant-charge approach, the preamplifier has to buffer the microphone output voltage, eventually introducing some gain, providing a suitable signal, with low output impedance, to the subsequent ADC, as shown in Fig. 9.4a. In this case, the input impedance of the preamplifier has to be extremely high (larger than 10 GΩ), in order to guarantee that the charge stored on the microphone capacitance is maintained, while providing, at the same time, a suitable DC bias voltage at the preamplifier input node. The biasing network at the preamplifier input is, therefore, very critical and represents typically the most challenging part of the preamplifier design. The solutions usually adopted to implement R_B are based on inversely biased diodes or switched networks. Resistor R_B introduces a high-pass filter with cut-off frequency $f_{HP} \approx 1/(2\pi R_B C_0)$, which has to be lower than 20 Hz to avoid loss of signal. The parasitic capacitance at the preamplifier input (C_{PA}) is also particularly important, considering that the output voltage of the microphone ΔV,

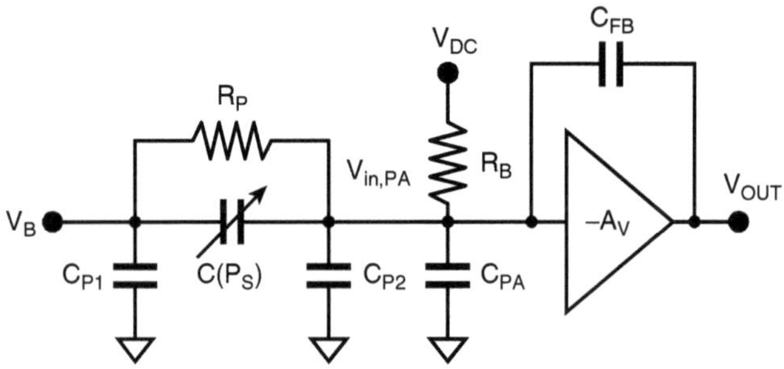

Fig. 9.5 Block diagram of the preamplifier for the constant-voltage approach

given by (9.2), in the presence of parasitic capacitances (both C_{P2} and C_{PA}), is actually attenuated, leading to

$$V_{in,PA} = \Delta V \frac{C_0}{C_0 + C_{P2} + C_{PA}} = -\Delta P_S \frac{\kappa_V C_0}{C_0 + C_{P2} + C_{PA}}. \tag{9.4}$$

This attenuation can often be quite substantial, thus leading to a degradation of the actual microphone sensitivity and, hence, of the *SNR*. This problem can be mitigated by bootstrapping C_{P2} and, eventually, also C_{PA}, as shown in Fig. 9.4b. In this case, the voltage across the parasitic capacitances is maintained constant, independently of the signal, and, therefore, $V_{in,\ PA} \approx \Delta V$. In order to achieve proper bootstrapping, the gain of the preamplifier (or, at least, of the preamplifier first stage) has to be unitary and, hence,

$$V_{OUT} = -\kappa_V \Delta P_S. \tag{9.5}$$

9.3.1.2 Constant-Voltage Approach

With the constant-voltage approach, the preamplifier has transform the charge ΔQ provided by the microphone, given by (9.3), into a voltage signal, with low output impedance, suitable to drive the subsequent ADC, as shown in Fig. 9.5. In this case, therefore, the preamplifier is actually a charge amplifier, whose output voltage is given by

$$V_{OUT} = \frac{\Delta Q}{C_{FB}} = \frac{\kappa C_0^2 V_B \Delta P_S}{\epsilon_0 A_V C_{FB}} = \frac{\kappa_Q \Delta P_S}{C_{FB}} = \kappa_V \frac{C_0}{C_{FB}} \Delta P_S. \tag{9.6}$$

The voltage swing at node $V_{in,\ PA}$ with the constant-voltage approach is quite small, thus making the effect of the parasitic capacitances C_{P2} and C_{PA} negligible.

Moreover, since $V_{in,\ PA}$, in this case, is a low-impedance node, the DC biasing of the preamplifier input is not critical and the value of resistance R_B can be in the tens or hundreds of MΩ range. Indeed, the high-pass filter introduced by R_B has cut-off frequency $f_{HP} \approx 1/(2\pi R_B C_{FB} A_V)$, which, for a given value of R_B, is much lower than with the constant-charge approach. Apparently, the constant-voltage approach looks more attractive than the constant-charge approach, since both the parasitic capacitance effect and the biasing of the input node are less critical. However, in practice, the constant-charge approach is almost always used [1, 4, 9, 13, 15, 18]. This is fundamentally due to three reasons.

Firstly, according to (9.5) and (9.6), the output voltage V_{OUT} in the constant-voltage approach does not depend only on κ_V, as in the constant-charge approach, but also on the ratio C_0/C_{FB}, which is not well controlled (C_0 and C_{FB} are realized with different technologies and depend on different process parameters). Therefore, the overall sensitivity of microphone and preamplifier ($V_{OUT}/\Delta P_S$) features a much larger variation with the process in the constant-voltage approach, where the spread is due to κ_V, C_0 and C_{FB}, than in the constant-charge approach, where the spread is only due to κ_V.

Secondly, in the constant-charge approach, the charge pump that generates voltage V_B does not have to provide any current, besides at system startup, whereas in the constant-voltage approach the charge pump has to deliver the signal charge ΔQ also during normal operation. As a consequence, in the constant-voltage approach the charge pump specifications are mode stringent, especially in terms of output impedance, in order to avoid distortion of the signal, due to dynamic variations of voltage V_B correlated with the sound pressure ΔP_S.

Finally, the constant-charge approach is the only available solutions for reading-out electret microphones, which are still dominating the market. Therefore, many research groups, historically working on microphone interface circuits, developed know-how on this approach, which can now be reused for MEMS microphones.

9.3.1.3 Force-Feedback Approach

Force-feedback has been commonly used to minimize the impact of mechanical imperfections and inherent non-linearities in MEMS capacitive sensors (e.g. accelerometers) [25, 26]. With the force-feedback approach the MEMS sensor is enclosed in an electromechanical feedback loop. The capacitive variations of the sensor are read-out by the interface circuit, whose output signal, either analog or digital, is fed back to the sensor as an electrostatic force, which counter-balances the incident force. The counter-balancing feedback reduces the movement of the mobile electrode, thereby, reducing non-linear effects. The same principle can be applied to MEMS microphones, modulating voltage V_B in the constant-voltage approach, in order to produce an electrostatic force that counterbalances the movement of the mobile electrode of the microphone. If the loop gain of the electromechanical feedback loop is large, any non-idealities of the microphone are strongly attenuated and the overall sensitivity of microphone and interface circuit is determined only by the components used to produce the force feedback (i.e. it does not depend on the microphone sensitivity κ_V), thus reducing the spread due to process variations. This approach is quite attractive and it is being investigated by several research groups [10, 27].

9.3.2 A/D Converter

The large majority of ADCs for audio applications are realized with ΣΔ modulators, in view of their inherent linearity and low power consumption. The main reason that makes ΣΔ modulators particularly suited for audio applications is the relatively small bandwidth of audio signals (20 { Hz} ÷ 20 { kHz}), which allows fairly large oversampling ratios to be achieved, while maintaining the clock frequency at acceptable values (few MHz). By trading accuracy with speed, ΣΔ modulators achieve *SNR* values larger than 60 dB with simple hardware and small area, considering that the *SNR* of a ΣΔ modulator of order L with N-bit quantizer and oversampling ratio M, is ideally given by Temes et al. [28], Maloberti [29] and Malcovati et al. [30]

$$SNR = \frac{2^{2N}3(2L+1)M^{2L+1}}{2\pi^{2L}}. \quad (9.7)$$

Following this trend, ΣΔ modulators represent the dominant solution for implementing the ADC also in the interface circuits for MEMS microphones [3, 4, 7, 9, 13, 14, 18, 25, 26]. Audio ΣΔ modulators for microphone applications are in most cases requiring single-bit output, in order to minimize the number of interconnections at PCB level, while delegating decimation and filtering to the digital signal processor (DSP) in charge of audio processing at system level.

Audio ΣΔ modulators can be implemented using either continuous-time or discrete-time architectures [28, 29].Continuous-time ΣΔ modulators can ideally achieve a lower power consumption than their discrete-time counterparts for the same *SNR*. However, they are more sensitive to clock jitter and process variations. Therefore, discrete-time ΣΔ modulators are still the dominant solution for MEMS microphone interface circuits, although the use of continuous-time architectures is spreading significantly.

9.3.3 Charge Pump

The bias voltage V_B of a MEMS microphone represents an important design parameter, since it affects directly the sensitivity κ_V. Therefore, the charge pump used to generate V_B represents an important part of a MEMS microphone interface circuit.

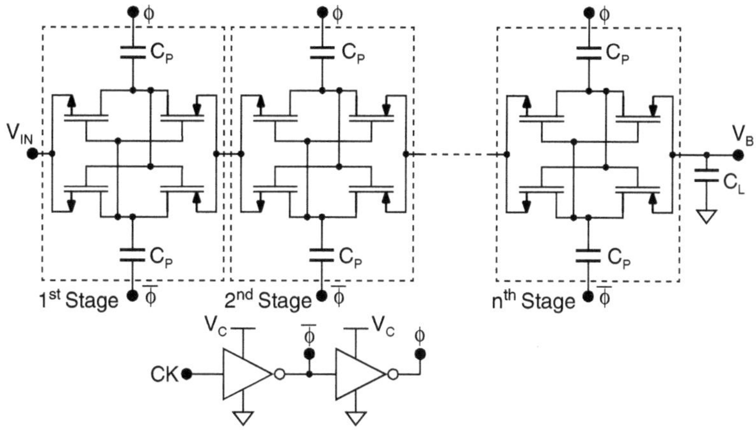

Fig. 9.6 Schematic of the charge pump typically used in MEMS microphone interface circuits

The schematic of the charge pump most widely used in MEMS microphone interface circuits is shown in Fig. 9.6. The circuit consists of a chain of cross-coupled latch stages. Each stage increases the output voltage of the previous one by the supply voltage V_C of the inverters used to generate the clock phases Φ and $\overline{\Phi}$, which drive the latches through capacitors C_P. The load capacitor C_L is used to reduce the ripple on the output voltage V_B. This charge pump topology alleviates the problem of reduced pumping gain that typically affects conventional Dickson charge pumps [31], thus achieving better efficiency. If required, closed-loop control of the output voltage can be easily achieved, by measuring the value of V_B and adjusting V_C accordingly.

9.4 Design Example: Constant-Charge Approach

The first design example is an interface circuit for MEMS microphones based on the constant-charge approach [12, 13]. Following the general block diagram shown in Fig. 9.3, the circuit consists of a buffer followed by a $\Sigma\Delta$ modulator. The input buffer, whose gain is unitary, is required in order to preserve the charge across the MEMS microphone capacitor, guaranteeing a linear conversion of the sound pressure into an electrical signal, according to (9.2) and (9.5). The unity gain buffer is followed by a fourth-order $\Sigma\Delta$ modulator, which transforms the analog input signal into a single-bit output stream, with a throughput frequency of 2.048 MHz.

9.4.1 Input Buffer

The main function of the unity-gain buffer is to provide a high input impedance, thus avoiding loss of charge on the MEMS condenser microphone. However, the

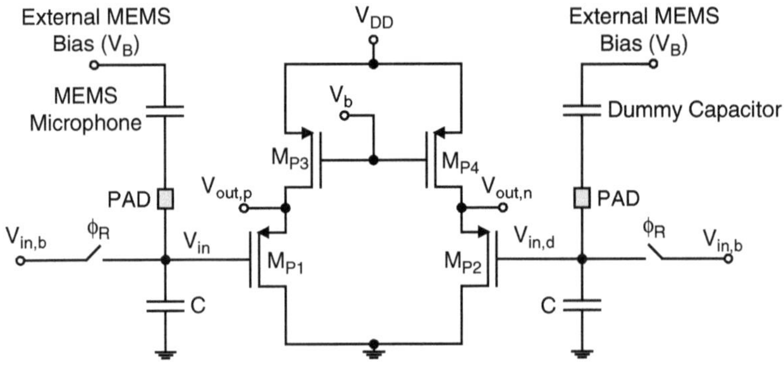

Fig. 9.7 Schematic of the high input impedance unity gain buffer

unity-gain buffer has also two additional functions: it has to provide sufficient driving capability for interfacing the subsequent switched-capacitor (SC) ΣΔ modulator, and it has to adapt the voltage on the lower plate of the microphone capacitor to the ΣΔ modulator input common-mode voltage. A unity-gain buffer as input stage is sufficient to achieve the required *SNR* (no additional gain is required). Moreover, the output of the unity-gain buffer can be used for bootstrapping the parasitic capacitance of the microphone, thus improving the system performance. The use of an amplifier with gain larger than one as input stage would, indeed, prevent this possibility.

The schematic of the unity-gain buffer used is shown in Fig. 9.7. The circuit is a pseudo-differential source follower. The main branch (M_{P1} and M_{P3}) is connected to the MEMS microphone, while a second identical dummy branch (M_{P2} and M_{P4}) is connected to a capacitor with the same nominal capacitance as the microphone. During the initial input bias and reset phase (Φ_R), the input nodes of both branches are charged to the bias voltage $V_{in,b}$. Therefore, both the microphone and the dummy capacitor are charged to $V_B - V_{in,b}$, V_B being the voltage used to bias the MEMS microphone.Voltage V_{in} is, therefore, proportional to the sound, while $V_{in,d}$ remains constant. The advantage of exploiting a pseudo-differential architecture is that the buffer output signal $V_{out,\,p} - V_{out,\,n}$ is, at first order, independent of the mean value of $V_{in,b}$, as well as of any noise coming from the power supply rail V_{DD} or from the bias voltage V_B, actually leading to a larger power supply rejection ratio, with respect to a single-ended solution. The noise contribution of the input buffer, after sampling, considering aliasing, is about 60 μV_{RMS} with white spectrum, which corresponds to 8.5 μV_{RMS}in the audio band. Considering the maximum input signal with a peak-to-peak amplitude of 650 mV, this leads to a *SNR* of about 88 dB.

9.4.2 *ΣΔ Modulator*

In the design of this interface circuit, we opted for a discrete-time ΣΔ modulator, which is more robust and predictable, although a continuous-time ΣΔ modulator could potentially achieve a slightly lower power consumption.

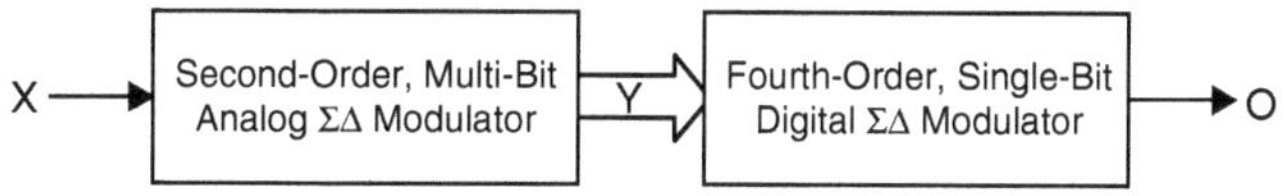

Fig. 9.8 Block diagram of the ΣΔ modulator

Considering a sampling frequency $f_s = 2.048$ MHz, with a signal bandwidth $B = 20$ kHz, and hence an oversampling ratio $M = f_s/(2B) = 51$, according to (9.7), the required $SNR \geq 80$ dB and a single-bit output stream can be achieved, for example, with a single-bit quantizer ($N = 1$) and a fourth-order noise shaping ($L = 4$). However, this solution suffers of instability for large input signals, thus requiring watch-dog circuits in order to guarantee saturation recovery. Moreover, at least four operational amplifiers (typically five) have to be used to design the loop filter.

Another possible solution is to use a 2–2 MASH ΣΔ modulator [28, 29] to achieve the required *SNR*, while overcoming instability issues. However, this solution does not provide a single-bit output stream, because of the additional digital filter required to combine the outputs of the cascaded modulators, and suffers of quantization noise leakage problems, due to mismatches between the analog integrators and the digital filter. Moreover, it still requires four operational amplifiers.

According to (9.7), the required *SNR* is also obtained with $L = 2$ and $3 < N < 4$ (e.g.12-level quantizer). This solution can be easily designed to be stable even for a large input signal and requires only two operational amplifiers to implement the loop filter. Moreover, multi-bit feedback alleviates the slew-rate requirements of the operational amplifiers. However, this solution does not provide fourth-order noise shaping nor single-bit output stream. These drawbacks can be solved by connecting at the output of the multi-bit, second-order, analog ΣΔ modulator a single-bit, fourth-order, digital ΣΔ modulator, operated at the same sampling frequency f_s, which truncates the multi-bit output down to a single bit and shapes the resulting truncation error with a fourth-order transfer function. The digital, fourth-order ΣΔ modulator is less critical than its analog counterpart, since it can be easily verified under any operating conditions, and, by using sufficiently large word-length in the integrators and a suitable noise transfer function, instability can be avoided. This solution, whose block diagram is shown in Fig. 9.8, is very promising to achieve the specifications of power consumption and resolution of the system. In order to verify the achievable performance with the used ΣΔ modulator architecture and derive the specifications for the building blocks, we performed behavioral simulations, including most of the non-idealities (kT/C noise, jitter, operational amplifier noise, gain, bandwidth and slew rate), using a dedicated toolbox in Simulink [30].The achieved *SNR* is 82.4 dB, which corresponds to an effective number of bits (*ENOB*) of 13.4.

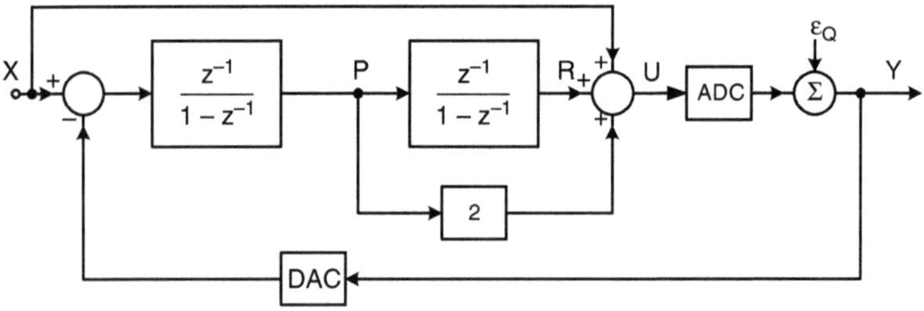

Fig. 9.9 Block diagram of the second-order, analog ΣΔ modulator

9.4.2.1 Second Order, Multi-Bit, Analog ΣΔ Modulator

Several solutions are available in literature to obtain a second-order, analog, SC ΣΔ modulator [32–36]. Among them, the second-order ΣΔ modulator architecture, whose block diagram is shown in Fig. 9.9 [37], is particularly suited for the considered application, since, thanks to the feed-forward paths from the input of the integrators to the input of the quantizer, the output of the integrators consists of quantization noise only, thus allowing low-performance (and hence low-power) operational amplifiers to be used.

The analog ΣΔ modulator consists of two integrators, one adder, a flash ADC, and a multi-bit DAC. The circuit features a unitary signal transfer function, $STF = 1$, and a noise transfer function,

$$NTF = \left(1 - z^{-1}\right)^2, \tag{9.8}$$

with second-order noise shaping.Both the integrator outputs consist of quantization noise only, whose maximum amplitude is equal to $V_{ref}/(k+1)$, where V_{ref} is the reference voltage (i.e. the full scale value) and $k = 2^N$ is the number of levels in the quantizer.

Figure 9.10 shows the SC implementation of the analog, second-order ΣΔ modulator. The circuit is actually fully-differential, although, for simplicity, Fig. 9.10 shows a single-ended version. An active block has been used to implement the adder before the quantizer, in order to reduce the capacitive load for the two integrators, thus reducing the power consumption. This solution requires an additional operational amplifier but, thanks to the reduced capacitive load, it consumes anyway less power than a solution based on a passive adder. The operational amplifiers used for the integrators and the adder are based on a folded-cascode topology. The common-mode feedback is realized with a SC network.

The quantizer (flash ADC) consists of $k = 11$ comparators, thus leading to a 12-level output code. The comparator used in the flash ADC consists of a pre-amplifier followed by a clock-driven regenerative latch. The fully-differential comparison between the input signals and the threshold voltages is performed before the pre-amplification stage by a SC network.

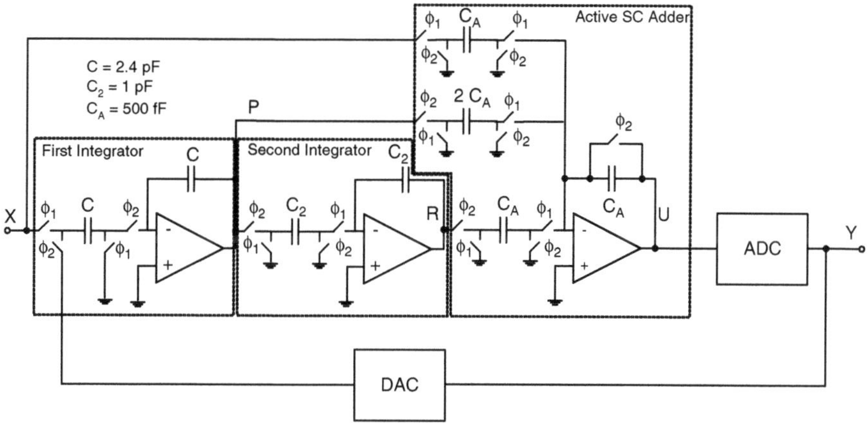

Fig. 9.10 Schematic of the SC implementation of the second-order, analog ΣΔ modulator

The DAC is realized by splitting the input capacitance C of the first integrator into 12 identical parts, which are alternately connected to $V_{ref,p}$, $V_{ref,n}$ or V_{agnd}, according to the quantizer output.

9.4.2.2 Fourth-Order, Single-Bit, Digital ΣΔ Modulator

The block diagram of the digital fourth-order, single-bit ΣΔ modulator is shown in Fig. 9.11. Denoting with Y and $\in_Q$ the modulator input and the quantization noise, respectively, the modulator output signal O is given by

$$O(z) = Y(z) + \epsilon_Q(z)\frac{(z-1)^2(z^2-1.99z+0.99)}{(z^2-1.079z+0.3014)(z^2-1.794z+0.8294)}, \tag{9.9}$$

thus leading to a unitary *STF* in the audio band and a *NTF* with fourth-order noise shaping. The coefficients of the ΣΔ modulator are implemented as the sum of no more than two terms, each expressed as a power of 2, thus avoiding the use of multipliers. The word-length in the internal registers is 8 bits for the first integrator, 10 bits for the second integrator, 15 bits for the third integrator, 16 bits for the fourth integrator, and 6 bits for the final adder, in order to avoid saturation and truncation, under any operating conditions.

9.4.3 Experimental Results

A prototype of the interface circuit has been fabricated using a 0.35-μm CMOS technology with four metal and two polysilicon layers. The circuit consumes

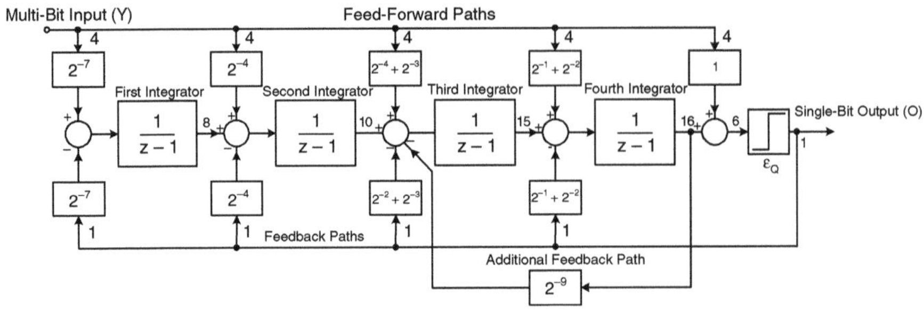

Fig. 9.11 Block diagram of the fourth-order, digital ΣΔ modulator

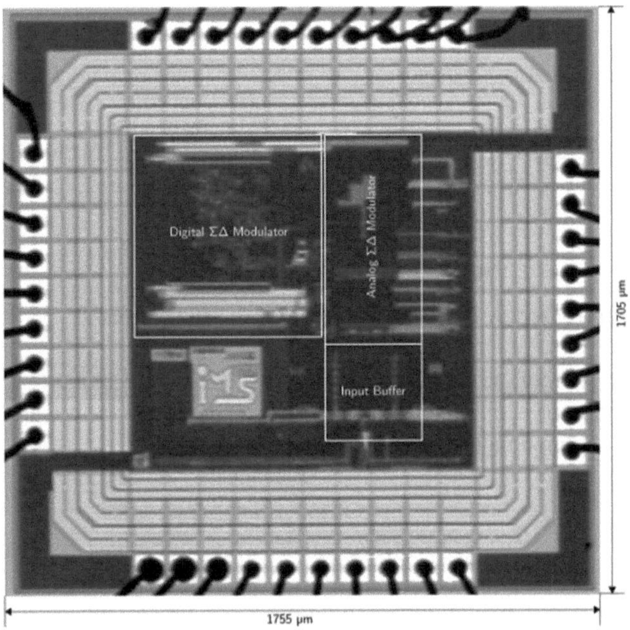

Fig. 9.12 Chip microphotograph of the interface circuit for MEMS microphones based on the constant-charge approach

215 μA for the analog section and 95 μA for the digital section, respectively, leading to an overall power consumption of 1.0 mW with a clock frequency of 2.048 MHz and a power supply voltage of 3.3 V.

Figure 9.12 shows the microphotograph of the fabricated chip, which occupies an area of 3 mm^2 (1,755 × 1,705 μm), including pads. The full-scale input signal amplitude is equal to the DAC reference voltage ($V_{ref} = V_{ref,p} - V_{ref,n}$), which has been set to ± 400 mV, i.e. V_{in} = 800 mV peak-to-peak, which, for the considered MEMS microphone, corresponds to about 106 dB_{SPL}.

Figure 9.13 shows the spectrum obtained at the output of the single-bit, digital ΣΔ modulator, by applying at the interface circuit input a differential sinusoidal

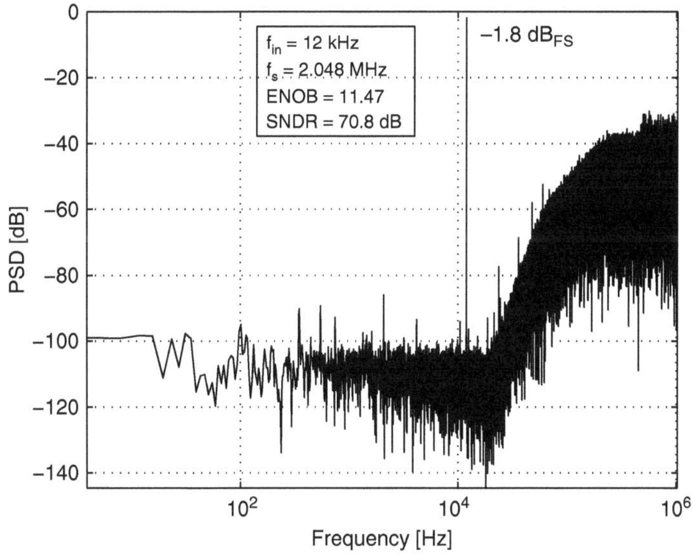

Fig. 9.13 Spectrum of the digital, fourth-order modulator output signal, applying at the interface circuit input a 12-kHz, − 1. 8-dB$_{FS}$ sinusoidal signal

signal with a frequency of 12 kHz and a peak-to-peak amplitude of 650 mV (i.e. − 1.8 dB$_{FS}$). The obtained signal-to-noise and distortion ratio (*SNDR*) is equal to 71 dB, while the quantization noise is fourth-order shaped. The*SNDR*value achieved is lower than what expected from behavioral simulations due to additional noise sources in the measurement (bias and reference voltages), but fulfills the specifications. By considering both noise and distortion contributions, the achieved *ENOB* is equal to 11.5. Neglecting the distortion and spurious components, we achieve instead a *SNR* of 80 dB, corresponding to 13.2 bits. The use of a feed-forward path in the analog, second-order $\Sigma\Delta$ modulator allows us to achieve the peak *SNDR* for an input signal amplitude as large as − 1.8 dB$_{FS}$, as shown in Fig. 9.14. Finally, Table 9.1 summarizes the most important measured performance of the interface circuit for MEMS microphones based on the constant-charge approach.

9.5 Design Example: Force-Feedback Approach

The second design example is an interface circuit for MEMS microphones based on the force-feedback approach. The circuit, whose block diagram is shown in Fig. 9.15 [8, 10], consists of a preamplifier, a third-order $\Sigma\Delta$ modulator and the force-feedback logic. The digital output bitstream of the $\Sigma\Delta$ modulator at $f_s = 2.52$ MHz is used to

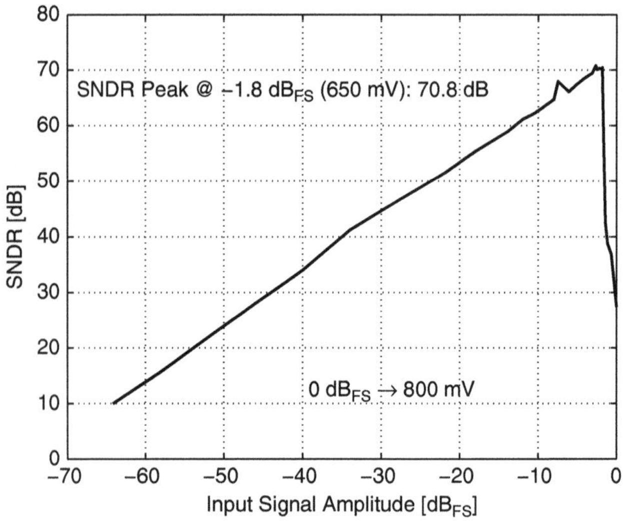

Fig. 9.14 Measured *SNDR* as a function of the input signal level

Table 9.1 Measured performance of the interface circuit for MEMS microphones based on the constant-charge approach

Parameter	Value
Bandwidth (B)	20 kHz
Signal-to-noise ratio (*SNR*)	80 dB
Signal-to-noise and distortion ratio (*SNDR*)	71 dB
Effective number of bits (*ENOB*)	11.5
Power supply voltage	3.3 V
Input buffer current consumption	80 μA
ΣΔ modulator current consumption	230 μA
Total current consumption	310 μA

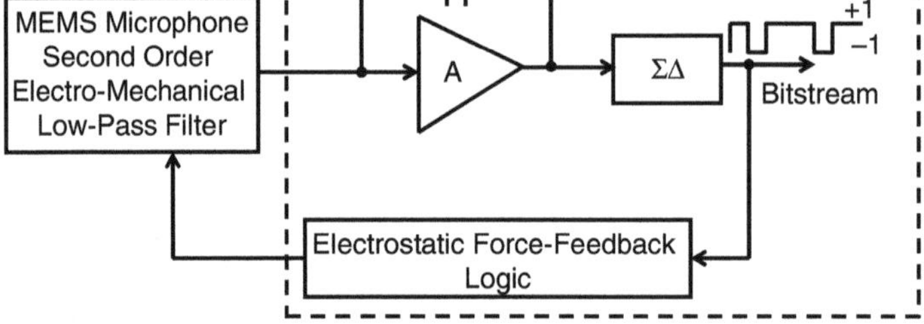

Fig. 9.15 Block diagram of the interface circuit for MEMS microphones based on the force-feedback approach

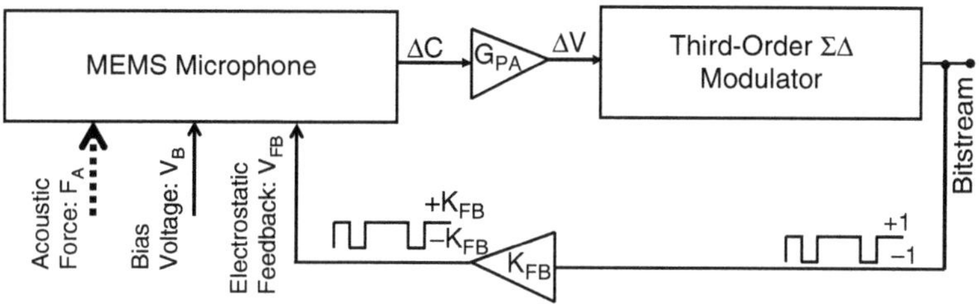

Fig. 9.16 Simulink model of the interface circuit for MEMS microphones based on the force-feedback approach

modulate the bias voltage V_B of MEMS microphone, to apply the counterbalancing electrostatic force feedback. A dummy capacitive branch is used to convert the single-ended input of the MEMS microphone into a differential output, exploiting a dynamic matching logic to equalize the dummy branch and the MEMS microphone.

9.5.1 System Modeling

The behavioral simulations for the force-balanced MEMS microphone are performed in Simulink, using the block diagram shown in Fig. 9.16. The preamplifier is replaced by a gain G_{PA}, which translates the capacitive variation into a corresponding voltage signal, which is provided to the third-order, single-loop ΣΔ modulator. Parameter K_{FB} represents the gain applied to the output of the ΣΔ modulator to control the magnitude of the feedback electrostatic force. The MEMS microphone is modeled as a second-order mass-spring-dumping system, according to

$$m\ddot{x} + b\dot{x} + k(x - x_0) = \frac{\epsilon_0 A (V_B - V_{FB})^2}{x^2} + F_A, \qquad (9.10)$$

where m is the mass, k the spring constant, and b the damping factor. Brownian noise of the sensor is also included in the model along with a $1/f$ noise component.

Figure 9.17 shows the simulated *SNDR* (including harmonic distortion) of the system as a function of the input acoustic pressure for different values of K_{FB}, with an oversampling ratio (M) for the ΣΔ modulator equal to 63. Force balancing can provide a maximum enhancement of 25 dB in the *SNDR* when $K_{FB} = 0.7$, especially close to the acoustic overload.

The closed-loop system can be considered as a hybrid ΣΔ modulator, in which the MEMS microphone also serves as an extra second-order loop filter, thus adding two more zeros in the *NTF*. Therefore, stability issues may arise, due to the additional poles in the loop transfer function. However, behavioral simulations, performed with the model shown in Fig. 9.16, demonstrate that, if the third-order

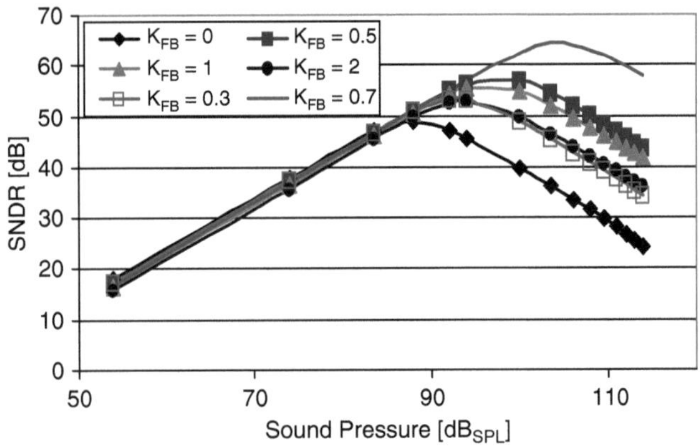

Fig. 9.17 Simulated *SNDR* of the closed-loop system as a function of the input acoustic pressure for different values of K_{FB}

$\Sigma\Delta$ modulator is by itself stable, stability of the closed-loop system is guaranteed over the whole sound pressure range (up to 114 dB_{SPL}) for $K_{FB} < 32$, which is well above the optimal value $K_{FB} = 0.7$.

9.5.2 *Preamplifier*

The preamplifier is a capacitive gain-stage based on the charge-amplifier topology, as shown in Fig. 9.18. The parasitic capacitors at the sensing node $V_{in,PA}$ have negligible effect, since the voltage swing at node $V_{in,PA}$ is quite small. The preamplifier exploits a dummy capacitive branch to convert the single-ended output of the MEMS microphone into a fully-differential signal. The preamplifier gain depends on the ratio between C_0 and C_{FB}, according to (9.6).

The bias resistance R_B introduces a high-pass filter in the preamplifier transfer function, with cutoff frequency $f_{HP} \approx 1/(2\pi A_V R_B C_{FB})$, which is below 20 Hz for $R_B > 10$ MΩ. Resistor R_B is implemented using a *p*-MOS transistor [38], exploiting the small voltage swing at node $V_{in,PA}$. The operational amplifier used is based on a mirrored cascode topology with SC common-mode feedback circuit.

The noise contribution of the preamplifier in the audio band, considering aliasing, leads to a *SNR* of about 80 dB.

9.5.3 *ΣΔ Modulator*

In this interface circuit, considering the sampling frequency $f_s = 2.52$ MHz and, hence, the oversampling ratio $M = 63$, according to (9.7), a third-order ($L = 3$),

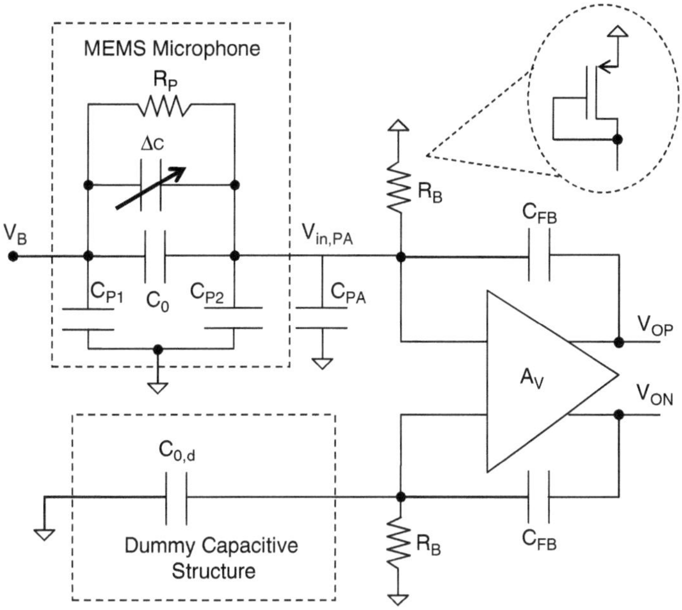

Fig. 9.18 Schematic of the preamplifier

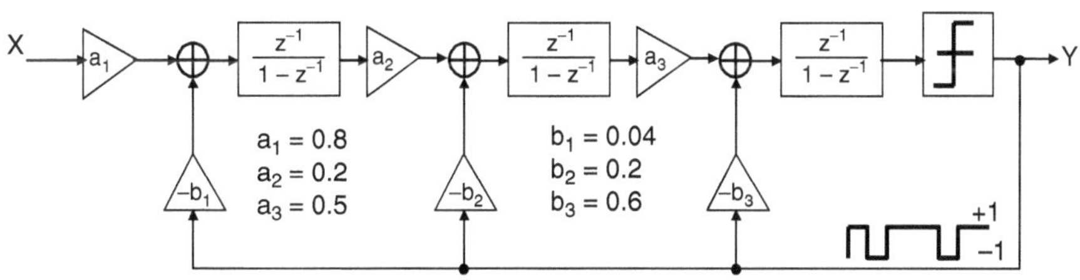

Fig. 9.19 Block diagram of the third-order ΣΔ modulator

single-bit ($N = 1$) ΣΔ modulator is sufficient to achieve the required $SNR \geq 60$ dB. The block diagram of the third-order ΣΔ modulator is shown in Fig. 9.19. The *STF* and the *NTF* are given by

$$STF = \frac{0.06z}{(z - 0.92)(z^2 - 1.47z + 0.55)}, \tag{9.11}$$

$$NTF = \frac{(z - 1)^3}{(z - 0.92)(z^2 - 1.47z + 0.55)}, \tag{9.12}$$

respectively.

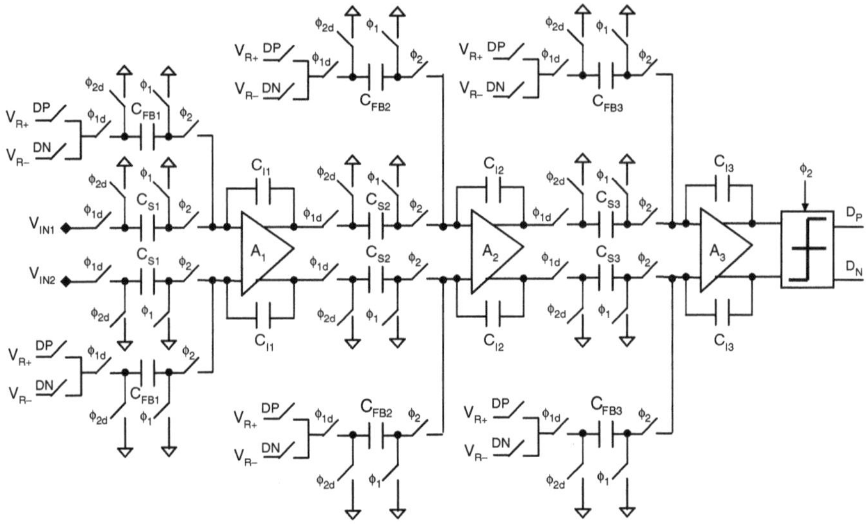

Fig. 9.20 Schematic of the SC implementation of the third-order ΣΔ modulator

Figure 9.20 shows the SC implementation of the third-order ΣΔ modulator. The feedforward and feedback paths are implemented using separate capacitors, thus relaxing the settling requirements of the operational amplifiers. The feedback path contains an extra switch, to select between positive and negative reference voltage (V_{R+} or V_{R-}). The first integrator has reduced output swing but the capacitors are large to keep the kT/C noise low, while the second and third integrator use smaller capacitors but the output swing is large. Therefore, all the integrators have almost the same settling requirements for the operational amplifiers. Bottom-plate sampling is used in the whole ΣΔ modulator to minimize the distortion due to charge-injection from switches.

The operational amplifiers used for the integrators are based on a telescopic-cascode topology. The common-mode feedback is realized with a SC network. The comparator used consists of a differential stage with regenerative load, followed by a set-reset flip-flop.

9.5.4 Force-Feedback Logic

The schematic of the circuit used to apply the force feedback to the MEMS microphone is shown in Fig. 9.21. The output of the ΣΔ modulator is applied to the backplate nodes of both the microphone and the dummy branch with feedback capacitors C_{FFB1} and C_{FFB2}, while the bias voltage V_B, provided by the charge pump, is applied through resistors R_{B2}.

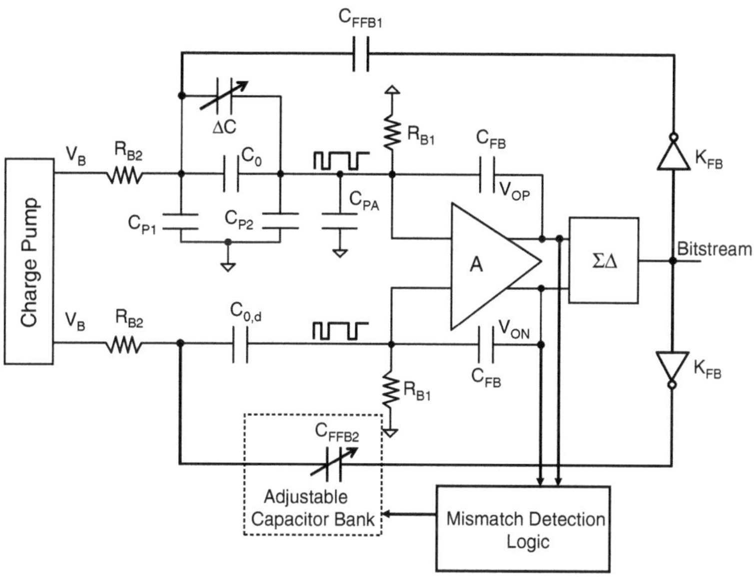

Fig. 9.21 Schematic of the force-feedback logic

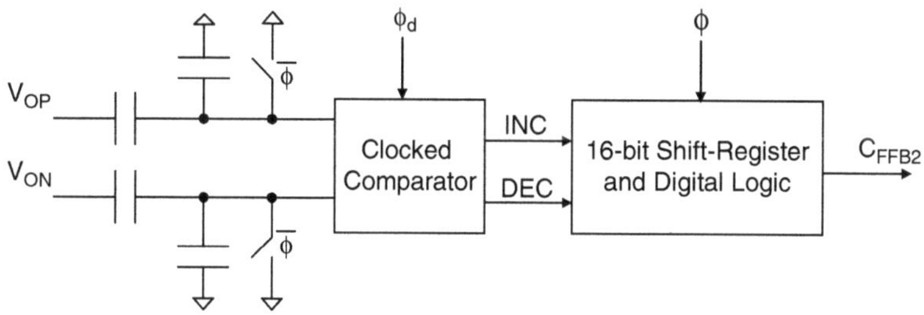

Fig. 9.22 Schematic of the mismatch minimization logic

The applied pulses, besides producing the desired force-feedback, reach also the input of the preamplifier through C_0 and $C_{0,d}$. If $C_0 = C_{0,d}$, these pulses represent a common-mode signal for the preamplifier and, therefore, their effect is negligible. However, if $C_0 \neq C_{0,d}$, a differential signal arises, thus degrading the performance of the interface circuit. To avoid this problem, the amplitude of the feedback pulse applied to the dummy branch can be controlled by varying C_{FFB2}, in order to compensate the mismatch between C_0 and $C_{0,d}$. This is achieved by the mismatch minimization logic, whose block diagram is shown in Fig. 9.22. This circuit consists of a high-pass filter and a 16-bit shift register. The outputs of the preamplifier V_{OP} and V_{ON} are applied to a comparator through a capacitive divider circuit. The input nodes of the comparator are reset to analog ground during clock phase $\overline{\Phi}$, while, during clock phase Φ, the comparator decides whether $V_{OP} > V_{ON}$ or V_{OP}

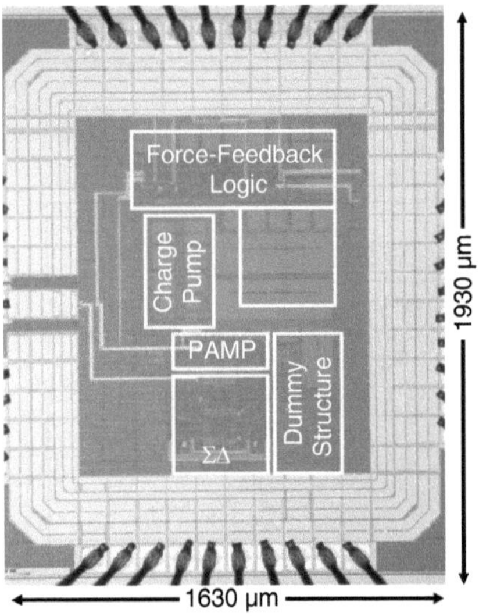

Fig. 9.23 Chip microphotograph of the interface circuit for MEMS microphones based on the force-feedback approach

$< V_{ON}$. Depending on this decision, the shift register is either incremented (*INC*) or decremented (*DEC*), thus adding or removing one element in the C_{FFB2} capacitor bank. The force-feedback pulses are applied to C_{FFB1} and C_{FFB2} through inverters with adjustable power supply, thus allowing to control parameter K_{FB}.

9.5.5 Experimental Results

A prototype of the interface circuit has been fabricated using a 0.35-µm CMOS technology with four metal and two polysilicon layers. The circuit consumes 210 µA for the analog section and 90 µA for the force-feedback logic, respectively, leading to an overall power consumption of 1.0 mW with a clock frequency of 2.52 MHz and a power supply voltage of 3.3 V.

Figure 9.23 shows the microphotograph of the fabricated chip, which occupies an area of 3.15 mm^2 (1,930 µm × 1,630 µm), including pads. The spectrum of the ΣΔ modulator output bitstream in the absence of input signal, with and without force feedback, is shown in Fig. 9.24. The force feedback lowers the noise floor and increases the noise-shaping order, as expected. Figure 9.25 shows the achieved *SNDR* as a function of the input signal amplitude with an input signal frequency of 1 kHz. The peak *SNDR* equal to 61 dB is achieved with an input signal amplitude of − 13 dB_{FS}, corresponding to a sound pressure of 104 dB_{SPL} for the considered MEMS microphone. By considering both noise and distortion contributions, the

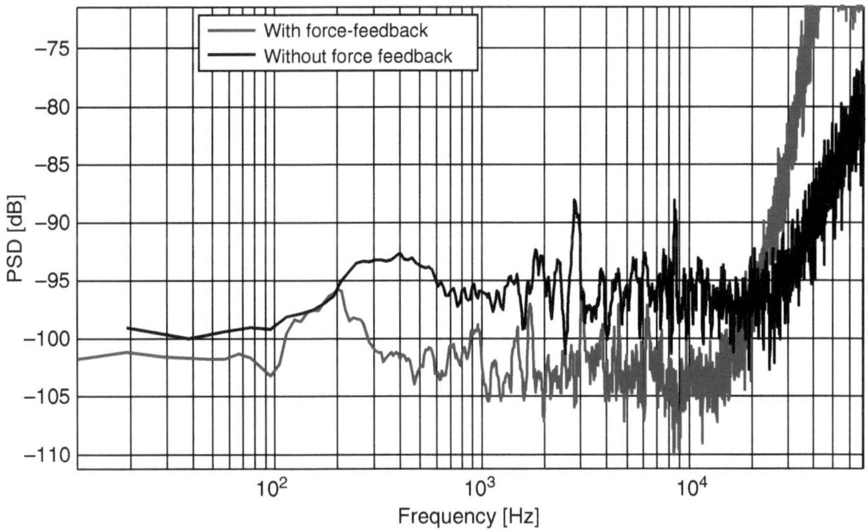

Fig. 9.24 Spectrum of the single-bit digital output signal with and without force-feedback in the absence of input signal

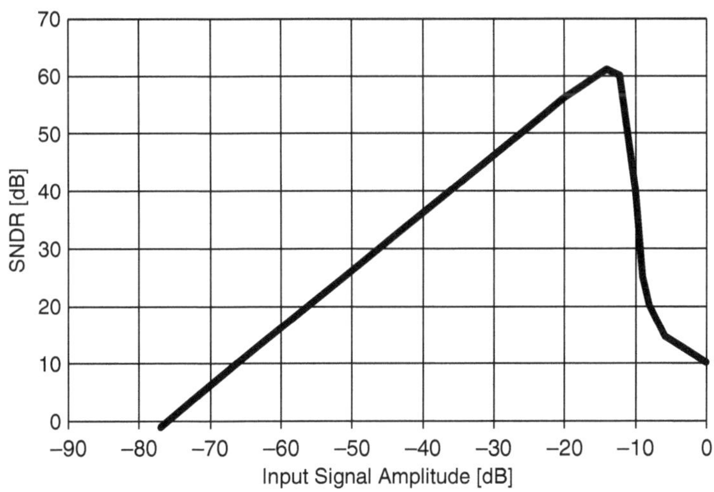

Fig. 9.25 Measured *SNDR* as a function of the input signal level

achieved *ENOB* is equal to 9.8. Neglecting the distortion and spurious components, we achieve instead a *SNR* of 72 dB, corresponding to 11.6 bits. Finally, Table 9.2 summarizes the most important measured performance of the interface circuit for MEMS microphones based on the force-feedback approach.

Table 9.2 Measured performance of the interface circuit for MEMS microphones based on the force-feedback approach

Parameter	Value
Bandwidth (*B*)	20 kHz
Signal-to-noise ratio (*SNR*)	72 dB
Signal-to-noise and distortion ratio (*SNDR*)	61 dB
Effective number of bits (*ENOB*)	9.8
Power supply voltage	3.3 V
Preamplifier current consumption	100 μA
ΣΔ modulator current consumption	110 μA
Force-feedback logic current consumption	90 μA
Total current consumption	300 μA

9.6 Conclusions

In this paper we presented an overview of the most important aspects in the design of interface circuits for MEMS microphones. The different interface circuit architectures and the main design trade-offs have been analyzed in detail. Moreover, we reported two design examples, actually implemented on silicon with a 0.35-μm CMOS technology. The literature overview and the experimental results reported demonstrate that MEMS microphones are already a valid alternative to electret devices in the field of consumer electronics, in terms of area, cost and performance. The next challenge in the development of MEMS microphones and interface circuits is to target high-end applications, where performance is a key issue. To achieve this goal an improvement of about 20 dB in terms of *SNDR* is required (the target *SNDR* in this case is of the order of 100 dB), thus requiring significant developments both in the MEMS devices and in the electronics.

References

1. Hsu YC, Chen JY, Wang CH, Liao LP, Chou WC, Wu CY, Mukherjee T (2008) Issues in path toward integrated acoustic sensor system on chip. In: Proceedings of IEEE Sensors. IEEE, Piscataway, pp 585–588
2. Malcovati P, Maloberti F (2005) Interface circuitry and microsystems. In: Korvink J, Paul O (eds) MEMS: a practical guide to design, analysis and applications. Springer, Dordrecht, pp 901–942
3. Bajdechi O, Huijsing JH (2002) A 1.8-V ΔΣ modulator interface for an electret microphone with on-chip reference. IEEE J Solid-State Circuits 37:279–285
4. Chiang CT, Huang YC (2009) A 14-bit oversampled delta-sigma modulator for silicon condenser microphones. In: Proceedings of IEEE IMTC. IEEE, Piscataway, pp 1055–1058
5. Pernici S, Stevenazzi F, Nicollini G (2004) Fully integrated voiceband codec in a standard digital CMOS technology. IEEE J Solid-State Circuits 39:1331–1334

6. van der Zwan EJ, Dijkmans EC (1996) A 0.2-mW CMOS ΣΔ modulator for speech coding with 80 dB dynamic range. IEEE J Solid-State Circuits 31:1873–1880
7. Zare-Hoseini H, Kale I, Richard CSM (2010) A low-power continuous-time ΔΣ modulator for electret microphone applications. In: Proceedings of IEEE ASSCC. IEEE, Piscataway, pp 1–4
8. Jawed SA (2009) CMOS readout interfaces for MEMS capacitive microphones. Ph.D. dissertation, University of Trento
9. Jawed SA, Cattin D, Gottardi M, Massari N, Baschirotto A, Simoni A (2008) A 828-μW 1.8-V 80-dB dynamic-range readout interface for a MEMS capacitive microphone. In: Proceedings of ESSCIRC. IEEE, Piscataway, pp 442–445
10. Jawed SA, Cattin D, Massari N, Gottardi M, Baschirotto A (2008) A MEMS microphone interface with force-balancing and charge-control. In: Proceedings of IEEE PRIME. IEEE, Piscataway, pp 97–100
11. Jawed SA, Nielsen JH, Gottardi M, Baschirotto A, Bruun E (2009) A multifunction low-power preamplifier for MEMS capacitive microphones. In: Proceedings of ESSCIRC. IEEE, Piscataway, pp 292–295
12. Picolli L, Grassi M, Rosson L, Malcovati P, Fornasari A (2009) A 1.0-mW, 71-dB SNDR, − 1. 8-dB_{FS} input swing, fourth-order sigma-delta interface circuit for MEMS microphones. In: Proceedings of ESSCIRC. IEEE, Piscataway, pp 324–327
13. Picolli L, Grassi M, Fornasari A, Malcovati P (2011) A 1.0-mW, 71-dB SNDR, fourth-order ΣΔ iinterface circuit for MEMS microphones. Analog Integr Circuits Signal Process 66:223–233
14. Le HB, Lee SG, Ryu ST (2010) A regulator-free 84-dB DR audio-band ADC for compact digital microphones. In: Proceedings of IEEE ASSCC. IEEE, Piscataway, pp 1–4
15. Citakovic J, Hovesten PF, Rocca G, van Halteren A, Rombach P, Stenberg LJ, Andreani P, Bruun E (2009) A compact CMOS MEMS microphone with 66-dB SNR. In: IEEE ISSCC digest of technical papers. IEEE, Piscataway, pp 350–351
16. Je SS, Kim JH, Kozicki MN, Chae JS (2009) A directional capacitive MEMS microphone using nano-electrodeposits. In: Proceedings of IEEE MEMS. IEEE, Piscataway, pp 96–99
17. Weigold JW, Brosnihan TJ, Bergeron J, Zhang X (2006) A MEMS condenser microphone for consumer applications. In: Proceedings of IEEE MEMS. IEEE, Piscataway, pp 86–89
18. Deligoz I, Naqvi SR, Copani T, Kiaei S, Bakkaloglu B, Je SS, Chae JS (2011) A MEMS-based power-scalable hearing aid analog front-end. IEEE Trans Biomed Circuits Syst 5(3):201–213
19. Scheeper PR, van der Donk AGH, Olthuis W, Bergveld P (1994) A review of silicon microphones. Sens Actuators A 44(1):1–11
20. Bergqvist J, Gobet J (1994) Capacitive microphone with a surface micromachined backplate using electroplating technology. J Microelectromech Syst 3(2):69–75
21. Kasai T, Sato S, Conti S, Padovani I, David F, Uchida Y, Takahashi T, Nishio H (2011) Novel concept for a MEMS microphone with dual channels for an ultrawide dynamic range. In: Proceedings of IEEE MEMS. IEEE, Piscataway, pp 605–608
22. Leinenbach C, van Teeffelen K, Laermer F, Seidel H (2010) A new capacitive type MEMS microphone. In: Proceedings of IEEE MEMS. IEEE, Piscataway, pp 659–662
23. Martin DT, Liu J, Kadirvel K, Fox RM, Sheplak M, Nishida T (2007) A micromachined dual-backplate capacitive microphone for aeroacoustic measurements. J Microelectromech Syst 16(6):1289–1302
24. Zou QB, Li ZJ, Liu LT (1996) Design and fabrication of silicon condenser microphone using corrugated diaphragm technique. J Microelectromech Syst 5(3):197–204
25. Lu C, Lemkin M, Boser BE (1995) A monolithic surface micromachined accelerometer with digital output. IEEE J Solid-State Circuits 30(12):1367–1373
26. Wu JF, Carley LR (2006) Electromechanical ΔΣ modulation with high-Q micromechanical accelerometers and pulse density modulated force feedback. IEEE Trans Circuits Syst I 53 (2):274–287

27. van der Donk AGH, Sprenkels AJ, Olthuis W, Bergveld P (1991) Preliminary results of a silicon condenser microphone with internal feedback. In: IEEE transducers digest of technical papers. IEEE, Piscataway, pp 262–265
28. Temes GC, Schreier R, Norsworthy SR (1996) Delta-sigma data converters. Wiley-IEEE Press, New York
29. Maloberti F (2007) Data converters. Springer, Dordrecht
30. Malcovati P, Brigati S, Francesconi F, Maloberti F, Cusinato P, Baschirotto A (2003) Behavioral modeling of switched-capacitor sigma-delta modulators. IEEE Trans Circuits Syst I 50:352–364
31. Dickson JF (1976) On-chip high-voltage generation in MNOS integrated circuits using an improved voltage multiplier technique. IEEE J Solid-State Circuits 11(3):374–378
32. Boser BE, Wooley BA (1988) The design of sigma-delta modulation analog-to-digital converters. IEEE J Solid-State Circuits 23:1298–1308
33. Matsuya Y, Yamada Y (1994) 1-V power supply, low-power consumption A/D conversion technique with swing-suppression noise shaping. IEEE J Solid-State Circuits 29:1524–1530
34. Ahn GC, Chang DY, Brown ME, Ozaki N, Youra H, Yamamura K, Hamashita K, Takasuka K, Temes GC, Moon UK (2005) A 0.6-V 82-dB delta-sigma audio ADC using switched-RC integrators. IEEE J Solid-State Circuits 40:2398–2407
35. Silva J, Moon UK, Steensgaard J, Temes GC (2001) Wideband low-distortion delta-sigma ADC topology. Electron Lett 37:737–738
36. Nam KY, Lee SM, Su DK, Wooley BA (2005) A low-voltage low-power sigma-delta modulator for broadband analog-to-digital conversion. IEEE J Solid-State Circuits 40:1855–1864
37. Kwon S, Maloberti F (2006) A 14 mW multi-bit ΣΔ modulator with 82 dB SNR and 86 dB DR for ADSL2+. In: IEEE ISSCC digest of technical papers. IEEE, Piscataway, pp 68–69
38. Harrison RR (2002) A low-power, low-noise CMOS amplifier for neural recording applications. In: Proceedings of IEEE ISCAS, vol 5. IEEE, Piscataway, pp 197–200

Chapter 10
Front End Electronics for Solid State Detectors in Today and Future High Energy Physics Experiments

Jan Kaplon and Pierre Jarron

Abstract We present circuit design techniques currently employed for the development of analog front end electronics dedicated to the readout of radiation semiconductor sensors used in tracking detectors for High Energy Physics (HEP) experiments, where the channel counts can be very large. It is shown that for very large numbers of channels, power consumption turns out to be a critical issue in the design of the analog front end. In general, Signal-to-Noise-Ratio (SNR) and speed requirements have to be optimized together with the permitted power consumption. A selection of amplifier circuits are discussed in the context of the evolution of the CMOS technologies that impose the adaptation of design techniques to the new properties of deep scaled MOS transistors.

10.1 Basics of Solid State Detectors

Semiconductor sensors have been present in HEP experiments for almost 30 years. Developed originally for the positioning (tracking) of the high energy particles in physics experiments, they found applications in other areas like nuclear medicine and industrial X-ray scanners.

Almost all present-day semiconductor sensors are based on the reverse biased p-n junction where the space charge region forms the detection volume, which usually extends to the physical sensor thickness. The primary interaction of incoming charged particles or gamma photons with the sensor material causes the creation of electron hole pairs. The electric field existing in the depleted structure provides drifting of the carriers towards electrodes inducing fast, of the order of

J. Kaplon (✉)
CERN, CH-1211 Genève 23, Switzerland
e-mail: Jan.Kaplon@cern.ch

P. Jarron
INFN, Torino, Italy

A.H.M. van Roermund et al. (eds.), *Nyquist AD Converters, Sensor Interfaces, and Robustness: Advances in Analog Circuit Design, 2012*,
DOI 10.1007/978-1-4614-4587-6_10,

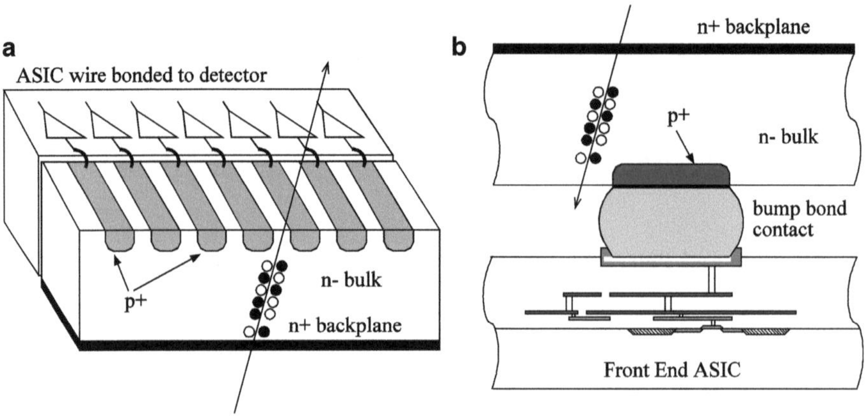

Fig. 10.1 Construction of single sided silicon strip detector (**a**) and hybrid pixel detector (**b**). The positive bias voltage is applied on the detector backplane. Each front end channel of the ASIC is wire bonded to the detector strip (**a**) or bump bonded to the detector pixel (**b**)

nanoseconds, current pulses. The spatial resolution is provided by the segmentation of the semiconductor crystal into several diodes, strips or pixels, read out separately by the front end electronics. Consequently the semiconductor detector can be modeled by an ideal current source representing the signal pulse and a parasitic capacitive network loading the front end inputs. For this last reason semiconductor detectors are classified as capacitive sensors.

The variety of the material used for the manufacturing of semiconductor detectors depends on the application. The positioning detectors in HEP experiments should detect the high energy particles with the minimum elastic scattering to avoid perturbation of the particle track. In this case low density materials will be preferable since they will be more transparent for the traversing particles. One of the most popular materials is Silicon, having excellent sensor properties and the advantage of being fully compatible with the monolithic technologies developed by the microelectronics industry.

Two examples of simple, single sided, silicon detectors are shown in Fig. 10.1. The p + strips or pixels forming diodes are implanted on n- silicon substrate. The positive bias voltage is applied on the backplane of the detector. The resistivity of the silicon material used for the sensors is of the order of a few kΩ-cm. For a typical 300 μm thick silicon detector, the bias voltages providing full depletion are of the order of 20–90 V. Each segment of the detector, a p + on n- silicon diode, is connected to a separate front end channel reading out signal pulses generated by the incoming charged particles or gamma photons. The signal created by the high-energy particle is proportional to the length of the interaction track. For a 300 μm thick, fully depleted, silicon detector, the Landau distribution has its most probable value around 3.6 fC (22,500 electrons). The collection times i.e. widths of the signal pulses from the detector are of the order of a few nano-seconds. In practice, depending on the required signal levels and acceptable collection times, which are also proportional to the track length, the thicknesses of the silicon detectors for tracking applications can vary between 100 and 500 μm.

Another field of applications for solid state detectors is X-ray detection. Materials of higher density offering higher stopping power and higher conversion efficiency are preferable in this domain. For medical imaging in 100 keV range, Cadmium-Zinc-Telluride (CdZnTe) is now the standard material for the construction of heads for Computer Tomography (CT) machines working in single photon counting mode. The number of electron hole (e-h) pairs generated in the detector by photoelectric absorption effect, when an X-ray photon deposits its full energy, will be proportional to the energy of the photon divided by the pair creation energy. For 100 keV photons the signals deposited in the CdZnTe detectors (pair creation energy 4.64 eV) fall in the same range as those for the relativistic Minimum Ionizing Particles (MIP) traversing 300 μm silicon detectors for tracking applications. It will be shown in one of the front end examples that a single architecture can be made to fulfill the requirements for both HEP and CT applications.

10.2 From Sensor Crystal to Detector System

Standard photolithographic processes allow for segmentation of the detecting areas into several diodes of different shapes of very fine pitch down to the micrometer scale. In general we can distinguish two arrangements of the diodes implanted on the common substrate: diodes implanted in the form of strips where the readout electronics can be placed next to the detector and connected to the sensor using standard wire bonding techniques (see Fig. 10.1a) or diodes implanted in the form of a pixel matrix where the silicon sensor and readout ASICs are connected using bump bonding methods, Fig. 10.1b.

Single sided silicon strip detectors provide one-dimensional information. In order to obtain a second coordinate, the detector module consists of two sensors with perpendicular or tilted strip axes (the latter a so-called stereo module). The example of the stereo module built for the ATLAS detector at CERN LHC (Geneva) is shown in Fig. 10.2. The module consists of four, $6.5 \times 6.5\ \text{cm}^2$ silicon detectors (two daisy-chained back to back) and it is equipped with 12 (6 on each side) 128 channel front end ASICs. The whole ATLAS SemiConductor Tracker (SCT) is built with 4,088 detector modules in the form of concentric barrels with enclosing disks placed at both ends, see Fig. 10.3. The total area covered by the silicon sensors is 61 m^2, which currently makes this the second largest silicon tracker in the world.

Figure 10.1b shows assembly of pixel detector connected to the front end electronics using bump bonding. The area of the front end ASIC completely covers the silicon sensor making together with electrical and mechanical hybrid a sandwich called a Hybrid Pixel Detector (HPD). The dimensions of one pixel cell might be in the micrometer range. For example, the ATLAS pixel detector uses pixels of

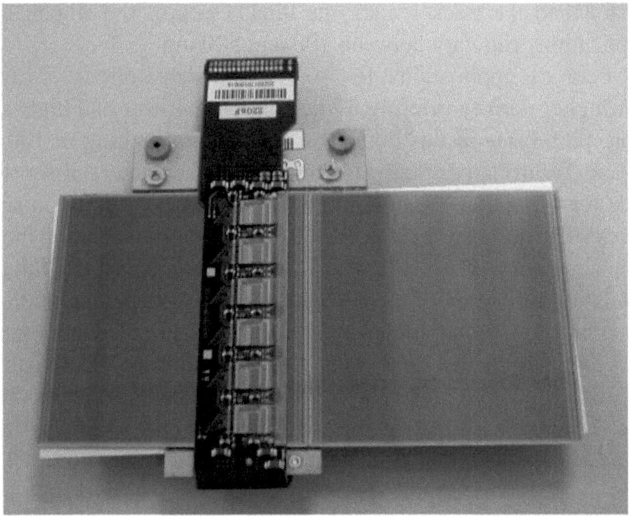

Fig. 10.2 Example of double sided silicon strip detector module of the ATLAS experiment at CERN (Geneva). The module consists of four, 6.5 by 6.5 cm^2 silicon detectors (two daisy-chained back to back) and it is equipped with 12 (6 on each side) 128 channel front end ASICs

Fig. 10.3 Photograph of one section of the semiconductor tracker of the ATLAS experiment. The full detector consists of 4 concentric barrels (2,112 modules) and 18 disks (1,976 modules). The total area of silicon is approximately 61 m^2

dimensions of $400 \times 50\ \mu m^2$. The number of pixels in one sensor is roughly 46,000 with the whole pixel detector comprising 1,744 modules, giving approximately 80 million readout channels. The total area covered by the pixel tracker in the ATLAS experiment is approximately 1.7 m^2.

The area of one silicon diode in the case of pixel and strip detector differs significantly and it is obvious that pixel detectors are more efficient when the number of tracked particles (occupancy) is higher. On the other hand the strip detectors can offer much higher area coverage providing lower mass and better transparency. Therefore in HEP trackers both sensor types are used simultaneously: closer to the interaction points, where the occupancy is higher pixel detectors with very fine granularity are used, whereas in the outer radii of the trackers, strip detectors are used.

In view of the millions of readout channels presented in the systems, it is clear that apart from the reception and amplification of signals from the semiconductor sensors, the front end electronics must provide additional processing of data including efficient data compression and transfer. Present-day front end ASICs consist of not only front end amplifiers with noise filtration circuits, but also digital circuitry providing data conversion (ADC or TDC), data buffering (RAM memories), data compression units and fast serializers. A modern front end ASIC is a typical example of a mixed mode VLSI circuit combining, on the same chip die, very sensitive amplifiers with the digital logic working with full swing CMOS levels. Providing a good separation between analog and digital circuits and keeping good Power Supply Rejection Ratio (PSRR) is one of the primary concerns. Another key issue for the designer of front end electronics is lowering the power consumption and overall mass of the detector whilst keeping noise, speed and dynamic range at the required levels.

10.3 Reception of Signals from Semiconductor Detectors

The basic function of the input amplifier is to interface the external sensor element to the on-chip circuit with the best possible charge transfer efficiency and signal amplification with minimum noise. There are three possible amplifier configurations which are based on common source transistor, common gate transistor and source follower. The common source stage is mainly used in charge sensitive amplifiers (CSA), the common gate circuit for high speed low input impedance readout channels [1], and the source follower circuit is used e.g. for monolithic radiation sensors [2]. The choice of the source follower is restricted to very low sensor capacitances, i.e. 10 fF, for which there is no need of voltage amplification.

In segmented semiconductor detectors the parasitic capacitance is dominated by the strip-to-strip or pixel-to-pixel fringing capacitance (c_{is}, Fig. 10.4). The primary readout configuration, resolving the problem of signal crosstalk and providing good efficiency of signal collection is the charge sensitive amplifier presented in Fig. 10.4.

The basic parameter of the preamplifier stage is the input impedance which should be low enough in comparison to the impedances constituted by the parasitic capacitances of the semiconductor detector. The feedback capacitor C_F, providing the integration of detector signals, is discharged with the feedback resistance R_F

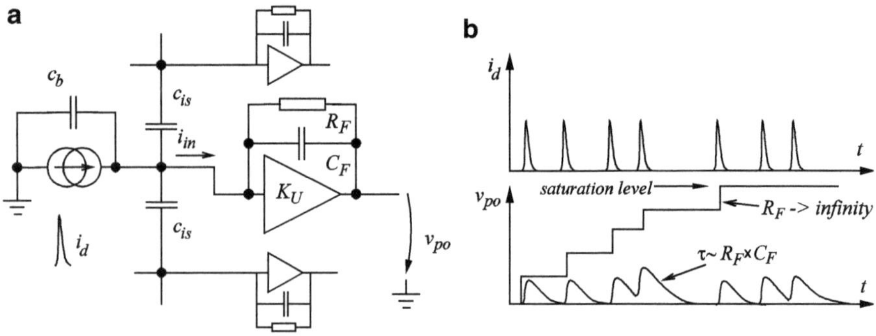

Fig. 10.4 Basic configuration of charge sensitive amplifier connected to semiconductor detector modeled as a current source in parallel with its total parasitic capacitance consists of sum of backplane capacitance c_b and fringing capacitances c_{is} (**a**). Responses to detector signal pulses, i_d, for two modes of operation, charge preamplifier and transimpedance preamplifier (**b**)

preventing saturation of this stage. The time constant of the feedback network τ_f will influence the input impedance characteristic of the amplifier impacting the charge collection efficiency and also the amplitude of the crosstalk signals. The input impedance of the preamplifier can be easily calculated in the operator domain:

$$Z_{in}(s) = \frac{Z_F(s)}{1 + K_U(s)} \approx \frac{Z_F(s)}{K_U(s)} \tag{10.1}$$

where $K_U(s)$ is the open loop gain of the preamplifier stage and Z_F is the feedback impedance. Assuming that preamplifier has one dominant pole τ_{PO} we can calculate the modulus of the input impedance in the frequency domain as:

$$|Z_{in}| = \frac{R_F}{K_U} \cdot \frac{\sqrt{1 + \omega^2 \cdot \tau_{PO}^2}}{\sqrt{1 + \omega^2 \cdot \tau_f^2}} \tag{10.2}$$

An example of two input impedance characteristics of the preamplifier working with two different feedback resistances is shown in Fig. 10.5. The preamplifier has been implemented in a commercial CMOS 250 nm process and optimized for a detector capacitance of the order of 20 pF. The cascode amplifier has an open loop gain of about 83 dB and dominant pole at around 200 ns. In the real circuit the equivalent feedback impedance is around 100 kΩ which, together with the 120 fF feedback capacitor, establishes a high-pass filter stage. This forms a part of the shaper, included in order to optimize the signal to noise ratio of the whole readout chain.

For very high frequencies the difference between the two characteristics is negligible. In that case the input impedance is limited by the position of the dominant pole of the amplifier and the feedback time constants (10.2). The input impedance can be lowered by either enlarging the amplifier bandwidth or by increasing the feedback capacitor. The use of a relatively low feedback resistance (100 kΩ) has a

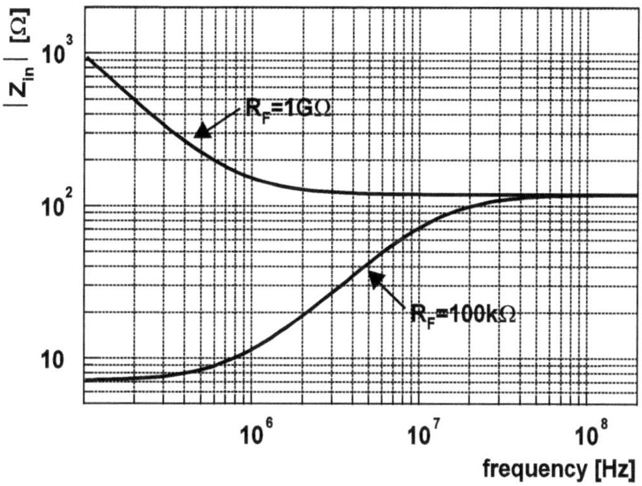

Fig. 10.5 Input impedance for the preamplifier working in charge mode ($R_F = 1$ GΩ) and transimpedance mode ($R_F = 100$ kΩ). The open loop gain of the amplifier is 83 dB with τ_{P0} equal to 200 ns, the feedback capacitance $C_F = 120$ fF (preamplifier optimized for detector capacitance of the order of 20 pF)

visible impact on the characteristic for low and medium frequencies where we can see a big improvement. We will say later that preamplifier works in transimpedance mode if its input impedance characteristic at low and medium frequencies is characterized by the feedback resistance. If the input impedance is increased at lower frequencies we will say that preamplifier works in charge integration mode.

From the point of view of the signal formation in charge sensitive preamplifier one would like to use smaller values of the feedback capacitor providing higher signal gain but not optimal from the point of view of lowering the input impedance. Still, one has to keep in mind that the signal charges from these semiconductor sensors are in the femto-coulomb range. In order to limit the noise contribution to the input stage only, one should ensure that the response amplitude of the preamplifier is sufficiently higher than the noise floor. In order to obtain a signal gain of the order of milli-volts per femto-coulomb one has to use feedback capacitances in the range of 10–100 fF. A way to satisfy both requirements, high signal gain and low input impedance, is to maximize the open loop gain and the bandwidth of the input amplifier. In practice, the open loop gain of the input stages should be of the order of 45 dB and 85 dB for pixel and strip detectors respectively.

The amplitude of crosstalk signals can be estimated using the model of the strip detector shown in Fig. 10.6, where the capacitance seen by the input of one channel consists of two parasitic capacitances to the neighbors (c_{is}) and one backplane capacitance (c_b). The amplitude of the crosstalk signals in frequency domain is simply the ratio of input impedance to the sum of the input impedance and the impedance of the inter-strip capacitance c_{is} (see Fig. 10.6).

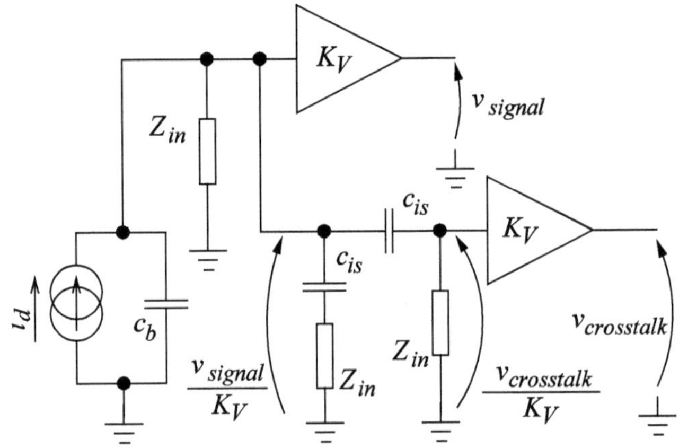

Fig. 10.6 Model of charge sharing between preamplifiers connected to silicon strip detector with a signal on one channel

$$|CrossTalk(j\omega)| = \left|\frac{v_{crosstalk}(j\omega)}{v_{signal}(j\omega)}\right| = \left|\frac{Z_{in}(j\omega)}{Z_{in}(j\omega) + Z_{is}(j\omega)}\right| \tag{10.3}$$

Where

$$Z_{in}(j\omega) = \frac{R_F\ (1 + j\ \omega\ \tau_{P0})}{K_U\ (1 + j\ \omega\ \tau_f)} \quad \text{and} \quad Z_{is}(j\omega) = \frac{1}{j\ \omega\ c_{is}} \tag{10.4}$$

The parameters of the preamplifier used as an example for crosstalk calculation are already described in the example from Fig. 10.5. The amplitudes of the crosstalk signals in frequency domain are plotted in Fig. 10.7. The differences in the crosstalk behavior for charge and transimpedance preamplifiers reflect their input impedance characteristics. The amplitude of the crosstalk in the time domain depends on the frequency characteristic of band pass filters applied after the preamplifier for signal to noise optimization and they are more pronounced in case of faster front end electronics. In practice, it is important to limit the amplitude of the crosstalk to below a few per cent. The transimpedance preamplifier shown in the example has been optimized for a detector of about 20 pF (c_{is} around 8 pF) and in the real ASIC is followed by a CR-RC^2 filter providing 22 ns peaking time response. The amplitude of the crosstalk signal in the time domain for this preamplifier is below 5%. The same preamplifier working in charge mode will exhibit crosstalk signals with twice the amplitude.

In general, transimpedance preamplifiers appear more interesting, however one should keep in mind that the feedback resistor, which improves the input impedance characteristic, might at the same time degrade the phase margin, which is an important parameter for the electronics implemented in huge systems such as the LHC tracking detectors. In order to ensure the stability and robustness of the front end electronics for tracking detectors the required phase margin is higher than

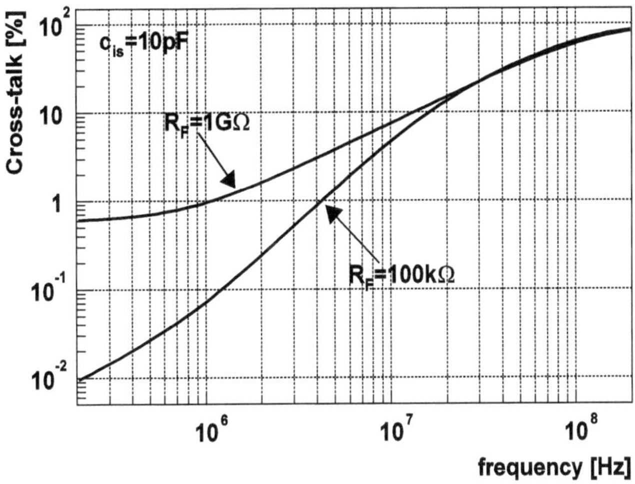

Fig. 10.7 Normalized cross-talk signals versus the frequency for charge and transimpedance preamplifiers loaded with 10 pF c_{is} capacitance

usual, being in the range of 85°–90°. To provide a fully asymptotic response of a transimpedance preamplifier having one dominant pole, it is sufficient to satisfy following inequality [3] pp. 6–7:

$$\tau_f \, C_F \, K_U > 4 \, c_d \, \tau_{P0} \tag{10.5}$$

where c_d is the total detector capacitance including all parasitic components. As we can see, for the case of transimpedance preamplifiers loaded with the given input capacitance, maximizing the open loop gain and bandwidth is also important to obtain a good phase margin.

Increasing the open loop gain and bandwidth of the core amplifier also serves to increase the PSRR which is especially important in large systems. Although the choice of the single ended architecture for the front end amplifier is dictated by the minimization of power consumption, one should keep in mind that all planar semiconductor detectors are intrinsically single ended. Therefore the single ended architecture of the front end amplifier is also a consequence of the detector construction. It can be shown that for feedback amplifiers its intrinsic PSRR parameter is modified by the feedback gain ($\beta \, K_U$), which for the charge sensitive preamplifier is expressed by following formulas

$$PSRR = PSRR_i \, (1 + \beta \, K_U) \quad \text{where} \quad \beta = \frac{1 + s \, C_F \, R_F}{1 + s \, (C_F + c_d) \, R_F} \tag{10.6}$$

Figure 10.8 shows the simulation of the open loop gain and PSRR of the preamplifier designed for the fast pixel readout in the NA62 [4] experiment at CERN. The design has been implemented in a commercial CMOS 130 nm process

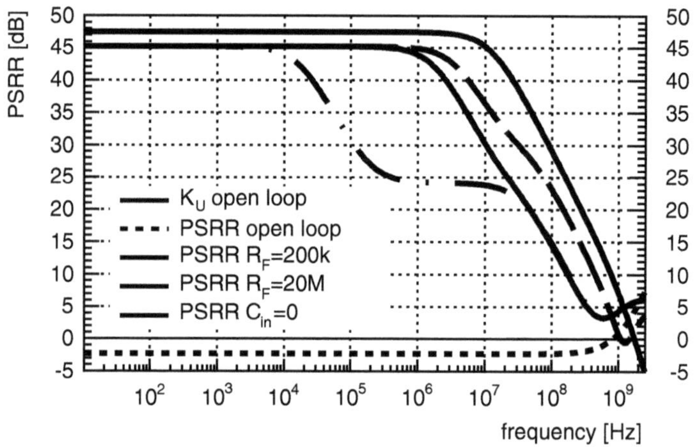

Fig. 10.8 The open loop gain and PSRR characteristics of the cascode amplifier designed for pixel GTK detector. The design has been optimized for 250 f. input capacitance. The open loop gain is 47 dB and the gain bandwidth product of the input cascode is 2 GHz. The PSRR behavior for preamplifier working in charge mode ($R_F = 20\ M\Omega$) and for open input ($C_{in} = 0$) has been shown for comparison purposes

and it is optimized for pixel capacitances of about 250 fF. The gain bandwidth product of the input cascode is approximately 2 GHz and the open loop gain is around 47 dB.

The intrinsic PSRR of the cascode in open loop configuration is −2 dB. As can be expected from the Eq. 10.6, the PSRR of the preamplifier working with the feedback ($R_F = 200\ k\Omega$, $C_F = 14$ fF) at low frequencies is 45 dB and towards higher frequencies follows the characteristic of the open loop gain.

A preamplifier loaded with a lower input capacitance would have a better PSRR at high frequencies (with $C_{in} = 0$ the feedback is loaded with gate capacitance of the input device and parasitic capacitance of the input pad only). It is worth noting that preamplifiers working in charge mode have significantly worse PSRR for middle frequency ranges due to lack of a R_F resistor. The overall PSRR characteristic of the input stage is actually modified by the band-pass filters following the preamplifier and used basically for SNR optimization. In order to preserve a good PSRR of the full readout chain at higher frequencies, it is important to design a preamplifier with bandwidth larger than the following filtering stages.

10.4 Types of Feedback in Charge Sensitive Amplifiers

Charge sensitive amplifiers must be reset to avoid saturation. In the past for LEP experiments, several tracking systems were equipped with a front end using a charge sensitive amplifier reset with a feedback CMOS switch controlled with the

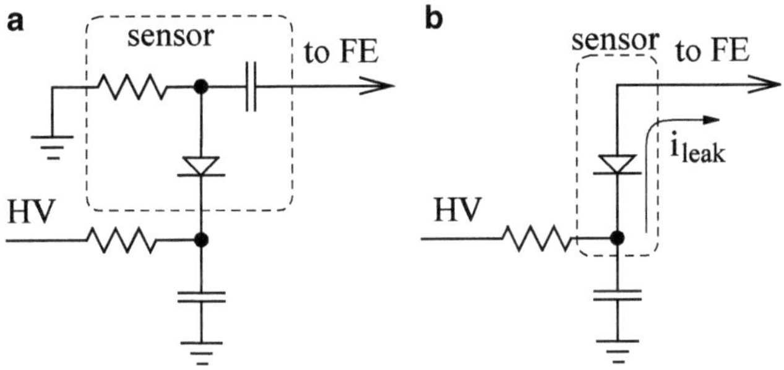

Fig. 10.9 Type of the coupling in the semiconductor detector, AC (**a**) and DC (**b**)

master clock of the collider machine [5]. However for colliders with a faster clock or asynchronous readout a continuous time reset is mostly used.

The basic function of the continuous time reset circuit is permanent discharging of the feedback capacitor with the given time constant. Depending on the type of the coupling between the detector and front end amplifier, the feedback circuitry has to deal in addition with the thermally generated DC leakage current which can flow in to the input of the amplifier. Two types of coupling are possible (see Fig. 10.9). For the strip detectors it is possible to integrate a capacitor in between the diode terminal and the readout pad together with a high value resistor (to avoid its noise contribution) providing the bias (Fig. 10.9a). The small area bias network is usually implemented on the edge of the detector. In the case of a pixel detector the situation is different. It is impractical or even impossible to integrate a coupling capacitor and bias resistor in each pixel. All hybrid pixel detectors use DC coupled detectors and front end electronics should be designed to tolerate the maximum expected leakage current which might be not negligible for sensors working in harsh radiation environments.

A limited selection of the feedback circuits used in front end electronics for semiconductor detectors is shown in Fig. 10.10. The simplest structure shown in Fig. 10.10a consists of a passive feedback resistor R_F. Depending on the technology and desired value of the feedback resistor, its implementation might be problematic. Another drawback is the lack of DC stabilization at the output of the preamplifier, which might create problems when DC coupling to the next stage. This is a simple design, and for relatively low values of the feedback resistance, it will tolerate reasonable values of detector leakage current, which will result in a change in the DC operating point only with no change to the feedback time constant.

Instead of a passive resistor in the feedback, an active device can be used. The active feedback amplifier shown in Fig. 10.10b was first described in [6]. The feedback transistor is biased into saturation by means of the input current source, denoted by i_f. The DC level at the preamplifier output can be set by the voltage applied on the gate of the feedback transistor which facilitates DC coupling to the next stage. Depending on the value of the feedback current we can distinguish three modes of operation [6]: square root compression, when the polarity of the signal

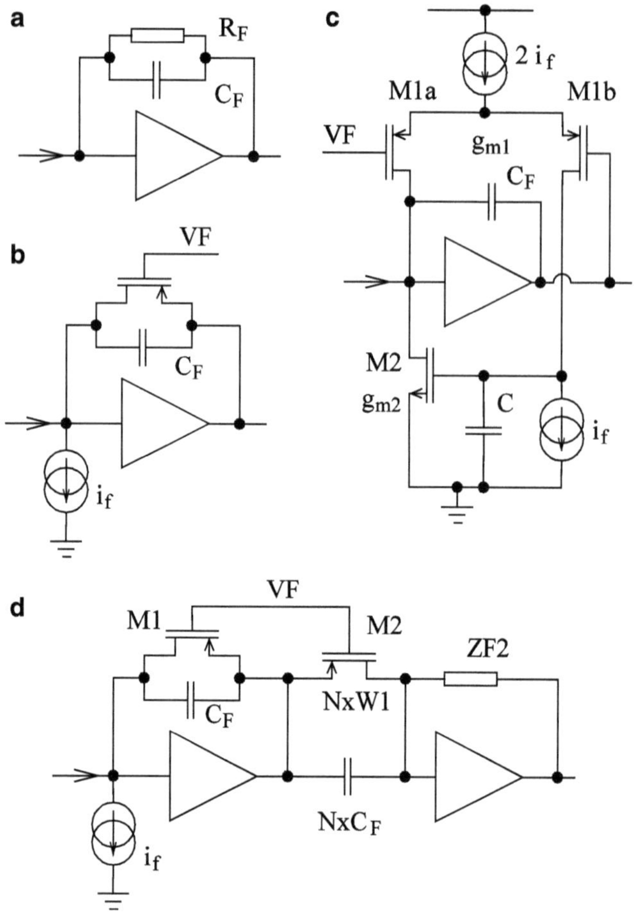

Fig. 10.10 Selection of the feedback circuits used in continuous time reset CSA

is negative and large compared to i_f, non-linear mode when the signal has the same polarity as i_f and dominates, switching OFF the feedback transistor, and the most frequently used linear mode when the input signal is significantly smaller that the feedback current (usually with $i_f > 300$ nA and signal < 12 fC). In this last mode, for small signal analysis, the feedback transistor can be represented by the resistance equal to $1/g_m$. In terms of noise performance, both the feedback transistor and the transistor in the current source, contribute to the equivalent parallel noise of the preamplifier. This parallel noise contribution is unfortunately higher than in the case of a passive feedback resistor of equivalent value (see analysis in [7]). One should stress that any current flowing into the amplifier input is adding (or subtracting) to the feedback current changing the transconductance of the feedback transistor i.e. the feedback time constant. Therefore the use of this configuration is restricted to AC coupled detectors or detectors where the expected leakage current is an order of magnitude less than the intended feedback current.

An active feedback configuration overcoming this problem is shown in Fig. 10.10c. The M1a transistor connected to the input of the preamplifier is equivalent to the feedback resistor of value $1/g_m$. Any excess current from M1b, generated by the leakage current attempting to flow into M1a, is integrated onto capacitor C and the resulting voltage controls the gate of M2, increasing the current through it. This slow feedback forces the DC signal (leakage) to flow into M2, working as a current source and keeping the g_m of M1a constant. In this configuration the leakage current might largely exceed the feedback current i_f without changing the feedback time constant. The detailed analysis of this configuration can be found in [8]. The drawback of this solution in comparison to the previous structure is the large number of transistors used, each of them contributing as a parallel noise source. As in the case of the previous architecture, the DC voltage at the preamplifier output is controlled by the feedback voltage VF applied to the gate of M1a. This configuration is used in many pixel front end designs [9–12].

A modification of the active feedback amplifier proposed initially for high energy X-Ray spectroscopy is shown in Fig. 10.10d. It addresses the problem of non-linear behavior for detector signals exceeding anticipated values of feedback current and also provides some tolerance to detector leakage current. This idea can also be realized for circuits designed for tracking applications where the intended feedback current is relatively low i.e. for preamplifiers working in charge mode. The square root behavior of the preamplifier output voltage with respect to detector current is compensated by means of a scaled transistor and capacitor sinking a multiplied detector current (by a factor of N) into the input of the transimpedance amplifier employed in the second stage. Together with the multiplied detector signal a multiplied leakage current has also to be sunk by the second stage transimpedance preamplifier. Although this configuration allows for minimization of feedback current and its noise contribution in the preamplifier, it demands a high performance second stage amplifier. Consequently, it is not very optimal from the point of view of power consumption, especially for very fast electronics where the preferred option is anyway a preamplifier working in transimpedance mode. Detailed analysis of this feedback configuration can be found in [13].

10.5 Architectures of the Input Stage

The choice of the architecture for the input stage always depends on the requirements concerning the AC parameters of the front end amplifier defined by the detector parameters and the technology in which the designed ASIC will be implemented. The demanding requirements of minimum noise and low power consumption usually lead to architectures for which the noise contribution is limited to the single ended input device. This architecture is compatible with the single ended architecture of the planar detectors under consideration.

Independently of the choice of optimum circuit topology, one has to solve the problem of the noise contribution from the devices employed in the active loads.

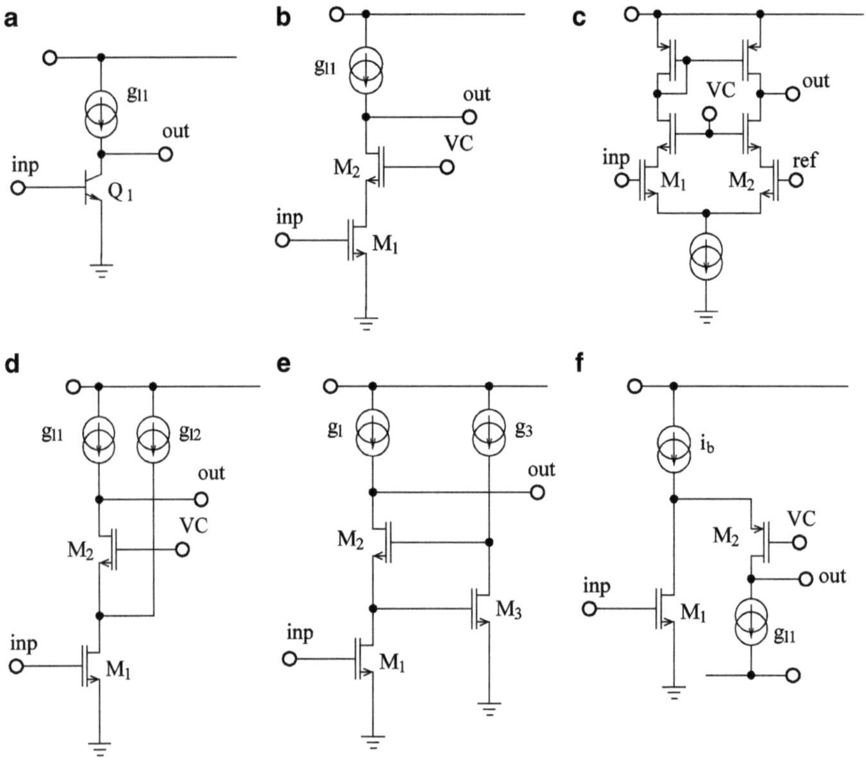

Fig. 10.11 Examples of the input stages of front end amplifiers for semiconductor detectors. (**a**) common-emitter amplifier; (**b**) telescopic cascode amplifier; (**c**) differential telescopic cascode; (**d**) telescopic cascode amplifier with bandwidth boosting; (**e**) regulated cascode amplifier; (**f**) folded cascode amplifier with symmetrical power supply allowing for power reduction

In the older CMOS technologies it was relatively easy to make this negligible by biasing those devices in the strong inversion region, to significantly limit the transconductance with respect to the transconductance of the active devices. In novel designs implemented in deep submicron processes, for many reasons, all transistors are biased in the moderate or weak inversion region i.e. the transconductance of the active load is close or equal to the transconductance of the active device. The problem is not new but already existed in old bipolar technologies where only bipolar transistors were available for the design. The resistive degeneration of the current sources solves this problem efficiently [14]. By providing a voltage drop across a degeneration resistor in the range of 120 mV, one can limit the noise contribution of the active load below 5% with respect to the contribution of the active device biased with the same current.

The simplest structure where the noise performance can be constrained by the parameters of the input transistor is a simple common-emitter amplifier with the bipolar transistor at the input and an active load (Fig. 10.11a).

This input stage can be implemented efficiently only in the processes offering fast bipolar transistors for two reasons. First, the intrinsic gain of CMOS devices is never high enough to provide sufficient open loop gain for optimizing the signal gain and input impedance for a given detector capacitance. Secondly, the Miller effect for CMOS transistors will be much more pronounced than in the case of small bipolar devices, degrading the preamplifier bandwidth and impacting negatively the input impedance characteristic as well as the PSRR for higher frequencies.

The basic architecture used in CMOS technologies for front end input stages is a cascode structure, which is a cascade of common-source and common-gate amplifiers. Figure 10.11b shows an example of a cascode amplifier with NMOS transistor at the input. The choice of the input transistor type, NMOS or PMOS, is determined by the noise optimization requirements. For high speed applications requiring higher input transistor transconductance, an NMOS device will be preferable. For the application where the noise performance will be limited by the flicker noise, the use of a PMOS device at the input will be favorable. The open loop gain of the cascode is defined by its enhanced output impedance boosted by the intrinsic gain of the cascode transistor M2. The Miller effect is avoided due to the very low voltage gain seen at the drain of the input transistor, which is loaded with the transconductance of the cascode transistor, M2. Because of the low gain of the input stage, one could logically expect a significant noise contribution from the cascode transistor, M2. Conveniently, this is not the case, and the noise contribution from the transistor M2 to the equivalent input noise of the cascode is reduced in proportion to the intrinsic gain of the input transistor, M1. Consequently, providing that the noise contribution from the active load is reduced using one of the previously described methods, the dominant noise source in the cascode amplifier is the input transistor. The gain bandwidth product of the cascode depends on the transconductance of the input transistor and the capacitance seen at its output, described by Eq. 10.7.

$$GBW = \frac{g_{m1}}{2\pi \, C_{OUT}} \tag{10.7}$$

This property of the cascode might be used to improve its characteristics at high frequencies. Figure 10.11d shows the cascode amplifier with an extra current source increasing the transconductance of the input transistor and, as a consequence, extending the bandwidth of the amplifier. This configuration is also beneficial from the perspective of possible noise contributions from the active load. By splitting the current source into two branches, one supplying the cascode and the other directly driving the input transistor we have two current sources providing equivalent bias of the input device that are not correlated with each other, i.e. the noise contribution before any noise reduction treatment is lower than in case of one single source supplying the same current. In practice the ratio between these two currents is set in the order of two to three which ensures the feasibility of the preamplifier compensation and keeps the noise contribution from M2 at a negligible level.

The folded cascode configuration (Fig. 10.11f) offers a similar possibility for bandwidth optimization. The bias of the input device can be optimized with respect to the noise requirements where the current in the cascode branch can be reduced allowing the dimensions of the transistors in the active load to be shrunk, which, in turn, limits the amplifier output capacitance. The drawback of this solution is the introduction of the extra current source biasing both the input transistor and the cascode branch. Some care must be taken to minimize the noise contribution from this extra current source. Without precautions, for all transistors biased in weak inversion region, this noise contribution would be higher than the noise coming from the input transistor. Historically, the single ended folded cascode stage has been used in front end amplifiers powered with symmetric voltage supplies. This allowed for connecting of the input transistor source to ground, lowering the overall power consumption. In the single power supply case, the overall power consumption of the folded cascode is higher than telescopic cascode discussed before.

The differential cascode amplifier shown in Fig. 10.11c is not optimal from the point of view of noise performance for a given power consumption. Nevertheless it has been used in a whole family of pixel front end chips [10] designed for very low input capacitances, which keeps the noise contributions from the differential structures at an acceptable level. The main motivation for this architecture is the fact that the differential input pair biased with the current source provides good isolation of the input transistors from the substrate noise in technologies where the guarding technique is not effective (low resistivity substrates).

In the deep submicron CMOS technologies, the designer faces the problem of lower intrinsic transistor gain, related to the short channel effects. For processes where the intrinsic transistor gain drops below 30 V/V a simple cascode configuration does not provide sufficient open loop gain for front end electronics designed for moderate detector capacitances, of the order of pico-farads. A solution for this problem might be to use a regulated cascode architecture in which the output impedance is enhanced by the gain of an extra amplifier built with the transistor M3, controlling the gate of the transistor M2 (Fig. 10.11e). The noise contribution of the transistor M2 is reduced proportionally to the intrinsic gain of the transistor, M1, multiplied by the gain of the boosting amplifier built with M3. This noise contribution can thus be completely neglected, in contrast to the noise contribution from the transistor M3. Although the noise from M3 is reduced proportionally to the intrinsic gain of the input transistor, M1, one should keep the M3 bias high enough to avoid seeing its noise contribution. This last constraint sets the limit for power consumption reduction in regulated cascode amplifiers implemented in these scaled CMOS processes with low intrinsic transistor gain.

It should be stressed that the gain enhancement in cascode and regulated cascode amplifiers is done by boosting the output impedance of these stages. In order to preserve the high gain of the input stage one must design the active load used in the amplifier to have an output impedance of the same magnitude. That is, for the cascode amplifier one has to use a cascode current source and for the regulated cascode one should employ a regulated cascode current source. In order not to deteriorate the output impedance characteristic of the structures working in a closed loop configuration, an extra buffering stage (usually a simple voltage follower) is mandatory.

The implementation of cascode architectures in deep submicron processes requiring relatively low supply voltages and degeneration of the active loads makes the biasing of all transistors in the saturation region challenging. Although operating all transistors in the weak inversion region makes this possible, with extra benefits in better noise performance exhibited by the transistors working in this regime, it results in a reduction of the transit frequency parameter and some degradation of the frequency characteristics.

The presented selection of structures is representative, though not at all exhaustive: there exist many other topologies which could be used as a front end input stage. Nevertheless, taking into account the noise, power consumption and frequency behavior, the authors consider cascode and regulated cascode architectures as optimal solutions for input stages implemented in deep submicron processes.

10.6 Noise Optimization

The optimization of the analog front end requires the trade-off between target performance and system constraints, in particular noise and power consumption. Power consumption is generally one of the key issues given the millions of channels required by the tracking systems of collider experiments. Noise and power consumption have opposite trends, which inevitably leads to compromise. This optimization is done for a given speed, usually imposed by the application. For binary front ends the decision is taken just after the front end amplifier and consequently a minimum signal to noise ratio is mandatory to keep the hit noise rate at a very low level , i.e. $<1\%$. From the Rice formula [15] a Signal to Noise Ratio (SNR) should be greater than 10:1 with the typical charge of a MIP approximately 22,500 e^-, the equivalent noise charge (ENC) should be lower than 1,500 e^- rms.

As was already mentioned, in order to improve the SNR of the readout chain, band-pass filters are usually used. A good compromise between performance, power consumption and simplicity, allowing for a very high scale of integration, is provided by CR-(RC)^n filters. They consist of one high-pass and n low-pass stages. Figure 10.12 shows the equivalent noise schematic diagram of a CSA connected to a detector represented by its parasitic capacitance c_d and signal source i_d. The electronic noise of the amplifier is represented by two equivalent noise sources, series v_{ns} and parallel i_{np}. It can be calculated [3] that the ENC for the electronics working with CR-(RC)^n type of the filters and connected to capacitive sensors is given by (10.8)

$$ENC[C] = \sqrt{\left(F_V \cdot \frac{\overline{v_{ns}}}{\Delta f} \cdot c_d / \sqrt{t_{peak}}\right)^2 + \left(F_i \cdot \frac{\overline{i_{np}}}{\Delta f} \cdot \sqrt{t_{peak}}\right)^2} \tag{10.8}$$

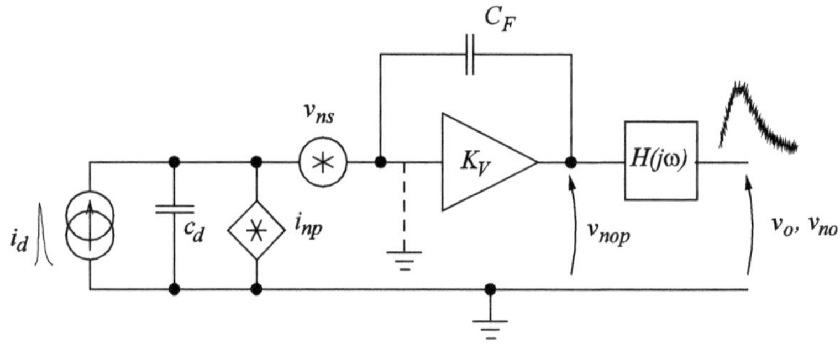

Fig. 10.12 Equivalent noise schematic of CSA connected to a semiconductor sensor. The electronic noise of a CSA is represented by series (v_{ns}) and parallel (i_{np}) noise sources. The frequency characteristic of the band-pass filter is represented by the transmittance H(jω)

where F_V and F_i are coefficients depending on filter order and t_{peak} is the peaking time of the response representing the frequency characteristic of the filter in the time domain. For most of the applications in LHC experiments, the optimization of t_{peak} is impossible; the required time response of front end electronics is defined by the experiment conditions and charge collection time from the sensor. In order to keep the ENC numbers at the specified level one has to optimize the noise source spectra, i_{np} and v_{ns}. In a properly designed preamplifier the dominant series noise source is the input transistor whereas the parallel noise source is determined by the feedback circuit and detector leakage current. Therefore the choice of the input transistor plays an important role not only in the speed and dynamic range but also in the noise performance. The choice between n-channel and p-channel is usually dictated by the importance of the contribution of the flicker noise. For slow shaping times, i.e. 1 μs, p-channel offers better performance because of its lower flicker noise whereas for fast shaping times, i.e. <100 ns, n-channel is preferable because of its high transconductance for a given drain current. The drain current should then be chosen according to the target noise because the device transconductance, which depends on the drain current, determines the thermal noise resistance.

The device dimensions, particularly the transistor width, should be optimized with regard to both the desired transconductance and the gate capacitance in parallel to the detector capacitance, since this latter quantity increases the series noise contribution (formula (10.8)). In practice gate length is kept as close as possible to the minimum gate length without increasing the excess noise. For submicron CMOS technologies this optimization leads in most cases to operating the input device in moderate or weak inversion. The main difficulty of such analytical optimization is the precise modeling of the transconductance and gate capacitance of the transistor for a wide range of drain current (I_{drain}) and transistor dimensions from strong to weak inversion. A fully analytical model providing continuity between the strong, moderate and weak inversion regions, proposed by

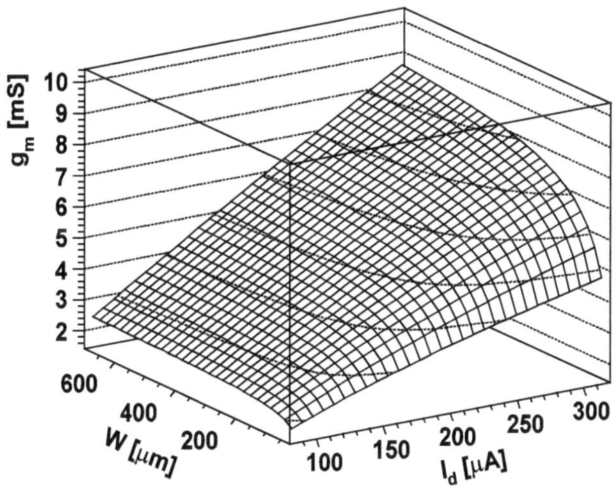

$$g_m = G(I_f) \cdot \frac{I_{drain}}{n \cdot U_T}$$

$$U_T = \frac{k \cdot T}{q} \qquad I_f = \frac{I_{drain}}{I_S}$$

$$G(I_f) = \frac{1}{\sqrt{I_f + \frac{1}{2} \cdot \sqrt{I_f} + 1}}$$

$$I_S = 2 \cdot n \cdot K_P \cdot \frac{W}{L} \cdot U_T^2$$

Fig. 10.13 The example of MOS transconductance plotted as a function of drain current and device width together with basic formulas of the EKV model. Example of NMOS transistor of gate length 500 nm from commercial CMOS 250 nm process

Enz, Krummenacher and Vittoz (EKV), is described in [16]. Figure 10.13 shows an example of the transconductance of a NMOS transistor from commercial CMOS 250 nm process plotted for a wide range of bias and transistor dimensions using the basic formulas of the EKV model. The U_T is the thermal voltage, n is the slope factor (parameter characterizing bulk transconductance) and K_p is transconductance parameter.

More detailed examples of noise optimization of preamplifiers built with bipolar and MOS transistors, taking into account complete noise models of the devices, can be found in [7] and [17].

10.7 Some Examples of Front End Designs

Figure 10.14 shows the front end of the ABCD3T chip designed for the ATLAS semiconductor tracker and implemented in the DMILL radiation hard BiCMOS process. It employs an input stage with a simple common-emitter amplifier. The design has been optimized for detector capacitances of about 20 pF. The peaking time of the amplifier response is of the order of 25 ns. The PMOS transistor loading the bipolar input device is biased in the strong inversion region and its transconductance is low enough to neglect its contribution to the noise. All current sources built with bipolar devices have been degenerated with resistors to limit their noise contribution and to improve device matching. The gain bandwidth product for the

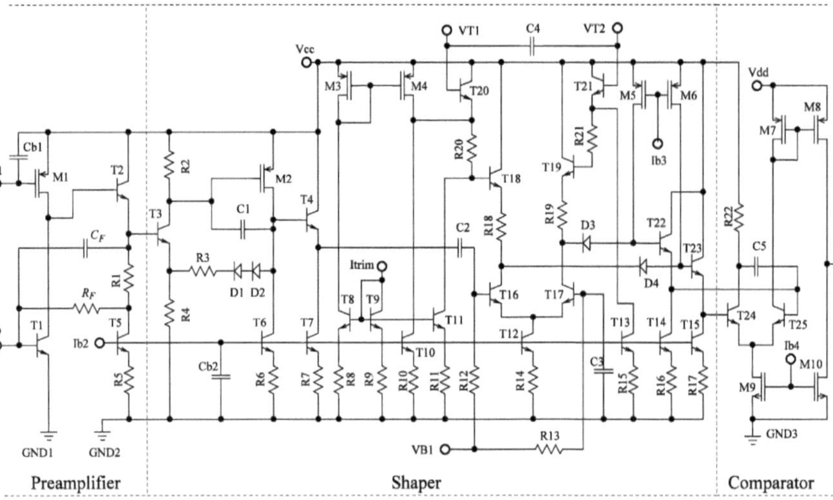

Fig. 10.14 Schematic of one channel of the front end of the ABCD3T chip for ATLAS semiconductor tracker (SCT) implemented in DMILL BiCMOS technology

nominal bias condition is approximately 1 GHz and the open loop gain is in the range of 65 dB, which restrict the noise contribution to the input device only. A consequence of the relatively small dimensions of the bipolar transistor imposed by the requirements of radiation hardness, is that base–collector capacitance is low enough to keep the Miller effect at an acceptable level. The transimpedance preamplifier is followed with a cascade of two common emitter (T3) and common-source (M2) amplifiers stabilized with resistive and capacitive feedback, providing both amplification and integration.

The feedback resistor R_F, equal to 80 kΩ, gives good tolerance to base current variations due to the spread of process parameters and radiation damage effects degrading β current gain of T1. This tolerance also translates to variations of detector leakage current. The PSRR of the readout chain is optimized using the previously described methodology i.e. the first and the second stages, being single ended amplifiers, work in feedback configuration, while the AC coupled discriminator works in open loop configuration and has a fully differential architecture. The ABCD3T front end chip provides binary information from the detector. At the end of the signal processing chain a leading edge comparator in Schmidt configuration has been used. The signal gain of the amplifier is roughly 50 mV/fC. As a result of the poor resistor matching exhibited by DMILL technology, a 4-bit trim DAC has been implemented in each channel to account for offset variations. The trimming current enters the T9/T8 current mirror causing a voltage drop on resistor R20, which adds to the threshold voltage applied differentially on transistors T20/T21, operating in emitter follower configuration. The overall power consumed by a single front end channel biased with nominal conditions is around

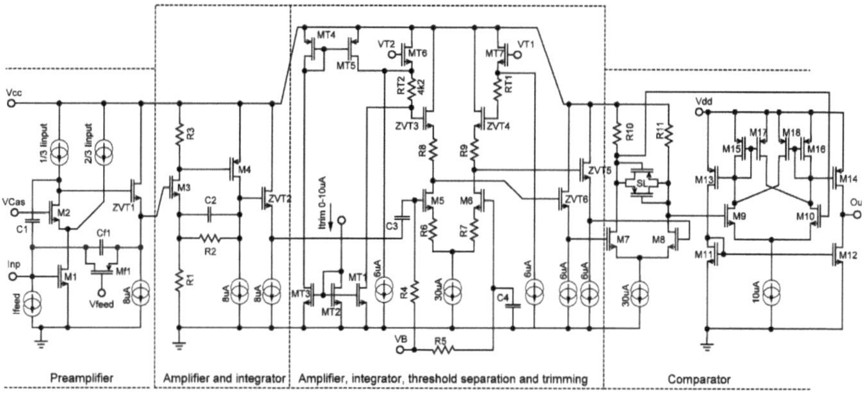

Fig. 10.15 A schematic of one channel of the fast transimpedance front end amplifier implemented in a commercial 250 nm CMOS technology

2 mW. The detailed description of the ABCD3T chip working in present ATLAS experiment can be found in [18].

The same architecture has been successfully implemented in a commercial 250 nm CMOS process yielding with a family of front end ASIC's for the readout of the silicon strip detectors of capacitances ranging from 1 to 30 pF [7, 19, 20] and CdZnTe sensors for Computer Tomography application [21]. Figure 10.15 shows the schematic of one channel of the front end chip implemented in quarter micron CMOS process.

The transimpedance preamplifier is built around telescopic cascode with a NMOS input transistor loaded with the cascode current source, providing an open loop gain higher than 80 dB. An additional cascode current source supplying the input transistor provides an improvement of the preamplifier bandwidth and helps in the reduction of noise from the active load (two uncorrelated noise sources instead of one). The gain bandwidth product of the cascode amplifier is around 600 MHz. The transconductance of the load transistors is reduced by the use of relatively long channel devices, reducing their noise contribution to a negligible level. The simple active feedback configuration used in the preamplifier allows it to be used with AC coupled detectors or with sensors exhibiting relatively low leakage current (limited to an order of magnitude less than feedback current, being in the range of 300 nA to 1 μA depending on the version). Similarly to the BiCMOS version, a second stage is built with the cascade of two common source amplifiers enclosed with resistive and capacitive feedback. The operating point of this stage is controlled by the DC level at the preamplifier output i.e. the voltage applied on the gate of the feedback transistor (V_{feed}) which is generated by the replica amplifier. The accurate control of this voltage allows for driving of the closed loop gain of this stage and lowering of the gain of the following differential amplifier, which is stabilized using the degeneration resistors. The architecture of the differential amplifier, used also for threshold interface, is similar to the BiCMOS version. Although the circuit is supplied with 2.5 V only, due to the use of the native

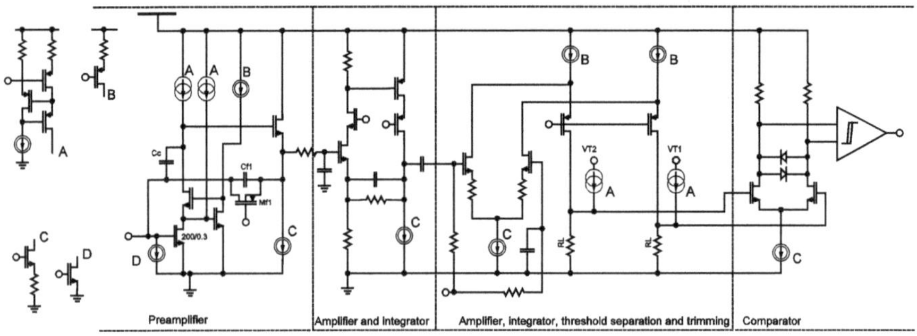

Fig. 10.16 A schematic of one channel of the front end prototype intended for super LHC ATLAS silicon strip tracker and implemented in commercial 130 nm CMOS process

NMOS transistors (0 V threshold) in the differential pair load, a dynamic range of better than 1 V is obtained. Together with the front end gain being around 100 mV/fC, it gives more than 10 fC of dynamic range, which is more than sufficient for the tracking application. The peaking time of the amplifier response is around 22 ns, which for LHC applications (requirement of unique association of event with a given Bunch Cross Over (BCO) clock equal to 25 ns) allows the use of a simple leading edge comparator. The DxCTA version of the front end optimized for CT applications is equipped with a window comparator providing a count rate in the range of 5 mega counts per second (MCps). The discriminator is equipped with 5-bit trim DAC with a selectable dynamic range, improving the circuit matching. The overall power consumption of a single channel powered with 2.5 V, depending on the version, is between 0.7 and 1.5 mW.

In response to the continuous demands of power reduction for the front end electronics intended for future experiments at the upgraded LHC collider, a similar architecture has been implemented in 130 nm commercial CMOS process. The schematic of one channel of front end amplifier optimized for detector capacitances up to 10 pF is shown in Fig. 10.16. As in the previous circuit the preamplifier works with an active feedback configuration in transimpedance mode, with the peaking time of the amplifier response tuned to 22 ns. A systematic use of the cascode and regulated cascode amplifiers resolved the problem related to the degradation of the intrinsic transistor gain in this technology (typically 30 V/V). The input stage built with the regulated cascode amplifier has an open loop gain of the order of 80 dB together with 2 GHz gain bandwidth product. The power consumption of the channel powered with 1.2 V supply is only 220 μW. The increase of the transconductance parameter in the technology used makes the biasing of the current sources in strong inversion impractical. The noise contribution from the active loads is reduced using the resistive degeneration method. The noise contribution from the cascode boosting amplifier is kept at negligible level by proper biasing of this stage. A relatively low voltage supply, 1.2 V, imposes a change on the discriminator architecture. A differential folded cascode with resistive loads serves both for amplification as well as for threshold interface, which is applied by means of two

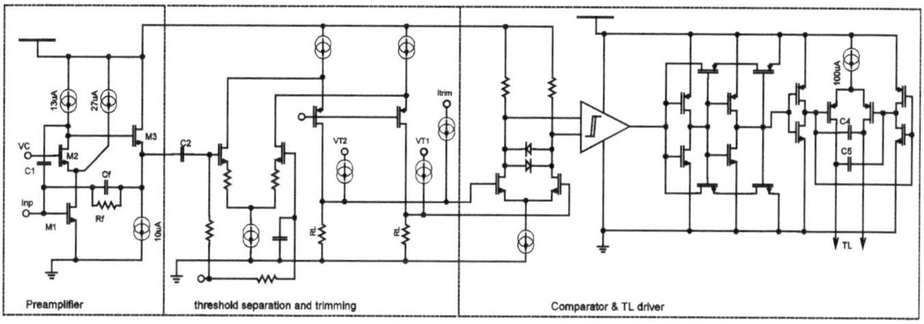

Fig. 10.17 A schematic of the NA62 GTK pixel front end amplifier implemented in commercial 130 nm CMOS process

high output impedance current sources. The dynamic range provided by this architecture is around 600 mV which, with a gain of 90 mV/fC, is sufficiently high for tracking applications. The front end channel ends with a differential comparator with a swing limiter at the input which improves the response time. More details about the performance of this front end prototype can be found in [22].

Another example of the front end amplifier designed for readout of silicon hybrid pixel detector at the NA62 [23] experiment is shown in Fig. 10.17. The front end circuit is intended for readout of 300 × 300 μm^2 silicon pixel detector with a RMS time resolution of better than 200 ps. The circuit, implemented in a standard 130 nm CMOS process, consists of a transimpedance preamplifier built with resistive feedback, differential folded cascode amplifier with comparator threshold interface followed by a three stage, fast discriminator with Schmidt trigger and transmission line driver with pre-emphasis. The input stage, built with telescopic cascode with a NMOS input transistor, and loaded with degenerated current sources provides 2 GHz GBP and 47 dB open loop gain. The rather moderate value of open loop gain with respect to previously described front ends is still sufficiently high for the intended detector capacitances, which are of the order of 250 fF: this combination permits the use of a relatively low value (14 fF) feedback capacitor (the simulated crosstalk signals are lower than 3%). A relatively low value of the feedback resistor, 200 kΩ, allows for detector leakage currents in the range of hundreds of nA, an order of magnitude higher than the expected maximum leakage at the end of the detector lifetime (20 nA).

The peaking time of the amplifier response is adjusted to 5.5 ns which is a compromise between requirements for discriminator jitter and charge collection efficiency. The equivalent noise charge of the front end connected to the detector is below 250 e^- and the power consumed by the analog part of the pixel channel is around 120 μW. More details about the performance of the prototype front end circuit [4] can be found in [24].

The presented front end amplifiers are examples of fast electronics exploiting the speed limits set by the process parameters, namely f_t. For this selection, a proper choice of architecture and mode of operation of the input stage i.e. charge or transimpedance preamplifier are especially important, and all problems related to the non-ideal frequency characteristic will be more pronounced than in the case of a front end amplifier designed for slower shaping times.

10.8 Conclusions

The design of a front end amplifier for a given application always implies a tradeoff between noise, speed and power requirements. As was already stressed, power consumption is a key issue in large systems for tracking detectors in HEP experiments. Therefore the architecture and performance of the front end amplifiers in terms of noise, speed and dynamic range result mainly from a compromise with the targeted power budget.

It has been shown that the performance of the input stage has an important impact on the overall performance of the front end channel not only on noise, charge collection efficiency, speed and crosstalk signals but also on the preamplifier phase margin and PSRR of the full chain. Increasing the open loop gain and the bandwidth of the core amplifier, while keeping an affordable power budget, is a key issue of the design of the input stage.

The evolution of CMOS technologies brings new design opportunities in terms of higher amplifier bandwidth and lower power consumption. However, the scaling of the transistor length causes degradation of the transistor intrinsic gain impacting the achievable open loop gain for a given topology. It has been shown that new design techniques, such as the regulated cascode, help to circumvent this issue however there are some limits imposed by possible excess noise contributions from the boosting stages. Although lower supply voltage accompanying the modern technologies helps reduce the power consumption, it also brings new challenges for the proper biasing of the transistors in cascode amplifiers as well as preserving the high dynamic range required in the output stages.

References

1. Anghinolfi F et al (2004) NINO: an ultra-fast and low-power front-end amplifier/discriminator ASIC designed for the multigap resistive plate chamber. Nucl Instrum Method A 533:183
2. Turchetta R et al (2003) Monolithic active pixel sensors (MAPS) in a VLSI CMOS technology. Nucl Instrum Method A 501:251
3. Kaplon J (2004) Fast bipolar and CMOS rad-hard front end electronics for silicon strip detectors. Ph.D. thesis, 2004 JINST TH 002
4. Martin E et al (2009) The 5ns peaking time transimpedance front end amplifier for the silicon pixel detector in the NA62 Gigatracker. In: Nuclear science symposium conference record (NSS/MIC), 2009 IEEE, pp 381–388
5. Bingefors N et al (1993) The DELPHI microvertex detector. Nucl Instrum Method A 328:447
6. Jarron P et al (1996) A transimpedance amplifier using a novel current mode feedback loop. Nucl Instrum Method A 377:435

7. Kaplon J, Dabrowski W (2005) Fast CMOS binary front end for silicon strip detectors at LHC experiments. IEEE Trans Nucl Sci 52(6):2713–2720
8. Krummenacher F (1991) Pixel detectors with local intelligence: an IC designer point of view. Nucl Instrum Method A 305:527
9. Anghinolfi F et al (1991) A 1006 element hybrid silicon pixel detector with strobed binary output. In: Nuclear science symposium conference record (NSS/MIC), 1991 IEEE, pp 255–262
10. Llopart X et al (2001) Medipix2, a 64k pixel read out chip with 55 μm square elements working in single photon counting mode. In: Nuclear science symposium conference record (NSS/MIC), 2001 IEEE, pp 1484–1488
11. Campbell M et al (1997) Readout for a 64×64 pixel matrix with 15-bit single photon counting. In: Nuclear science symposium conference record (NSS/MIC), 1997 IEEE, pp 189–191
12. Ballabriga R et al (2007) The Medipix3 prototype, a pixel readout chip working in single photon counting mode with improved spectrometric performance. IEEE Trans Nucl Sci 54(5):1824–1829
13. De Geronimo G, O'Connor P (1999) A CMOS detector leakage current self-adaptable continuous reset system: theoretical analysis. Nucl Instrum Method A 421:322
14. Bilotti A, Mariani E (1975) Noise characteristics of current mirror sinks/sources. IEEE J Solid-State Circuits 10(6):516–524
15. Rice S (1944) Mathematical analysis of random noise. Bell Syst Tech J 23:282–332, 24:46–156
16. Enz C, Krummenacher F, Vittoz E (1995) An analytical MOS transistor model valid in all regions of operation and dedicated to low-voltage and low-current applications. J Analog Integr Circuit Signal Process 8:83–114, Kluwer Academic Publishers
17. Kaplon J, Dabrowski W (2006) Experience with bipolar front-end amplifiers and perspectives for LHC upgrade. Nucl Instrum Method A 568:877
18. Dabrowski W et al (2005) Design and performance of the ABCD3TA ASIC for readout of silicon strip detectors in the ATLAS semiconductor tracker. Nucl Instrum Method A 552:292
19. Aspell P et al (2008) VFAT2: a front-end "system on chip" providing fast trigger information and digitized data storage for the charge sensitive readout of multi-channel silicon and gas particle detectors. In: Nuclear science symposium conference record (NSS/MIC), 2008 IEEE, pp 1489–1494
20. Dabrowski W et al (2009) Design and performance of the ABCN-25 readout chip for ATLAS inner detector upgrade. In: Nuclear science symposium conference record (NSS/MIC), 2009 IEEE, pp 373–380
21. Moraes D et al (2008) CERN_DxCTA counting mode chip. Nucl Instrum Method A 591:167
22. Kaplon J, Noy M (2012) Front end electronics for SLHC semiconductor trackers in CMOS 90 nm and 130 nm processes. IEEE Trans Nucl Sci 59(4):1611–1620
23. Fiorini M et al (2011) The Gigatracker: an ultra-fast and low-mass silicon pixel detector for the NA62 experiment. Nucl Instrum Method A 628:292
24. Noy M et al (2011) Characterisation of the NA62 GigaTracker end of column demonstrator hybrid pixel detector. 2011 JINST 6 C11025

Part III
Robustness

Robustness in integrated circuits was and still is a very important concern. For that, in this part an overview and some robustness issues are discussed. Among them we have the radiation hardness, which will be discussed in the first two contributions. Of course transistor matching properties are also limiting factors which have become nowadays even design parameters. However the effect of different technologies such as organic or even nano-meter technologies are effected or even dominated by matching effects. As a result those issues and technology choices and their effect are discussed in the next three contributions.

The first Chapter presents an overview in the world of radiation hard design. A problem that is often underestimated, but very important if high robustness is required. Even for daily live circuits, due to the mass employment, very high robustness is required and the performance can be effected even under natural radiation.

The second Chapter presents the design of radiation hard CMOS Time-to-Digital-Convertors (TDC). By a proper topology selection high performances can be achieved even under high radiation doses. Measurements and radiation performance degradation and solutions are discussed.

The third Chapter is an overview of the matching properties and requirements for analog circuits. The effect on circuit performances and the effect of technology choices are discussed. Some design techniques and effects are described in detail

The fourth Chapter addresses the design and effect of matching in organic integrated circuits. Those emerging technologies still suffer from parameter variability and as such design techniques to minimize the effect are becoming part of the circuits. An example of an integrated AD converter and several amplifier topologies are described.

The last Chapter deals with the search towards nanometer technologies. Different technology choices such as bulk, SOI or fin-FET technologies are analyzed based on statistical variability analyses and simulation tools. Due to the different function slopes between voltage and currents of the different technologies, variability can have sometimes reverse worse effects towards the design corners. As a result the importance of variability analysis and simulation tools in the technology choices are becoming more and more important.

Michiel Steyaert

Chapter 11
How Can Chips Live Under Radiation?

Erik H.M. Heijne

Abstract Interactions of different types of radiation in silicon are discussed together with effects on devices. Long-term irradiations cause 'Total-Ionization-Dose' degradation and 'Single Event Effects' occur when dense ionization upsets a small area in a chip. At the CERN Large Hadron Collider LHC we expect a severe radiation environment, yet sophisticated chips are needed. Some remedies against radiation effects are illustrated. One can use changes in technology, in device geometry, in circuit design or in layout. At system level one can recover loss of functions or data. Trends in CMOS technology call for continuous study of behaviour of new devices under radiation. The increased use of chips for critical functions everywhere imposes study of rare effects of radiation, not only in extreme conditions. With large areas of silicon in operation worldwide, low probabilities do result in real incidents.

11.1 Introduction

Radiation of all kinds generates free charge carriers in semiconductors, as long as the energy of the incident quantum exceeds the bandgap energy. This can be desirable, as in an imager or a solar cell, or very bad if the radiation upsets the normal device functions. Semiconductor devices have been introduced just a couple of years after the second world war, and military and space applications soon suffered from this problematic sensitivity to radiation. A series of yearly conferences started, the Nuclear and Space Radiation Effects Conference NSREC, reported in the Transactions on Nuclear Science of the IRE, later IEEE, and these continue today. A specialized industry developed for the supply of adequately hardened circuits for

E.H.M. Heijne (✉)
CERN PH Department, CH1211 Genève 23, Switzerland

IEAP-CTU, Prague, Czech Republic

NIKHEF, Amsterdam, The Netherlands
e-mail: erik.heijne@cern.ch

A.H.M. van Roermund et al. (eds.), *Nyquist AD Converters, Sensor Interfaces, and Robustness: Advances in Analog Circuit Design, 2012*,
DOI 10.1007/978-1-4614-4587-6_11,

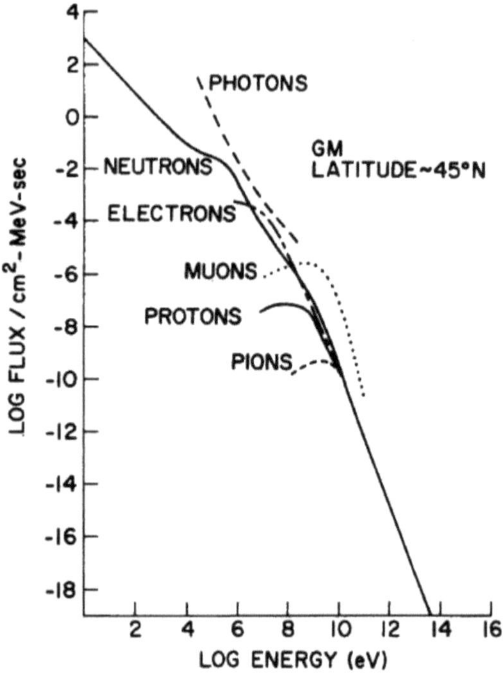

Fig. 11.1 Average values of cosmic ray particles at sea level, 45° geomagnetic latitude, from Ziegler [2]. Large variations with time and place are possible

this class of users. The earliest spectacular, radiation–related, public incident was the failure of the TELSTAR communication satellite, soon after its launch in 1962. Civilian applications of devices and integrated circuits at first appeared unaffected, until ~1975 it was discovered that alpha particle impact could change the state of semiconductor memory cells [1]. In 1979 James Ziegler at IBM showed that also sea level cosmic rays could influence silicon devices [2]. He concluded that the technology had just gone over a threshold in miniaturization, beyond which the radiation-generated charge becomes comparable with the charges that determine the function of the device. Ever since, radiation has, or should have been a concern in the operation of integrated circuits for critical applications, and for example, the aircraft industry has taken this seriously [3]. In recent years, with ever more widespread use, also for several consumer applications one has to be aware of radiation effects. Chips in cars and in medical devices must be free from such interferences. Moreover, one has to take into account the statistics of radiation impacts for a very large area of deployed silicon. Figure 11.1 shows the intensity of the cosmics-induced radiation, after Ziegler [2]. The proportion of neutrons at sea level is ~95 % of the cosmic-ray-induced particles.

This presentation will focus in Sect. 11.2 on interactions in silicon devices by various types of radiation and the effects on device operation. In Sect. 11.3 the characteristics of some specific environments will be discussed, including the extreme example of the CERN Large Hadron Collider LHC. An overview of remedies is given in Sect. 11.4. Circuit design and layout techniques are mentioned. One might think that there is not much to be added anymore to the knowledge about radiation effects, but the trends in CMOS technology towards smaller dimensions and new materials require continuous research. For different technology choices the radiation hardness of deep submicron IC evolves in opposite directions and trade-offs have to be found. One still may encounter unexpected and maybe even new effects, as the one discussed in Sect. 11.6. Finally, some conclusions are presented.

11.2 Radiation: Interactions in Silicon and Device Effects

Both natural and man-made environments present a variety of radiation quanta that interact with integrated circuit materials. Some specific examples are discussed in Sect. 11.3. Here we first look at the main effects in silicon devices.

11.2.1 Total Ionizing Dose TID and Threshold Shifts

All kinds of ionizing radiation loose energy and create free carriers in silicon as well as in the isolating oxide or nitride layers. The mean energy loss per generated e^-h^+ pair is 3.64 eV in Si, and ~17 eV in SiO_2. The charge moves to interface layers, and induces a shift of the threshold voltage in the MOS structures, gate oxides as well as field oxides. The shift is found to be proportional to the square of the adjacent oxide thickness, except for very thin gate oxides where it vanishes. Saks and coworkers studied this behaviour already in 1982 [4] and their results are shown in Fig. 11.2. For comparison, some later data measured at CERN in commercial CMOS technologies are shown as well. In advanced CMOS (below 0.3 μm with $t_{ox} < 8$ nm) the V_T of gates becomes virtually insensitive to ionizing radiation. This work led us at CERN to use for LHC circuits a commercial 0.25 μm CMOS technology with gate oxide of 5.5 nm [5].

However, the interfaces under the thick field oxide remain very sensitive and leakage paths can be created between transistors and under the bird's beak regions in LOCOS isolation, shorting source and drain. A remedy (Sect. 11.4.2) with some area penalty is the ancient form of enclosed transistor, already used at RCA ~1975 [6].

11.2.2 Electro-magnetic Pulse (EMP) or Flash

For early military applications an Electro-Magnetic Pulse EMP created by a nuclear explosion should not prevent the critical circuits to function. The dense flux of photons and photo-generated electrons flood all silicon with free carriers that immediately have to be evacuated. Today, similar situations seem to be rare, but

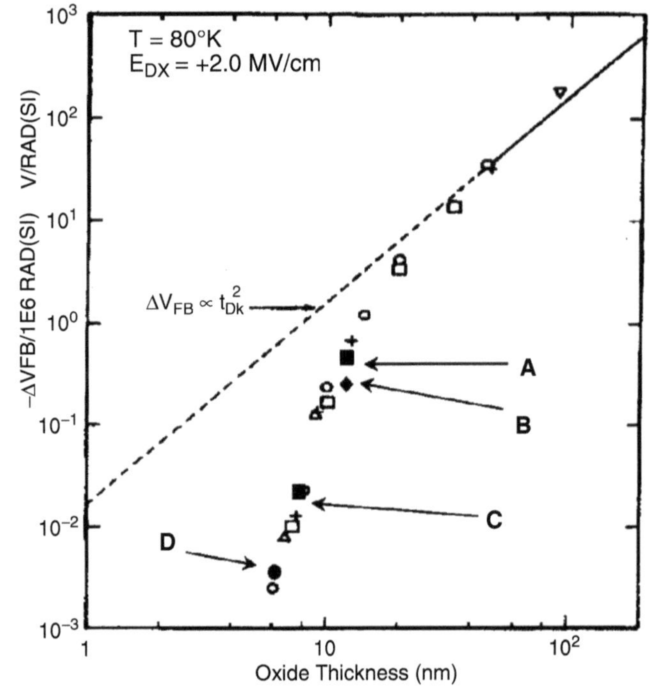

Fig. 11.2 Threshold shift ΔV_T in MOS capacitors at 80 K after 1 Mrad irradiation as a function of the gate oxide thickness published by Saks et al. [4]. Data points A to D later have been measured at CERN in test samples from commercial CMOS technologies

in some pulsed laser applications the flash may be very intense as well. This type of interaction will not be discussed further. The Wassenaar convention imposes restrictions on chips that can survive such environments.

11.2.3 Crystal Defects

Energetic particles, especially neutrons, can kick atoms in the Si crystal from their lattice sites, where the binding energy is ~20 eV, and create crystal defects. The initial Si lattice vacancy is mobile and eventually combines into a stable defect with some other defect or with a doping atom. Such defects have well-defined energy levels, many close to midgap, ~0.4 eV. Typical introduction rate is one defect per cubic centimeter per incident particle per square centimeter. Defects act as charge trapping site, reduce the carrier lifetime, influence the effective doping density and contribute to generation-recombination current in depletion regions. Crystal defects are an important damage mechanism for imaging devices, creating pixels with large dark currents. In bipolar devices the gain is reduced via the minority carrier lifetime.

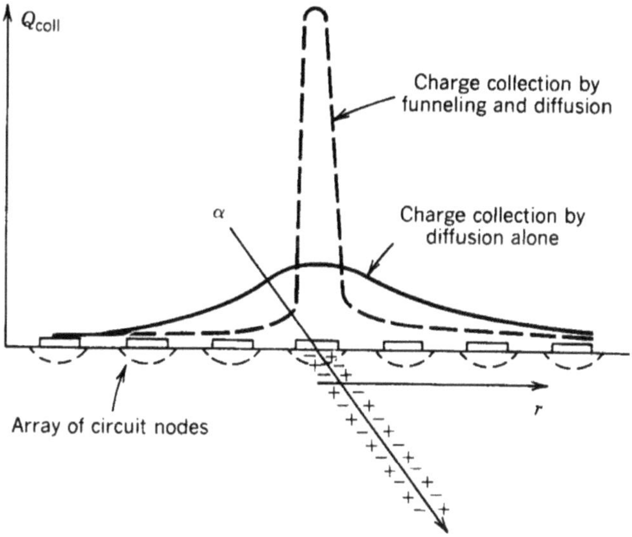

Fig. 11.3 Illustration of the funnel effect on an array of junctions following an MeV alpha-particle strike [7]

Ion implantation already creates a lot of crystal defects, so that in-process annealing may be needed on top of the thermal treatment for activation. Most defects can be annealed below ~500°C but for critical implant profiles only a limited thermal budget is available.

11.2.4 Single Event Effects SEE

Total dose and displacement effects accumulate gradually over longer periods and the chip functioning may degrade only after a fairly long time. On the contrary, Single Event Effects are immediately apparent when a particle hits a sensitive node and generates in that position a charge that exceeds some critical value. It was already mentioned that alpha particle impact can alter the content of a memory cell. A 5 MeV alpha particle deposits all its energy along a pathlength of ~20 μm, generating a charge of ~1.3×10^6 electrons or 220 fC. This charge cloud can be collected towards a node of the circuit, even if initially much of the charge is in the bulk and has some lateral extension. This process of 'funneling' was described by McLean and Oldham [7] and is illustrated in Figs. 11.3.

Precise modelling of the interaction between the energy deposition and the device geometry can now be done with tools such as Sentaurus Device (previously ISE) from Synopsys [8] as shown in Fig. 11.4. The carrier densities are calculated as a function of time, and compared with the normal device parameters. Experimentally, one has to determine the critical charge that leads to an upset of the function.

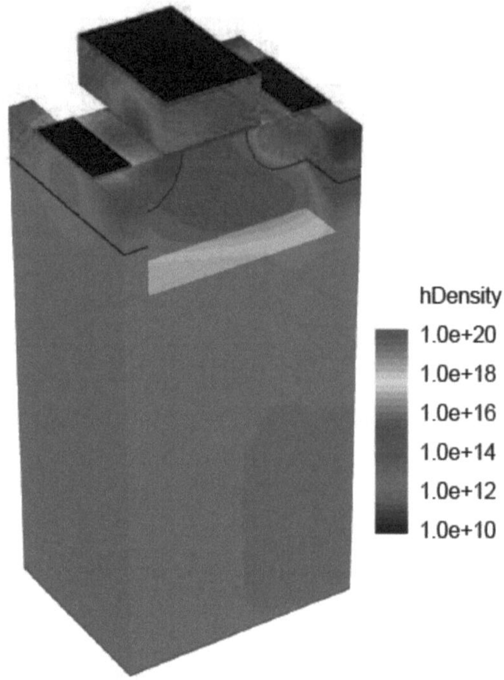

Fig. 11.4 Simulation of a heavy ion striking an NMOS transistor, using Sentaurus Device, Figure 6 from Datasheet Synopsys; www.synopsys.com/products/tcad/tcad.html [8]. The simulation result, which in color can be consulted in the original datasheet [8], illustrates the hole densities at different places in the device, as a function of time. The scale of the densities over 10 orders of magnitude is indicated on the right, but quite invisible in a B/W representation

Also one has to measure the effective cross sectional area, as only a small part of the chip may be sensitive. An example is the measurement by El-Mamouni et al. [9] of ion-induced charge collection in bulk FinFETs. If a 20 × 20 μm area is scanned with an ion beam (spot 1.5 × 0.7 μm, step 0.1 μm) only in the small drain region a charge signal is observed, as illustrated in Fig. 11.5. However, the sensitive area is much larger than both the 200 nm drain and the beam spot. Apparently some fraction of the charge signal is contributed from the bulk Si under the fin.

Single Event Effects are the most difficult to study, yet the most dangerous in practice as they may disrupt the functioning of systems quite unexpectedly. A stunning example was repeatedly occurring Single Event Burnout SEB of IGBT power supply components in the beginning of the German ICE train operations. Immediate studies [10] concluded that only a cosmic-ray-induced short could explain the failures. Voltage derating provided a temporary solution.

A variety of Effects have been discovered, and more are added as technology proceeds to smaller dimensions and smaller charges in the chip functions. More on this in Sect. 11.5. Besides SEB, another of the catastrophic things that may happen is Single Event Gate Rupture SEGR, which obviously also is destructive.

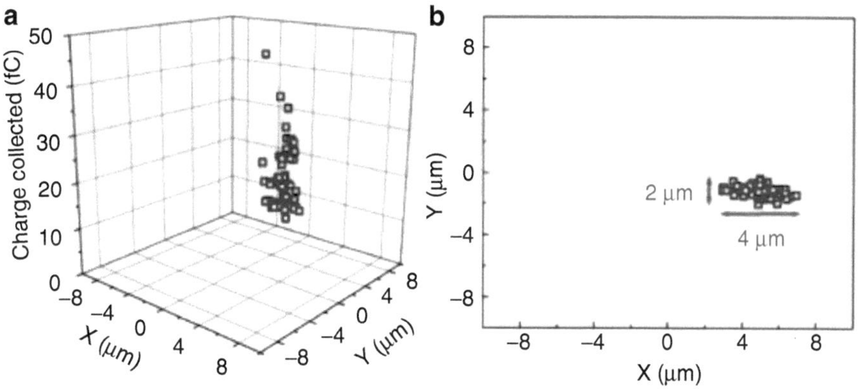

Fig. 11.5 (**a**) Charge signals in fC from a heavy ion beam in the drain terminal of a bulk FinFET with a 5 nm fin width and a gate length of 70 nm. (**b**) Projection of the measurement points on the (x, y) plane, showing the actual sensitive area [9]

However, most of the effects are interrupting the functioning or destroying the data but are non-destructive for the chip as such. In Single Event Latchup SEL a large charge deposit creates a conductive path through the Si bulk between the V_{DD} and V_{SS} power supply terminals. The ensuing large current may become destructive without latchup detection, if the chip is not reset quickly enough.

The proximity of nodes in the circuit also plays a role in the probability for a Single Event Effect to happen, due to the charge sharing between different collecting electrodes. A study by Atkinson et al. [11] shows a 70% reduction of the cross section for a grouped chain of inverters in a single well, compared with a single chain in an identical well, as illustrated in Fig. 11.6a, b.

The most widely known Single Event Upsets lead to loss of digital data in memories or in digital processors. Just as the earlier Effects mentioned, Single Effect Transients also occur in analog circuits, with a recent example in a bandgap reference, discussed by Zanchi et al. [12]. The transient due to an ion impact is shown in Fig. 11.7. Understanding of the origin of this transient allowed to practically eliminate the occurrence, using different approaches. One solution was a diode clamp on the n-well to + V_{CC}.

11.3 Practical Radiation Environments

Sometimes, chips are to be used in hostile environments and special studies are needed. However, even in normal circumstances the use of ICs can be affected by incident radiation, although failure rates are low. The widespread use of chips compensates for the low rate per unit and when many square meters of chips are exposed, a few incidents still may make it to the newspapers. Sometimes a very energetic ion or neutron from the cosmic ray background can cause a critical chip to

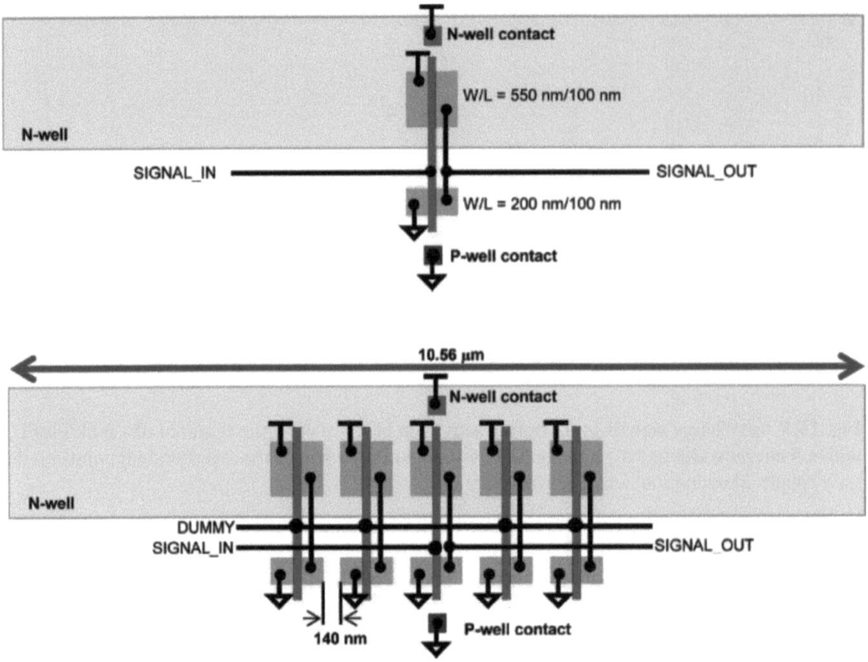

Fig. 11.6 (**a**) Single inverter in n-well (**b**) The same inverter surrounded by two more on either side in a chain, which reduces the collected charge [11]

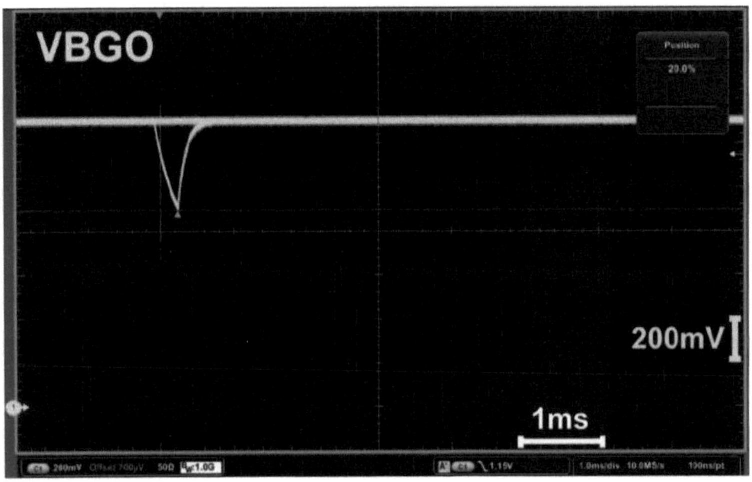

Fig. 11.7 A heavy ion induced long duration pulse (LDP) in a 'Brokaw' bandgap reference. The duration of ~ 600 μs impairs the function [12]

fail in operation, in a car or in an airplane.[1] It is essential to know the radiation environment by simulation and by measurements, and to understand the possible failures of circuits. An example can be the already mentioned cosmic-ray generated neutrons. Most of these are 'thermalized' with energy ~ .025 eV. Such neutrons can be efficiently captured by the isotope ^{10}B which breaks up with a local energy deposit of several MeV leading to a possible Single Event Effect. Boron is present in overglass chip protection and also as implanted B doping. One can reduce this effect of thermal neutrons by using pure ^{11}B, which is a cheap by-product of the manufacturing of pure ^{10}B for medical purposes.

As an example of extremely difficult situations, the LHC at CERN has an unprecedented radiation environment for such particle physics experiments. Simulations show for the central detector region, close to the collision point an intensity of $5 \times 10^{13}\,\mathrm{cm}^{-2}$ hadronic particles per year or ~50 kGy/year [13]. The high intensities of the colliding beams are needed to produce a sufficiently large number of the very rare interactions that one is looking for. Yet it is required to place in this region semiconductor particle detectors with readout electronics. The distribution of radiation is very inhomogeneous, with the intensities falling off radially, away from the interaction point. However, downstream also a dense radiation field is generated in the so-called calorimeters that serve to absorb and measure the energy of most of the outgoing particles. Moreover, these emit large numbers of neutrons all around. Overall, a variety of radiation situations exists, for which the electronics has to be optimized. Studies started already in the early 1980s. Most of the chip designs were made between 1999 and 2003 and have used the standard 0.25 μm CMOS technology of IBM, with enclosed transistor layout [5]. Some designs have been implemented in the radhard BiCMOS technology DMILL [14]. In general, the prototype designs have been tested for radiation hardness before manufacturing was undertaken. The commercial-off-the-shelf (COTS) components that have been used e.g. in power supplies, also were tested beforehand.

Probably even more severe conditions than in LHC will exist in nuclear fusion reactors, and for the ITER project numerous studies are underway.

The space-borne applications of microelectronics continue to present a challenge for chip design. The communication networks depend more and more on satellites for critical transmissions, using large numbers of processors and memories. In outer space, the proton component is the most important, partially of solar origin, while heavy ion impacts there cause the more complicated trouble. In low-earth orbits the Van Allen radiation belts also contain many lower energy electrons. Large solar eruptions can modify these environments even for longer periods.

The relevance of radiation resistance for avionics is illustrated in Fig. 11.8 using the Medipix imager developed at CERN for quantum imaging dosimetry. Intensities at flight altitudes of 10–14 km are 20–100 times higher than at sea level, as already was shown in Fig. 11.1.

---

[1] Failures due to inadequate computer programming in some cases seems to happen more frequently and obscure upsets by radiation impacts.

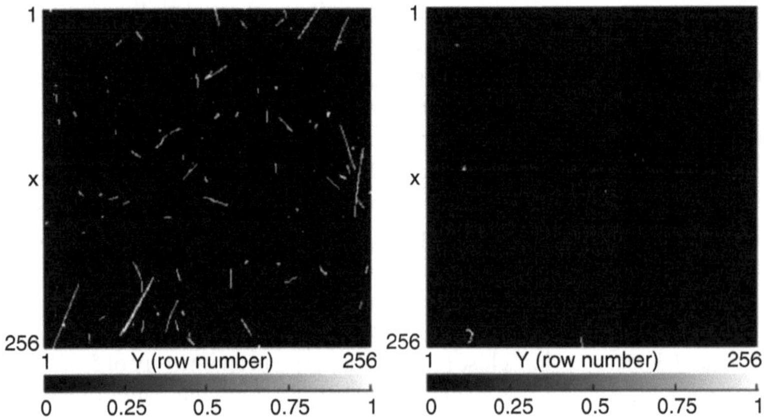

Fig. 11.8 (*left*) 60 s exposure of a medipix imager at 10 km altitude: 105 clusters of which 36 muons dose rate 0.9 + 0.8 μSv/h (electrons, resp. muons) (*right*) similar 60 s exposure at sea level: nine clusters. Dose rate 0.1 μSv/h

11.4 Some Remedies Against Radiation Effects

Once the precise nature of a radiation-induced failure in the semiconductor device is discovered, one can engineer a remedy in one way or another. Unfortunately, in many cases the failed component or system is not available anymore, such as with spacecraft. Preliminary ground testing and dummy systems are necessary precautions to begin with. With extensive expertise one can pre-empt many potential failures. There exist different approaches to mitigation of radiation effects and a few remarks will be made here.

11.4.1 Technology and Devices

Several suppliers manufacture radiation hard chips of reasonable complexity in qualified, radhard technologies. Many aspects of the processing come into play. The use of insulating substrates such as sapphire, reduces the volume in which mobile charge carriers can be generated by the radiation, so that less charge is there to influence the chip functioning. Silicon-On-Insulator SOI has a similar effect, as the Buried Oxide (BOX) layer isolates the active Si from the carrier wafer volume. However, the Si-SiO_2 interface is a sensitive place where charge may accumulate and induce undesirable potentials in the circuit. Special technological steps have been invented to reduce or cancel out the fixed charge retention at the interfaces. One can try to immobilize the radiation-induced carriers so that they can not move

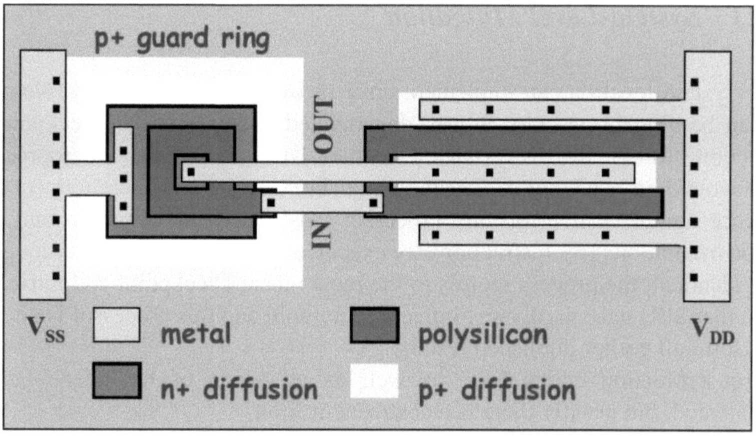

Fig. 11.9 A CMOS inverter using radiation tolerant layout rules

to sensitive nodes, or on the contrary one can create paths so that this charge is evacuated away from these sensitive nodes. Adding body ties or diode clamping are examples of the latter solution.

Modifications in technologies usually are costly, because in long evaluation sequences the manufacturability and reliability have to be established. If then the number of sold components is small, their unit cost will be very high.

At CERN a few designs have been implemented in commercial radhard technologies, and so far these circuits behave well. An example was discussed in the presentation by Jan Kaplon in this Workshop.

11.4.2 Circuit Design and Layout

The field oxide has already been mentioned as a source of leakage between transistors or between source and drain, for n-channel devices in particular, after accumulation of interface charges under TID. The use of Enclosed Layout Transistors ELT overcomes this problem to a large extent. Adding guard bands also may help improve the influence of radiation-induced charge. An example of the special layout is given in Fig. 11.9. for an inverter.

Sensitivity for radiation effects often may be reduced in the circuit design by increasing capacitance so that the impact of added charge is less.

In logic design one has various approaches, such as the Dual-Interlocked Cell (DICE) proposed by Calin et al. [15]. New ideas are proposed continuously.

11.4.3 System-Level Mitigation

At the system level on can implement error recovery and redundancy. Nowadays this can be already on-chip. A well-documented example is the checkpoint and Instruction Retry Recovery IRR in the IBM POWER6 microprocessor, manufactured in a 65 nm SOI technology. For accelerated measurement of its resilience against soft errors the processor was tested under proton and under neutron irradiation [16]. Instructions are executed in parallel, and results compared. If not identical, the process returns to the most recent checkpoint and retries. It is shown that SEU have hardly any influence anymore and this processor is quoted to outperform all earlier published results.

Error correction codes ECC as well as all kinds of redundancy can be implemented, but usually there is a cost/benefit limit.

11.5 Impact of Technology Trends

The trends in CMOS technology create all the time new challenges for implementing radiation resistant ICs, due to new materials and new processing. At the same time, some developments by themselves solve existing problems. The ever thinner gate oxide turned out to become nearly free of threshold shift below ~8 nm, which occurred for CMOS generations with linewidth below ~0.25 μm. Still, things may not always be that straightforward.

In Fig. 11.10 some measurements by Faccio and Cervelli [17] at CERN are shown for transistors of fixed, minimum length, and different widths, made in a commercial 0.13 μm CMOS technology. They find an edge effect for narrow transistors, leading to increased V_T which tends to disappear if the channel is made wider, or if the enclosed layout is used.

For future technologies, such as e.g. those using FinFETs there obviously is hardly any experience available, and much work has to be done to understand their performance under radiation. Song et al. [18] studied the threshold and subthreshold shifts due to total ionizing dose as a function of the fin width. Quite contrary to the findings for 'classical' devices studied by Faccio, here the wider fins experience larger shifts. A TEM (Transmission Electron Micrograph) picture of a device is shown in Fig. 11.11 and some of the results are shown in Fig. 11.12. The increase with fin width is attributed to the larger impact of the buried oxide layer (BOX) which is 200 nm thick, much more than the 4 nm thickness of the gate oxide itself, around the other three sides of the fin.

Only a glimpse can be given on work going on for new technologies. The amount of signal charge tends to become smaller for smaller dimensions. One expects radiation problems to become more serious, in particular in memories. However, countermeasures are found and amazingly enough, the sensitivity for SEU has decreased over several generations. Ziegler has an interesting website where one can find the diagram of Fig. 11.13.

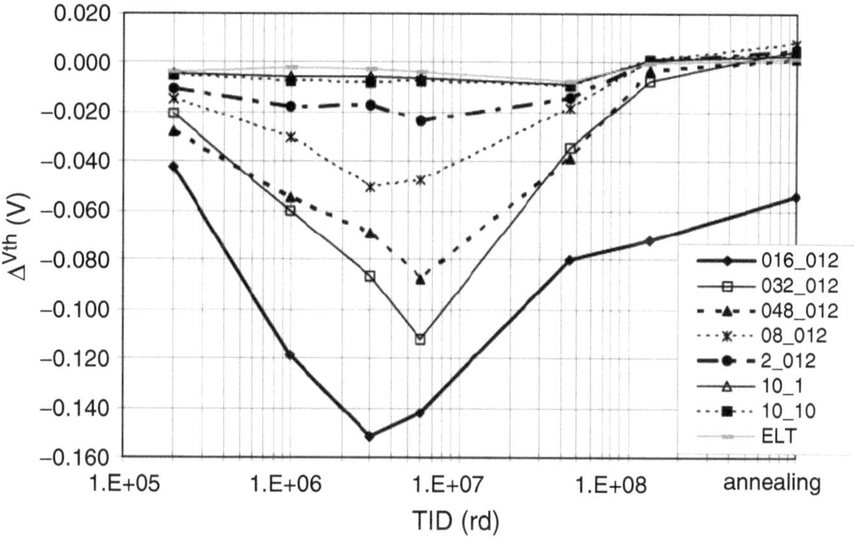

Fig. 11.10 Threshold shift for 0.13 μm transistors as a function of irradiation dose in rad(SiO_2) up to 136 Mrad. The measurements have been made with transistors with minimum length and different gate widths. *ELT* is enclosed layout. The last point at the right each time refers to full annealing at 100° C

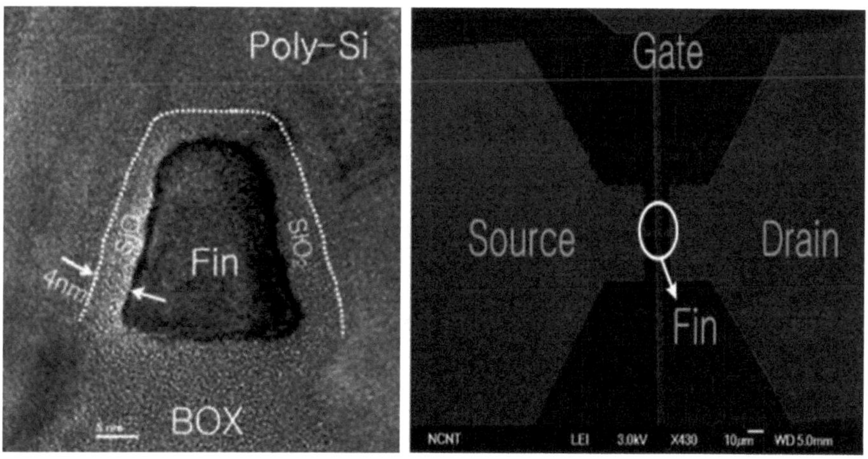

Fig. 11.11 (*left*) Cross-sectional TEM and (*right*) top-down SEM images of a triple-gate MOSFET. The TEM image shows the fin width, while the gate-length dimension is normal to the plane of the image (From [18])

Contrary to expectations, the failure rate per bit has decreased by 6 orders of magnitude, while the size of the memory chips over that period (until ~2005) increased by ~3 orders, from 200 kB to just under 1 GB. This has been achieved generally even without advanced ECC, which for most applications is not judged worthwhile.

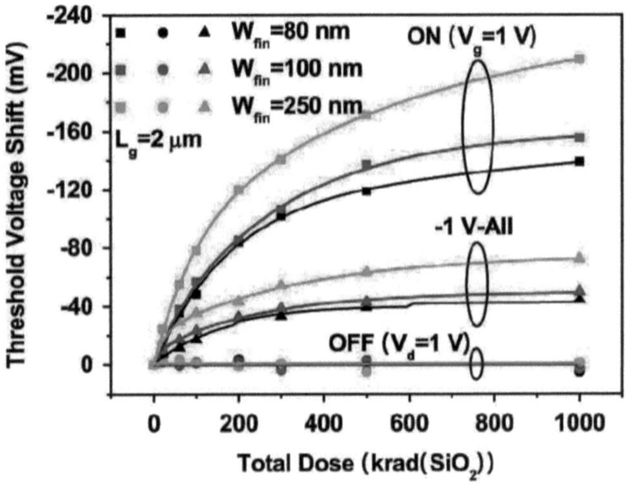

Fig. 11.12 Threshold voltage shift degradation versus dose for different fin widths. The transistors are biased in the "ON" (*top*) or "OFF" or "−1V-All" bias conditions during irradiation. *Lines* indicate general trends (From [18])

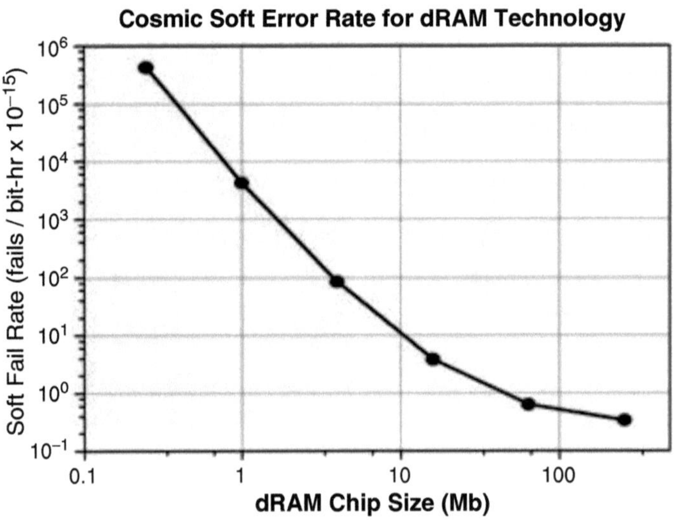

Fig. 11.13 Trend of soft failure rate as a function of chip size, and implicitly of the technology used. The diagram can be found on Ziegler's website [19]

11.6 Unexpected Effects Related to Radiation

Radiation may cause defects in devices, or it can interact with defects that were already there before. This leads to phenomena that may be difficult to explain, such as unexpected noise behaviour or even device failure. An example here is the observation of switching between different, but for each pixel well-defined levels of dark current in a CMOS image sensor. These levels appear as 'Random Telegraph Signals' RTS in the sensor images. The switching frequency can be in the mHz region, or even slower. RTS was reported for high performance JFET already ~1980, where the phenomenon was attributed by Kandiah to charging and discharging of single defects in or close to the channel [20, 21]. In a recent study, Virmontois et al. [22] could attribute RTS in pixels to metastable defects, either in the bulk or in the interface between the Trench Isolation oxide and the edge of the n-type implant of the photodiode, as shown in Figs. 11.14 and 11.15.

The CMOS imagers have been tested with neutrons that cause mainly displacement damage in the crystal (top row in Fig. 11.14) or with X-ray photons causing Total Ionizing Dose TID. After irradiation, many more pixels exhibit RTS behaviour due to increased dark current. TID creates many more defective pixels, but pixels with RTS after displacement damage show multiple RTS levels and

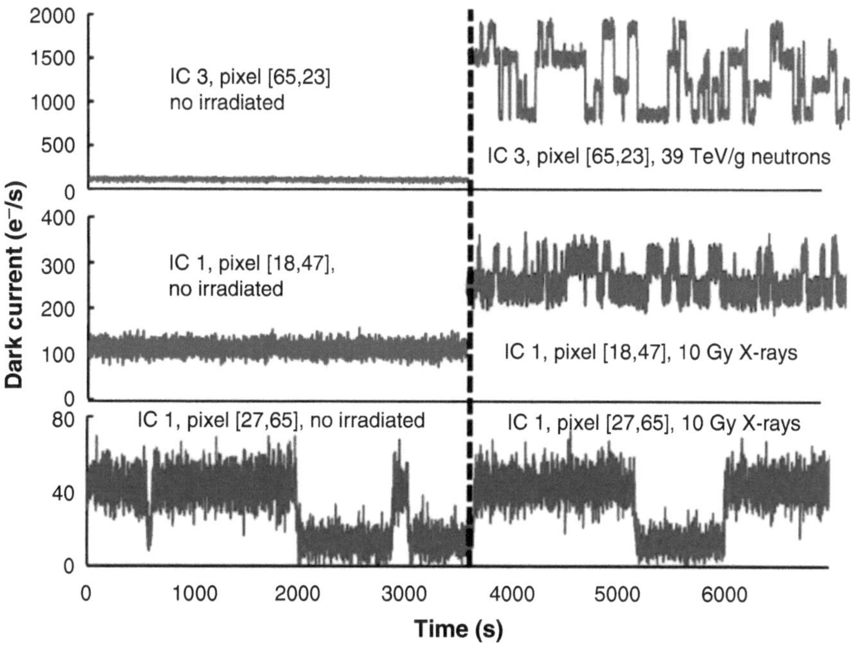

Fig. 11.14 Some pixel dark currents as a function of time, showing RTS behaviour. Measurements are shown before irradiations (*left*) and after 14 MeV neutron irradiation (*top right*) or after a dose of 10 Gy of 10 keV X-rays (*middle* and *bottom right*) (From [22])

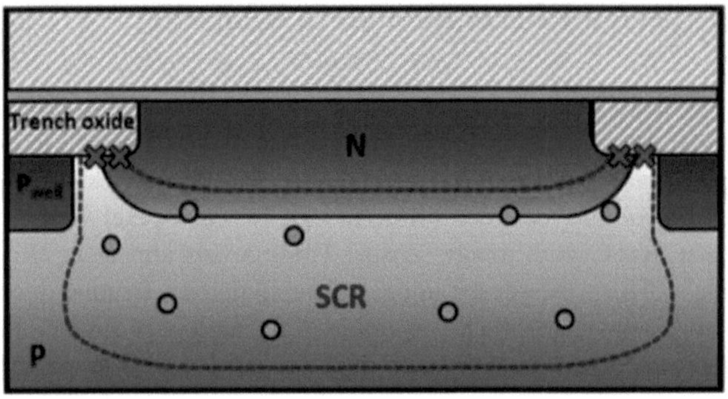

Fig. 11.15 Cross section of in-pixel photodiode. Defects responsible for RTS are indicated: *blue circles* in the bulk created by neutrons, *red crosses* are interface states from TID (Diagram from [22])

higher amplitudes (top row Fig. 11.14). Some pixels have initially already RTS. Also the devices studied by Kandiah were not irradiated and had such defects from the beginning.

With a modified sensor design as illustrated in Fig. 11.16, using a recessed trench oxide they could dramatically reduce the number of pixels with RTS noise due to TID. A more modest reduction in defective pixels due to crystal displacement damage could be achieved by a redesign of the space charge region in the photodiode.

11.7 Conclusions

In several scientific and industrial situations, the microelectronics circuits have to be radiation resistant. The CERN LHC experiments have served as an example for using custom-designed radhard chips in standard CMOS. In order to mitigate catastrophic effects, a number of points can be kept in mind.

- In the first place, do not simply forget to think about radiation effects; radiation is everywhere, even if it is of low intensity. A large energy deposit may occur.
- What is the expected radiation environment including the types of particles and their approximate intensities around the chip.
- Check if radiation data are available for the chip technology to be chosen.
- Which radiation effects may affect the planned chip design: considerations are very different for memory, an ADC, imager, processor, FPGA, etc.
- Which parts of the circuit might be sensitive for SEU, for threshold shifts, for carrier lifetime degradation?
- Can the chips in the application be shielded, or is an open packaging method better? Note that for some high energy particles, such as electrons or protons

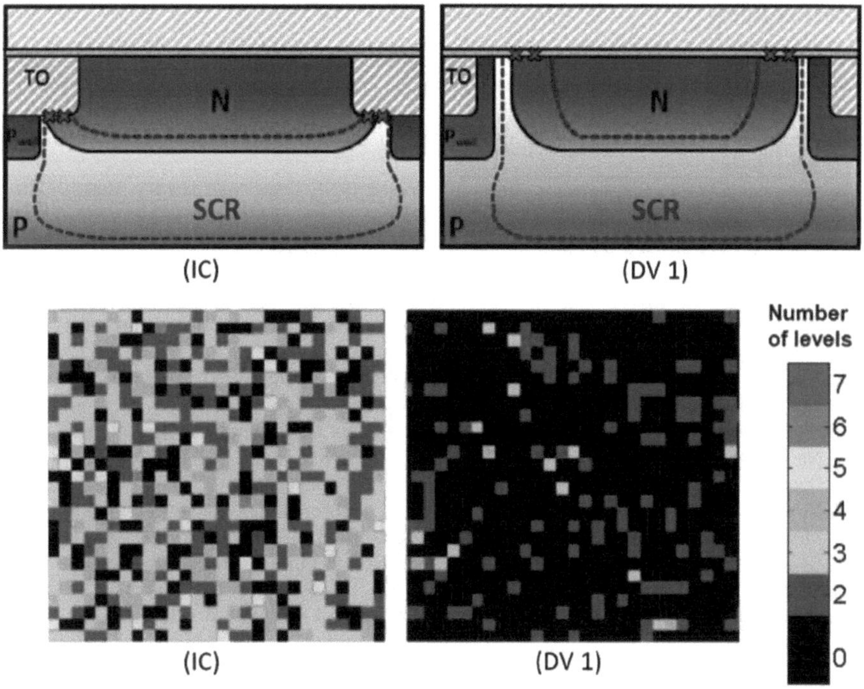

Fig. 11.16 *Top*: left standard design (*IC*) of the photodiode, *right* (*DV1*) improved design with recessed trench oxide. *Red crosses* indicate interface states. *Bottom*: Mapping of dark current RTS occurrence after 100 Gy of X-rays [22]

with hundreds of MeV, the shielding may increase the intensity through multiplication in showers.
- Possibilities for hardening by layout: body ties, supply clamps, enclosed transistors,. . .,etc.
- Is there need for system-level remedies; of course, 'system-level' also can be on-chip, in a 'system-on-chip' approach.
- Use of redundancy: multiple paths for processing, logic with triple voting or even higher? built-in ECC,. . ., etc.

Today, in many IC applications one must not ignore the influence of ambient radiation on the proper functioning of the circuit.

Acknowledgements Many people have contributed to the know-how at CERN in radiation effects in chips. Contrary to my original fears, it has been quite possible to meet with many scientists and learn about this subject in open scientific meetings. The IEEE has played an important role from the beginning. It is through these contacts and through the yearly NSREC and RADECS that progress continues to be made.

References

1. Binder D, Smith EC, Holman AB (1975) Satellite anomalies from galactic cosmic rays. IEEE Trans Nucl Sci NS-22:2675
2. Ziegler JF, Lanford WA (1979) Effects of cosmic rays on computer memories. Science 206:776–788
3. Normand E (1996) Single-event effects in avionics. IEEE Trans Nucl Sci NS-43:461
4. Saks NS, Ancona MG, Modolo JA (1984) Radiation effects in MOS capacitors with very thin oxides at 80K. IEEE Trans Nucl Sci NS-31:1249
5. Anelli G, Campbell M, Delmastro M, Faccio F, Florian S, Giraldo A, Heijne E, Jarron P, Kloukinas K, Marchioro A, Moreira P, Snoeys W (1999) Radiation tolerant VLSI circuits in standard deep submicron CMOS technologies for the LHC experiments: practical design aspects. IEEE Trans Nucl Sci NS-46:1690
6. Dingwall AGF, Stricker RE (1977) C2L: a new high-speed high-density bulk CMOS technology. IEEE J Solid-St Circ SC-12:344
7. McLean FB, Oldham TR (1982) Charge funneling in n- and p-type Si substrates. IEEE Trans Nucl Sci NS-29:2018
8. Datasheet Synopsys. www.synopsys.com/products/tcad/tcad.html
9. El-Mamouni F, Zhang EX, Pate ND, Hooten N, Schrimpf RD, Reed RA, Galloway KF, McMorrow D, Warner J, Simoen E, Claeys C, Griffoni A, Linten D, Vizkelethy G (2011) Laser- and heavy ion-induced charge collection in bulk FinFETs. IEEE Trans Nucl Sci NS-58:2563–2569
10. Kabza H, Schulze H-J, Gerstenmaier Y, Voss P, Wilhelmi J, Schmid W, Pfirsch F, Platzoder K (1994) Cosmic radiation as a cause for power device failure and possible countermeasures. In: IEEE Cat 94CH3377-9 Proceedings of 6th international symposium power semiconductor devices & IC, Davos, pp 9–12
11. Atkinson NM, Ahlbin JR, Witulski AF, Gaspard NJ, Holman WT, Bhuva BL, Zhang EX, Chen L, Massengill LW (2011) Effect of transistor density and charge sharing on single-event transients in 90-nm bulk CMOS. IEEE Trans Nucl Sci NS-58:2578–2584
12. Zanchi A, Buchner S, Hafer C, Hisano S, Kerwin DB (2011) Investigation and mitigation of analog SET on a bandgap reference in triple-well CMOS using pulsed laser techniques. IEEE Trans Nucl Sci NS-58:2570–2577
13. Aarnio PA, Huhtinen M (1993) Hadron fluxes in inner parts of LHC detectors. Nucl Instrum Methods A336:98
14. Dentan M, Abbon P, Borgeaud P, Delagnes E, Fourches N, Lachartre D, Lugiez F, Paul B, Rouger M, Truche R, Blanc JP, Leroux C, Delevoye-Orsier E, Pelloie JL, de Pontcharra J, Flament O, Guebhard JM, Leray J-L, Montaron J, Musseau O, Vitez A, Blanquart L, Aubert J-J, Bonzom V, Delpierre P, Habrard MC, Mekkaoui A, Potheau R, Ardelean J, Hrisoho A, Breton D (1996) DMILL, a mixed analog-digital radiation-hard BiCMOS technology for high energy physics electronics. IEEE Trans Nucl Sci NS-43:1763
15. Calin T, Nicolaidis M, Velazco R (1996) Upset hardened memory design for submicron CMOS technology. IEEE Trans Nucl Sci NS-43:2874–2878
16. Sanda PN, Kellington JW, Kudva P, Kalla R, McBeth RB, Ackaret J, Lockwood R, Schumann J, Jones CR (2007) Soft-error resilience of the IBM POWER6 processor. IBM J Res Dev 52:275
17. Faccio F, Cervelli G (2005) Radiation-induced edge effects in deep submicron CMOS transistors. IEEE Trans Nucl Sci NS-52:2413–2420
18. Song J-J, Choi BK, Zhang EX, Schrimpf RD, Fleetwood DM, Park C-H, Jeong Y-H, Kim O (2011) Fin width and bias dependence of the response of triple-gate MOSFETs to total dose irradiation. IEEE Trans Nucl Sci NS-58:2871–2875
19. Ziegler JF. Trends in electronic reliability – effects of terrestrial cosmic rays. http://www.srim.org/SER/SERTrends.htm
20. Kandiah K, Whiting FB (1978) Low frequency noise in junction field effect transistors. Solid-St Electron 21:1079

21. Kandiah K, Deighton MO, Whiting FB (1989) A physical model for random telegraph signal currents in semiconductor devices. J Appl Phys 66:937–946
22. Virmontois C, Goiffon V, Magnan P, Saint-Pe O, Girard S, Petit S, Rolland G, Bardoux A (2011) Total ionizing dose versus displacement damage dose induced dark current random telegraph signals in CMOS image sensors. IEEE Trans Nucl Sci NS-58:3085–3094

Books

Holme-Siedle A, Adams L (1993) Handbook of radiation effects. Oxford Science, Oxford. ISBN 0 19 856347 7

Ma TP, Dressendorfer PV (eds) (1989) Ionizing radiation effects in MOS devices & circuits. Wiley,New York. ISBN 0-471-84893-X

Messenger GC, Ash MS (1986) The effects of radiation on electronic systems. Van Nostrand Reinhold,New York. ISBN 0-442-25417-2

Pantelides ST (1986) Deep centers in semiconductors. Gordon and Breach,New York. ISBN 2-88124-109-3

van Lint VAJ et al (1980) Mechanisms of radiation effects in electronic materials. Wiley-Interscience (out of print)

Proceedings

Series of Proceedings of IEEE nuclear and space radiation effects conferences in December issues of IEEE Trans Nucl Sci NS – from 1965 till present

Series of Proceedings of European conference on radiation effects on components and systems RADECS 1991, 1993, 1995, 1997 etc., now also in IEEE TNS

Series of Proceedings of international conference on defects (and radiation effects) in semiconductors, Royaumont, Paris, 1964

Series of Proceedings of international conference on defects (and radiation effects) in semiconductors, Dunod, 1965

Series of Proceedings of international conference on defects (and radiation effects) in semiconductors, Reading, 1972. IOP Conf Series 16. ISBN 0 85498 106 3

Series of Proceedings of international conference on defects (and radiation effects) in semiconductors, Freiburg, 1974. IOP Conf Series 23. ISBN 0 85498 113 6

Series of Proceedings of international conference on defects (and radiation effects) in semiconductors, Dubrovnik, 1976. IOP Conf Series. ISBN 0 85498 xxxxx

Series of Proceedings of international conference on defects (and radiation effects) in semiconductors, Nice, 1978. IOP Conf Series 46. ISBN 0-85498-137-3

Series of Proceedings of international conference on defects (and radiation effects) in semiconductors, North Holland, Amsterdam, 1982. Reprint Physica 116 B+C, etc.

Series of Proceedings of international conference on defects (and radiation effects) in semiconductors, Berkeley, 1999. Elsevier Reprint Physica B 273–274

Series of Proceedings of international conference on defects (and radiation effects) in semiconductors, Berkeley, 2007. Elsevier Reprint Physica B 401–402

Chapter 12
Radiation-Tolerant MASH Delta-Sigma Time-to-Digital Converters

Ying Cao, Paul Leroux, Wouter De Cock, and Michiel Steyaert

Abstract Time-to-Digital Converters (TDCs) are key building blocks in time-based mixed-signal systems, used for the digitization of analog signals in time domain. A short survey on state-of-the-art TDCs is given. In order to realize a TDC with picoseconds time resolution as well as multi MGy gamma-dose radiation tolerance, a third-order time-domain $\Delta\Sigma$ TDC structure is proposed. The first prototyping TDC, implemented in 0.13 μm, consumes only 1.7 mW from a 1.2 V supply. It achieves a time resolution of 5.6 ps and an ENOB of 11 bits, when the oversampling ratio (OSR) is 250. The SNDR is mainly limited by the skew error introduced by the comparator delay, which can be mitigated by using a delay-line assisted calibration technique. It improves the ENOB of the TDC to 13 bits and achieves a wide input dynamic range of 100-ns. The TDC also exhibits enhanced radiation tolerance owing to the mismatch-insensitive nature of the $\Delta\Sigma$ structure. Even after a total dose of 3.4 MGy at a high dose rate of 30 kGy/h, the ENOB only drops by 1 bit and, for an OSR of 250, a 10.5 ps time resolution is still achieved.

Y. Cao (✉)
KU Leuven, ESAT-MICAS, Kasteelpark Arenberg 10, Room 91.21,
B-3001 Heverlee, Leuven, Belgium

SCK CEN, Mol, Belgium
e-mail: Ying.Cao@esat.kuleuven.be

P. Leroux
KU Leuven, ESAT-MICAS, Kasteelpark Arenberg 10, B-3001, Heverlee, Belgium

K.H. Kempen, Geel, Belgium

W. De Cock
SCK CEN, Mol, Belgium

M. Steyaert
KU Leuven, ESAT-MICAS, Kasteelpark Arenberg 10, Room 01.21,
B-3001, Heverlee, Belgium

A.H.M. van Roermund et al. (eds.), *Nyquist AD Converters, Sensor Interfaces, and Robustness: Advances in Analog Circuit Design, 2012*,
DOI 10.1007/978-1-4614-4587-6_12,

12.1 Introduction

Recently, high resolution time-to-digital converters (TDCs) have gained more and more interest due to their increasing implementation in digital PLLs, ADCs, jitter measurement and time-of-flight (TOF) rangefinders. Specifically, a laser TOF rangefinder which operates by measuring the flight time of a light pulse from a laser transmitter to the target and back to an optical detector, can be applied to many industrial purposes, e.g., positioning of vehicles, detecting particles in atmosphere, measuring level heights in radioactive environments, etc. The accuracy of a TOF rangefinder is mainly determined by the resolution of the TDC employed. Upcoming applications in nuclear fusion reactors like the International Thermonuclear Experimental Reactor (ITER), electronic components are required to stand for higher than 5 MGy total dose radiation level [1], where the threshold voltage, transconductance, and delay of a transistor undergo dramatic changes. In these cases, the high accuracy and robustness of the TDC need to be inherent to the design.

Conventional TDCs were built based on the CMOS gate-delay-line structure [2], whose highest achievable resolution is limited by the intrinsic delay of a CMOS inverter gate. In order to get sub-gate-delay resolution, the Vernier method [3, 4] was commonly used. However, the mismatch problem caused by process variation limits its effectiveness. Although calibration can be applied to compensate for the mismatch error [5], huge efforts are needed since each delay element in the TDC has to be tuned individually. Other methods to achieve sub-gate-delay resolution such as time amplification (TA) [6, 7], local passive interpolation (LPI) [8, 9], and successive approximation (SAR) [10], are vulnerable to changing transistor characteristics and temperature variations. Therefore, they are not suitable for applications in harsh environments.

By contrast, data converters which adopt the noise-shaping technique to achieve high resolution are highly immune to environmental noise and component mismatch. The technique has been successfully implemented in analog-to-digital conversion for years, which is well known as $\Delta\Sigma$ modulator. Some works brought the same principle into phase-domain data conversion, as in [11, 12], and achieved first-order and second-order noise-shaping, respectively. However, these phase-domain $\Delta\Sigma$ modulators are analog intensive approaches, and the phase information is converted back to the voltage-domain. As the technology scales down, this becomes less attractive due to the difficulty of achieving high-gain and wide-bandwidth analog blocks under strongly reduced supply but relatively unchanged threshold voltage. Moreover, their performances are highly relying on the linearity of the front-end phase detector, which practically limits the dynamic range of the input time signal.

The goal of this work is to design a digital intensive high-order noise-shaping TDC, which has radiation-tolerance up to 5 MGy. It is required to achieve better than 10 ps resolution and a wide input range of 10 ~ 100 ns. In this paper, a brief review on existing TDCs is first given in Sect. 12.2. Section 12.3 investigates the architecture of the MASH TDC. Section 12.4 describes implementation of the first prototyping MASH $\Delta\Sigma$ TDC (Chip I). In Sect. 12.5, the delay-line assisted calibration technique is introduced to solve the skew error problem caused by the comparator delay. The Pre-Rad measurement results are shown in Sect. 12.6, where Sect. 12.7 discusses the radiation assessment results at both low (1.2 kGy/h) and high (30 kGy/h) dose rate. A conclusion is drawn in Sect. 12.8.

12.2 A Short Survey on Time-to-Digital Converters

Quantizing the time interval between a start and stop signal, and representing it as a digital code, is the basic task of a TDC. The early first TDCs were actually performed in two steps: time-to-voltage conversion (TVC), followed by voltage-to-digital conversion (VDC), as in [13, 14]. The time signal is mapped into an analog voltage in the first phase, by using a charge pump. The amplitude of the voltage corresponds to the width of the time frame. In the second step, this voltage is translated into a digital code by a conventional analog-to-digital converter (ADC). A 30 ps resolution is the best that has ever been reported based on this configuration [13]. The performance is mainly limited by the nonlinearity in the TVC unit and the resolution of the ADC.

As opposed to the traditional analog method, a TDC could also be designed in time domain, where a circuit gains most profits from technology downscaling in terms of speed, power consumption and area. The simplest time domain TDC is a counter. By using a high-speed low-jitter reference clock, the counter can digitize a time signal with high resolution. However, when the requirement for the TDC resolution is increased to a few picoseconds, the reference clock frequency becomes unreasonably high ($>$100 GHz) with respect to power consumption and system complexity. Therefore, a real TDC employs phase-aligned parallel counting clocks to achieve high resolution rather than using one single external reference clock. A CMOS gate delay-line can serve this purpose. Many TDC architectures based on a delay-line core are reported in the last decade, and some achieve better than 10 ps resolution, as introduced in [2–12, 15]. They can be summarized into following categories:

- Flash TDC [2–5, 8, 9],
- Pipeline TDC [6, 7],
- Successive approximation TDC [10],
- Noise-shaping TDC [11, 12, 15].

Similar to their ADC counterparts, each type of TDC performs well in one area, e.g., resolution, bandwidth, robustness, or power consumption, and lacks in another, as will be discussed below.

12.2.1 Flash TDC

A Flash TDC uses a linear delay-cell ladder with a D flip-flop (DFF) at each rung of the ladder to compare the input time signal to successive reference time units [2], as shown in Fig. 12.1a. The advantages of this circuit are obvious: It employs a very

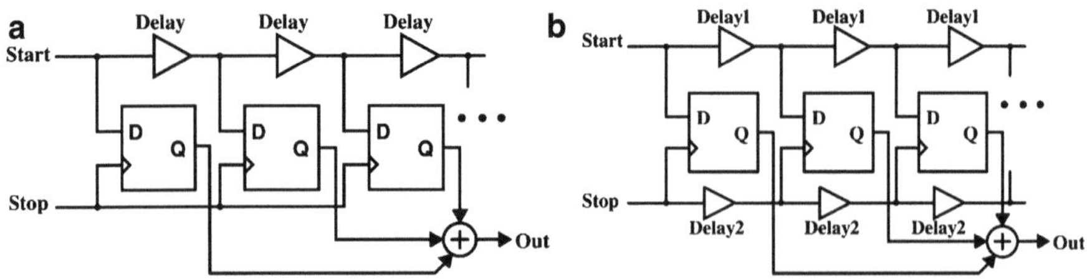

Fig. 12.1 (**a**) Basic structure of a delay-line based Flash TDC, (**b**) Basic structure of a Vernier delay-line TDC

simple structure with only delay cells and DFFs. Hence, it is very area efficient. The resolution is determined by the intrinsic CMOS gate delay, which scales according to the technology scaling factor. But practically, the TDC resolution does not continue to improve with technology downscaling, due to the worsened mismatch problem between delay cells. Therefore, the finest achievable resolution of the basic delay-line TDC is limited to around 20 ps. Moreover, the input range of the Flash TDC is limited by the length of the delay-line. The circuit's area and power consumption then increase linearly with the time interval to be measured.

In order to obtain sub-gate-delay resolution, the Vernier method [3] is commonly used, as shown in Fig. 12.1b. Instead of using only one delay chain, the Vernier TDC utilizes two independent delay lines on both start and stop signal paths to improve the time resolution. The resolution of the Vernier TDC is then given by $t_{Delay1} - t_{Delay2}$. However, along with the resolution, the sensitivity of the Vernier TDC to mismatch has also been amplified. Although calibration can be applied to compensate for this error [5], huge efforts are needed since each delay element in the TDC has to be corrected individually.

Another technique to improve TDC resolution below that of a gate-delay is to subdivide the coarse time interval given by an inverter delay line. This concept can be realized by placing a resistor divider between two nodes of an inverter, as presented in [9]. The improvement in resolution for the interpolation architecture over the gate-delay is similar to that of the Vernier architecture, and is practically limited by the non-linear impedance of the delay elements during signal transients.

12.2.2 Pipeline TDC

It is well-known that, a Pipeline ADC uses two or more steps of subranging and residue amplification technique to achieve high resolution. The same idea can also be realized in the design of TDCs, with the help of a time amplifier (TA). Figure 12.2 shows the conceptual diagram of a Pipeline TDC, which was proposed in [6]. First, the input time signal is digitized by a coarse Flash TDC. The conversion result is then converted back to a reference time and subtracted from the original input. The residue time is then applied to another Flash TDC after amplification. The effective

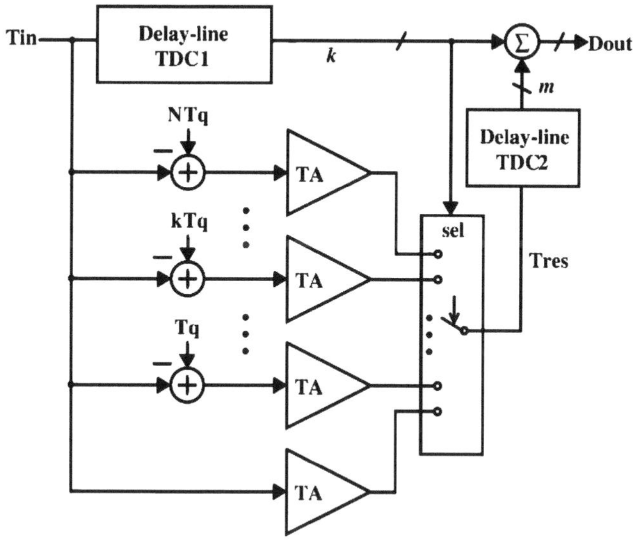

Fig. 12.2 Conceptual diagram of the TA-based Pipeline TDC

resolution of the second Flash TDC is thereby improved by a factor of the gain of the TA. However, unlike voltage, the residue time cannot be stored unless it is transformed to other forms such as voltage or current. Therefore, in a Pipeline TDC, every possible time residue must be created and amplified separately. This significantly increases the system latency, and limits the input range of the TDC, since the linear working region of the TA is quite restricted. Moreover, the gain of the TA is also very sensitive to its working environment, mismatch and process variation. Although a resolution as fine as 1 ps has been reported [7], TA-based Pipeline TDCs are not appropriate candidates for nuclear application.

12.2.3 Successive Approximation TDC

Successive approximation has been widely used in the design of ADCs to reach high resolution at the cost of conversion time. In the time domain, a successive approximation TDC [10] resolves the time difference between the start and stop signal one bit at a time in N cycles using binary search, as illustrated in Fig. 12.3. Due to the irretrievable nature of a time signal, the bidirectional adjustment required by the binary search is implemented by making both signal paths adjustable, rather than adjusting only one signal back and forth. The two delayed versions of input signals, fb1 and fb2, propagate cyclically in two separate loops formed by digital-to-time converters (DTCs), whose delays are controlled by the successive approximation register (SAR).

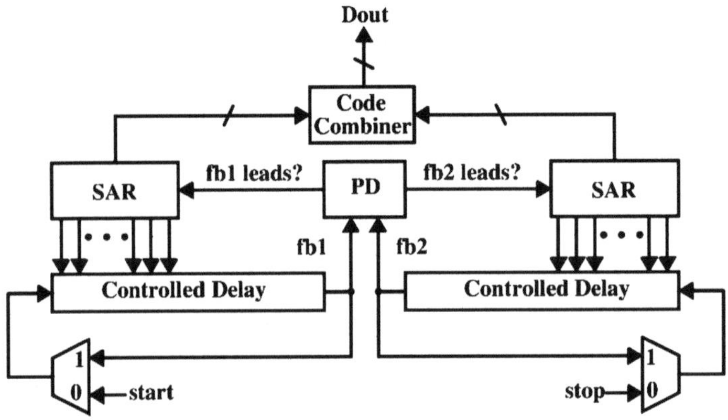

Fig. 12.3 Signal flowchart of the successive approximation TDC

At the beginning of the conversion, the DTC at the start path has a delay of $T_{REF}/2$. The relative timing of fb1 and fb2 is compared with a phase detector (PD) to determine which signal is leading. The SAR will adjust its value according the output of the PD. Whenever the signals fb1 and fb2 are aligned within 1 LSB, the conversion is complete. The fine resolution of the SAR TDC is obtained by interpolation. For an 8-bit operation, 128 unit interpolators (e.g. capacitors) are needed in one DTC, which occupies large area. In order to achieve a wide input range, the SAR TDC has to be configured as a coarse-fine architecture, which has more severe matching problems and consumes more power. Similar to the Vernier method, it also suffers from mismatch and process, temperature variation. The SAR TDC could work together with the DLL structure to have better PVT tolerance. However, this is not compatible with technology downscaling and also introduces system instability.

12.2.4 Noise-Shaping TDC

All TDCs described above, are following the Nyquist criterion. The TDC resolution relies on either a minimized CMOS gate-delay or good matching between delay cells. Looking to the evolution of ADCs, the $\Delta\Sigma$ ADC achieves high resolution by noise-shaping, and it is insensitive to component mismatch. This makes the $\Delta\Sigma$ TDC a preferable choice for applications in radiation environment.

A Gated-Ring-Oscillator (GRO) TDC with first-order noise-shaping has been reported in [15]. In the GRO TDC, the input time signal is used to enable/disable a ring oscillator, as illustrated in Fig. 12.4. One single measurement is done by counting all the phase transitions in the oscillator during the enabling phase. The quantization error, which refers to the intermediate state of the oscillator, is preserved between measurements. This results in a first-order noise-shaping on the quantization noise. One issue in the GRO TDC, which could completely disrupt the noise-shaping

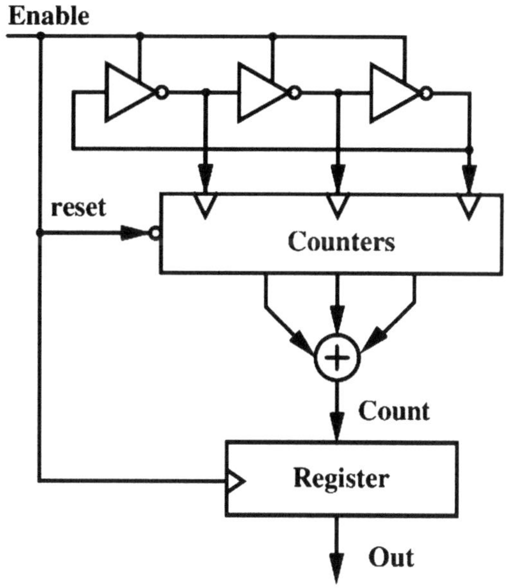

Fig. 12.4 Basic structure of the GRO TDC

behavior, is the existence of large skew error, caused by charge redistribution during the 'silent' phase of the TDC. The skew error results in imperfect preservation of the quantization error. It can only be suppressed by oversampling, thus forfeits the benefit gained from noise-shaping. Another drawback of the GRO TDC, is the difficulty of achieving high-order noise-shaping with this structure, which could principally further improve the TDC resolution and reduce the need for fast delay elements.

12.3 Architecture of the 1-1-1 MASH ΔΣ TDC

The first third-order ΔΣ TDC has been briefly reported in [16]. It achieves better than 10 ps time resolution when a coarse quantization step of 16 ns is used. The on-chip generated quantization reference clock has a frequency depending primarily on passive components, which shows intrinsic PVT variation tolerance. The data conversion is mainly being processed in the time-domain, which benefits most from technology downscaling.

The commonly used ΔΣ modulator structure which consists of integrators can not be directly adopted by a TDC, due to the difficulty of realizing a time integrator. The error-feedback structure is impractical for a ΔΣ ADC since its performance is limited by the inaccuracy of the analog subtractors. However, time subtraction can be easily realized by nand/nor operation, and moreover, the error-feedback structure does not require an explicit integrator in the loop. A first-order noise-shaping TDC can be built in an error-feedback manner, as shown in Fig. 12.5. The input time signal *tin* is first digitized by using a reference clock *tref*. The quantizer can

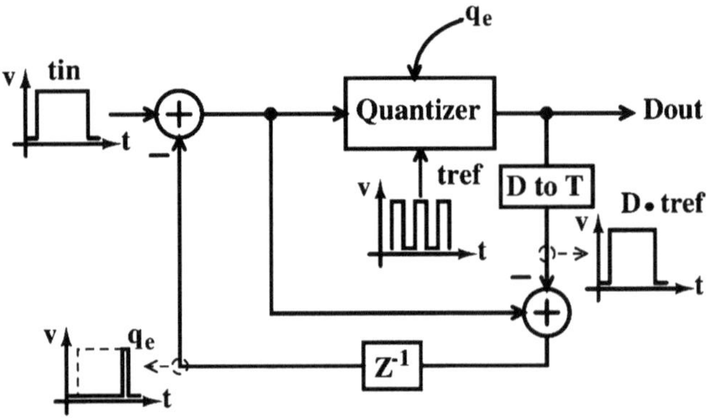

Fig. 12.5 Behavior model of the error-feedback structure based TDC

simply be a counter, which is enabled/disabled by the input signal. A quantization error is present at the output when the input signal is not an integer times of the reference clock period. It can be reproduced by subtracting the corresponding reference time of the digital output from the input signal. A memory element is inserted in the feedback loop to preserve the quantization error before it is being subtracted from the next input signal.

However, directly preserving the quantization error in the time regime is still impossible with current technologies: the time information has to be converted into another intermediate physical quantity such as voltage or charge. A relaxation oscillator can generate a clock by alternatively charging and discharging two capacitors. The phase of the clock corresponds to the voltage on each capacitor. The time can be measured by enabling the oscillator during the measurement interval and counting the number of periods of the generated clock. When the oscillation stops, the phase of the clock, which refers to the quantization error, can be stored on the capacitor as a residue voltage. First-order noise-shaping can be achieved by forwarding this error information into the next measurement phase, hence canceling the low frequency quantization noise. Principally, it is similar with the GRO [15], but the skew error caused by charge redistribution during the start and stop of the oscillator is negligible here due to the large capacitance C.

By cascading multiple error-feedback structures, a high-order noise-shaping TDC can be formed. In this work, a third-order MASH $\Delta\Sigma$ TDC is demonstrated. The system architecture of a 1-1-1 MASH $\Delta\Sigma$ TDC is shown in Fig. 12.6. All three stages have the same structure and are followed by a digital processing block. It is algebraically equivalent to a conventional 1-1-1 MASH $\Delta\Sigma$ modulator. Each stage works as a relaxation oscillator, controlled by the input time signal.

It works as follows: In the first stage, the time signal *tin* controls a current to charge one of the two capacitors during its active phase. For instance, *vinp1* starts rising when *vinn1* stays at *vlow*, as illustrated in Fig. 12.7. When *vinp1* exceeds the threshold voltage *VREF*, the comparator output becomes '1'. This reverses the state

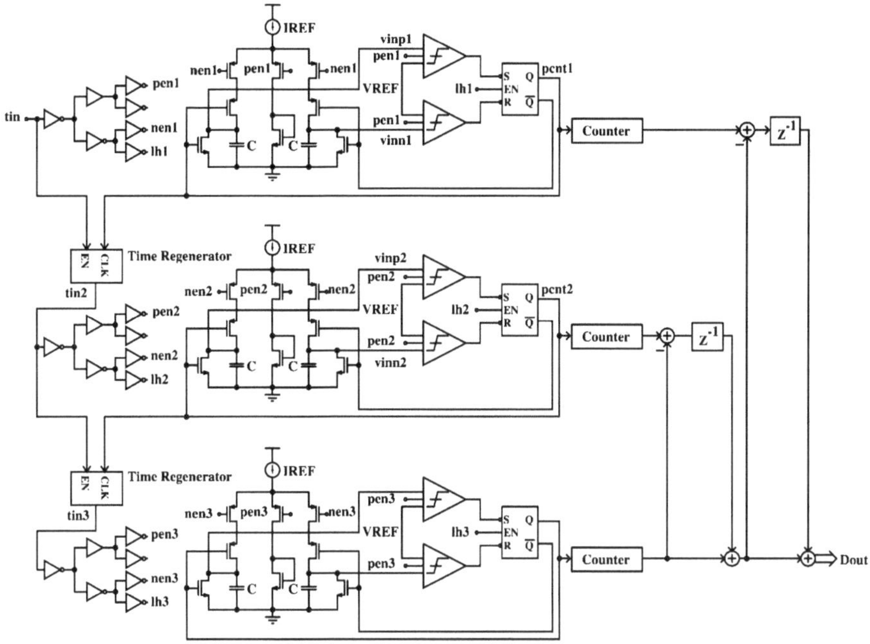

Fig. 12.6 System architecture of the 1-1-1 MASH ΔΣ TDC

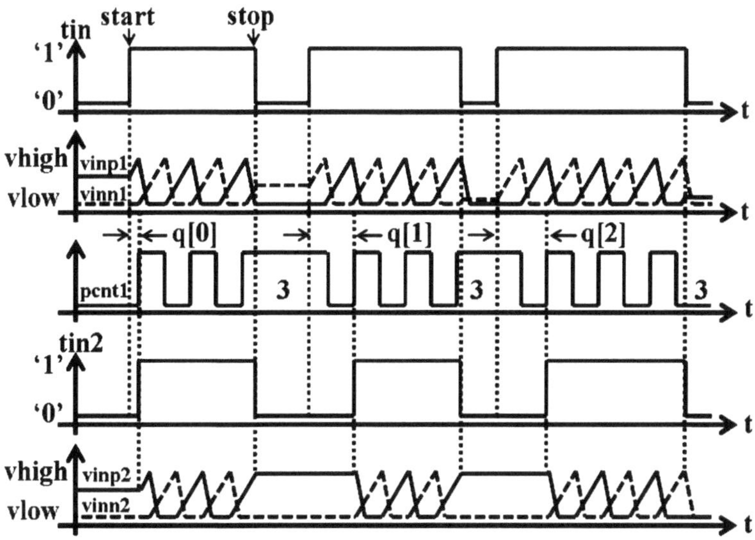

Fig. 12.7 Timing diagram of the MASH ΔΣ TDC

of the SR-latch, and triggers the oscillation. The output of the oscillator is connected to a 4-bit counter. The final result in the counter is a digitized copy of the input signal with large quantization error. After the stop signal arrives, the charging current is disconnected from the capacitors; the counter is first read out and then reset to 0. By preserving the residue voltage on the capacitor at the end of each measurement interval, the quantization error $q[k-1]$, which refers to the phase of the oscillator clock, is also preserved. The quantization error takes on a uniform distribution over the interval $[0, T_d)$, where T_d is the coarse quantization step size. When the next measurement is initiated, the previous quantization error will be subtracted from the next input, due to the factor that the counter is only driven by the rising edge of the clock. The overall quantization error introduced into this measurement can then be described as

$$q_{err}[k] = q[k] - q[k-1]. \tag{12.1}$$

The third-order noise-shaping is obtained by cascading all three stages. The time signal which feeds into a following stage is generated by subtracting the quantization error from the input of the previous stage. This is done by taking the first rising edge of the counting clock as the new start signal, and keeping the same stop signal as the TDC's initial input. The output of the 1-1-1 MASH TDC is given by

$$Dout = tin + \left(1 - z^{-1}\right)^3 \cdot q_{err3}. \tag{12.2}$$

where q_{err3} is the quantization error in the third stage. All digital blocks used for signal processing are synchronized by the falling edge of the input time signal. The theoretical *rms* value of the quantization noise power can be written as [17]

$$q_{err_{rms}} = \frac{T_d}{\sqrt{3}} \frac{\pi^L}{\sqrt{2L+1}} (OSR)^{-(L+1/2)}, \tag{12.3}$$

where OSR is the oversampling ratio, and L is the order of the noise-shaping function. An example TDC using a T_d as 16 ns, OSR as 25, and L as 3, will then ideally have an *rms* quantization error of 1.4 ps. If a higher OSR of 250 is employed, and the other parameters remain the same, the theoretical *rms* quantization error can be reduced to only 0.4 fs, which is far below the physical noise floor of the TDC.

In the MASH ΔΣ TDC, there is also a special relation between the OSR and full scale input range. The bandwidth of the input signal, BW, is set to 100 kHz in this design. The sampling clock of the TDC system is then equal to 2BW·OSR. Due to the input signal's timing nature, the peak-to-peak full scale input signal amplitude T_{fspp} has to be smaller than one period of the sampling clock. For instance, when the OSR is 25, T_{fspp} can not exceed 200 ns.

12.4 CHIP-I: First Prototyping of the MASH ΔΣ TDC

12.4.1 Design Consideration

The finest achievable resolution of the MASH TDC is in practice mainly limited by the thermal noise of switches, the phase noise in the relaxation oscillator, the skew error introduced by comparator delay, and the switching charge injection. Among them, the phase noise and skew error draw most attention, since they are at a similar power level as the quantization noise.

Phase noise, caused by timing jitter, is a commonly used specification for measuring the performance of an oscillator. The primary contributors to the relaxation oscillator's jitter are the noise current flowing into the capacitor and the noise voltage present in series with the threshold level. The latter is generally the dominant cause of jitter, due to its much larger contributing bandwidth [18]. The relation between the SNR and jitter can also be predicted from a simulation of the behavioral MASH TDC model with additive white noise in the reference voltage, as shown in Fig. 12.8. With a higher OSR, the SNR drops due to the decrease of the input range. But it has better tolerance to the timing jitter.

The skew error, e_{skew}, introduced by the comparator delay, occurs only when the oscillator needs to be started and stopped. When the relaxation oscillator is turned off, the comparator state may not be perfectly preserved due to the hysteresis. This will introduce extra noise into the preserved quantization error, and it can only be suppressed by oversampling. The skew error occurs when the stop signal arrives during the time the comparator enters its state-reversing phase, when *vinn* or *vinp* just exceeds *VREF*. A delay always exists before the comparator can make a final

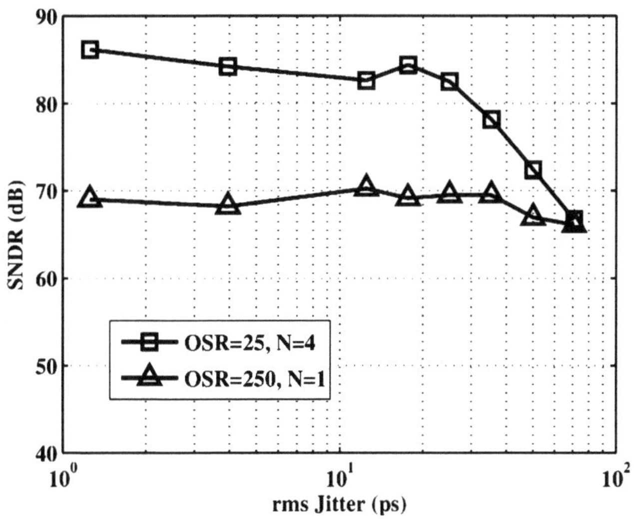

Fig. 12.8 Simulated SNR versus timing jitter

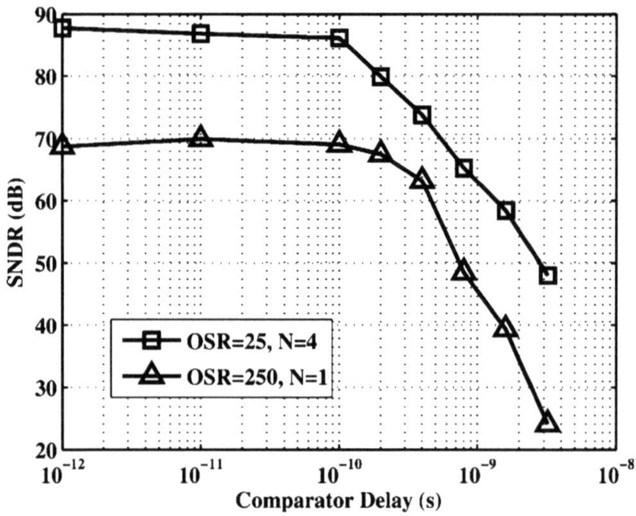

Fig. 12.9 Simulated SNR versus comparator delay

decision according to its input change. It is impractical to save all the intermediate states of the comparator when the system enters idle. So the output of the comparator will continue rising till its final state '1' even if the oscillation has been stopped. Therefore, when the next start signal arrives, the SR-latch will immediately reverse its state, and alternates the capacitor being charged. This will result in a change in the counting clock period and introduce extra skew error to the preserved quantization residue time.

The relation between the SNR and the comparator delay is shown in Fig. 12.9 by simulation. When increasing the OSR, the skew error can be reduced by the same order. However, the full scale input is decreasing at the same time. Eventually, the SNR of the TDC remains almost unchanged. But the time resolution ($T_{fs}/2^{ENOB} - 1$) can be improved at the cost of a smaller input time range. Comparing to the quantization noise and timing jitter, the skew error becomes the dominant noise source in the MASH TDC. In order to achieve a higher SNR, either the comparator delay has to be limited within a small amount or calibration needs to be applied.

12.4.2 Circuits Implementation

The 1-1-1 MASH ΔΣ TDC is implemented in 0.13 μm CMOS. The main circuit blocks in the TDC are the relaxation oscillator, the counter, the time regenerator, and digital processing units. Among them, the performance of the relaxation oscillator determines the highest achievable SNR of the TDC. The conventional two-capacitor relaxation oscillator structure is used in this design, due to its simplicity of controlling. The specification of the relaxation oscillator is derived

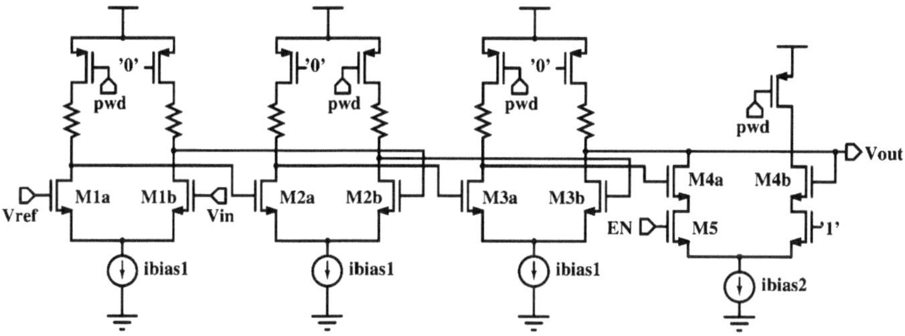

Fig. 12.10 Schematic of the four-stage threshold-detection comparator

first according to the noise analysis presented above, and then a trade-off has been made between the time resolution and power consumption.

The on-chip relaxation oscillator provides the reference clock for the ΔΣ TDC, whose frequency therefore needs to be stable over process and temperature. The period of the relaxation oscillator can be expressed as $(VREF{\cdot}2C)/IREF + 2t_{cmp}$. If the comparator delay t_{cmp} is small enough compare to the whole clock period, the oscillator frequency becomes $IREF/(VREF{\cdot}2C)$. By correlating *VREF* and *IREF* as $VREF = IREF{\cdot}R$, its frequency becomes depending only on passive components, which is $1/(2{\cdot}RC)$. Thus, it exhibits inherent PVT variation tolerance and the matching between stages is better than for its MASH ADC counterparts. In the conversion to time noise the local noise voltage is divided by the charging slope of the capacitor. Thus, a larger slope is mostly desirable. In this design, $IREF = 50$ μA, $VREF = 650$ mV, and $C = 0.64$ pF, which gives a charging slope of 80 μV/ps. According to simulation, the relaxation oscillator has a phase noise of −87 dBc/Hz at 100 kHz offset frequency, which is adequate to achieve better than 14 bits ENOB.

Figure 12.10 shows the schematic of a threshold-detection comparator used in the relaxation oscillator. It is built in a multi-stage structure, for high speed consideration. Each of the first three stages has a gain of 10 dB and consumes 40 -μA current. The last stage provides a higher gain of 20 dB with a power consumption of 80 μA. The total delay of the comparator is 0.8 ns. The switch M5 in the last stage, controlled by *EN*, is added to reduce the skew error caused by the comparator delay. A constant-g_m biasing circuit is used to provide biasing current for the comparator. It adjusts the biasing current adaptively according to the threshold voltage variation of the MOS transistor, and keeps the transconductance constant. It also generates the charging current *IREF* for each stage and the reference voltage *VREF* for the comparator.

The time regenerator produces the input time signal for a following stage. As explained earlier, it generates a timing pulse whose rising edge is aligned at the same moment as the first rising edge of the counting clock, and the falling edge comes when the stop signal arrives. This can be achieved by using a RS latch. In order to inactivate the second, the third, … rising edges of the counting clock, a D flip flop is placed at the *CLK* path before the RS latch, as shown in Fig. 12.11.

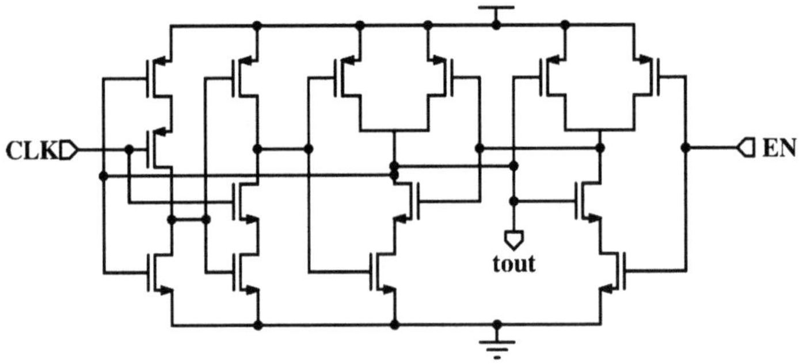

Fig. 12.11 Schematic of the time regenerator

12.5 Delay-Line Assisted Calibration

As explained in Sect. 12.4.1, the main origin of noise in the MASH TDC is the skew error introduced by the comparator delay. According to simulation results, the delay of the comparator has to be limited within 200 ps in order not to degrade the SNDR of the 1-1-1 MASH TDC, when a moderate OSR of 50 is adopted. A threshold-detection comparator at that high speed consumes huge power, and for some old technologies, it is even unrealistic to achieve that delay. Fortunately, this skew error can be calibrated by using a coarse delay-line [19].

One stage of the 1-1-1 MASH $\Delta\Sigma$ TDC with delay-line assisted calibration (TDC-CAL) is shown in Fig. 12.12. The RS-latch is controlled by the output of the calibration unit *EN* rather than *tin*, but both have the same 'stop' edge. *lh*, the complementary signal of *tin*, is sent to the delay-line, for which the detailed structure is shown in Fig. 12.13. Each delay cell has a delay of 150 ps. The delay-line calibration unit becomes active only after the stop signal arrives. When the comparator's output becomes '1' during the inactive phase of the TDC, the state of each delay cell will be sampled by its connected arbiter. For instance, if *lh* has passed through 3 delay cells, the states of all arbiters will be "1110...0". The switch connected to the third delay cell will also be closed, which charges *sel* to '1'. *sel* will be discharged only after the next start signal has also passed through the same delay cells as the stop signal, then activate *EN*. Since the rising edges of *lh* and *sel* have always passed through the same delay cells, the matching between those cells in the calibration unit becomes less important. The resulting waveform on the capacitor is shown in Fig. 12.14 by blue lines. In this way, the skew error caused by the comparator delay is compensated at an accuracy of 150 ps. Therefore, a large comparator delay is now tolerable by the design, which means the power consumed by the comparators can be minimized without deteriorating the TDC's performance.

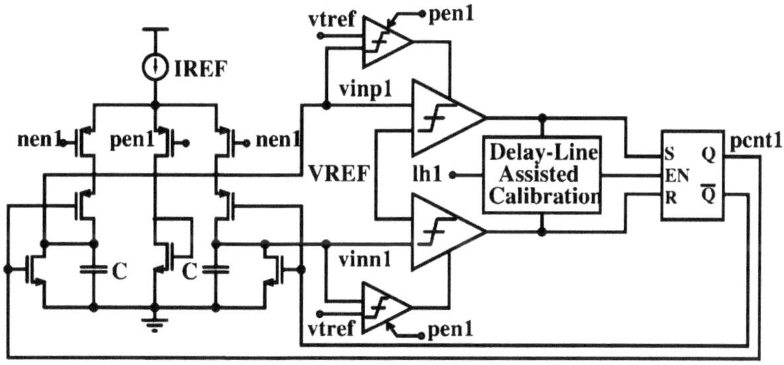

Fig. 12.12 One stage of the MASH TDC with delay-line assisted calibration

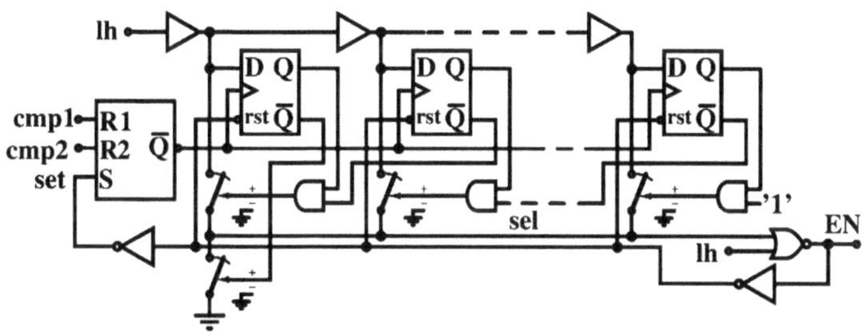

Fig. 12.13 Structure of the ten stages delay-line calibration unit

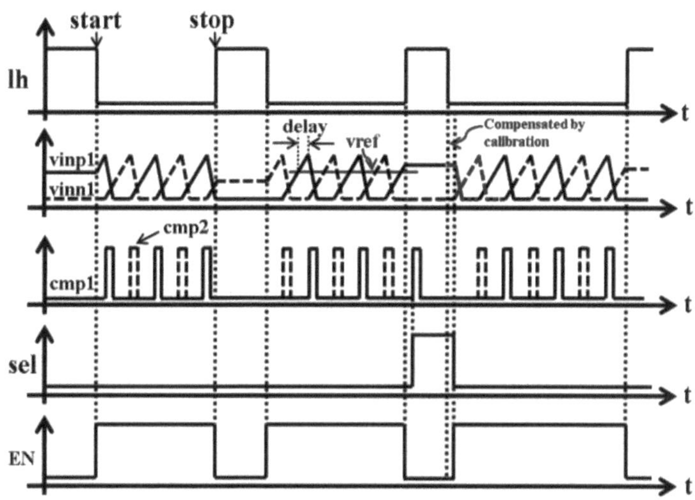

Fig. 12.14 Timing diagram of the TDC-CAL

12.6 Pre-rad Measurement

A pulse-width-modulation (PWM) signal, modulated by the sine-like time signal, is employed to evaluate the performance of the TDC (CHIP I). The conversion rate can be varied from 5 to 50 MHz. For a signal bandwidth of 100 kHz, this corresponds to an OSR of 25–250. The TDC is first configured in a low conversion rate (5 MHz) mode. The full scale input range determined by the conversion rate is then 200 ns. An 18 kHz–3 dBFS PWM signal is applied to the input of the TDC. The output spectrum is shown in Fig. 12.15a. It shows an SNDR of 60.3 dB. The resolution of the TDC is mainly limited by the skew error introduced by the comparator delay.

In the second measurement, the OSR is increased to 250. This requires a higher conversion rate, which is now 50 MHz. The full scale input range is hence reduced to 20 ns. One should notice that, although increasing the OSR can reduce the skew error in the comparator, and further improve the TDC's SNDR, the effective quantization bits of the counter are also reduced at the same time, since the quantization step size is remaining constant. Consequently, the full scale SNDR and ENOB of the TDC will stay nearly unchanged regardless of the value of OSR. Therefore, a trade-off between time resolution and input range exists in the TDC. The power spectrum of the TDC output shown in Fig. 12.15b, is carried out with a 22 kHz–40 dBFS input, which has a peak to peak amplitude of 200 ps. An ENOB of 11 bits and a resolution of 5.6 ps are achieved, respectively.

The same PWM signal described in the previous measurement setup, is again used here to evaluate the effectiveness of the delay-line calibration technique. The rising edge of the PWM pulse represents the start signal, where the stop signal is located at the falling edge. The carrier frequency is set to 10 MHz, which turns to a full scale peak-to-peak input range of 100 ns. Then for a bandwidth of 100 kHz, the OSR is 50. Figure 12.16 shows the output spectrum of the MASH TDC with an 18 kHz 10 ns (−20 dBFS) peak-to-peak input. It shows an SNDR of 55.2 dB and 6 ps time resolution. It can be clearly seen that, without calibration, the large skew error caused by the comparator delay will introduce distortion and increase the baseband noise.

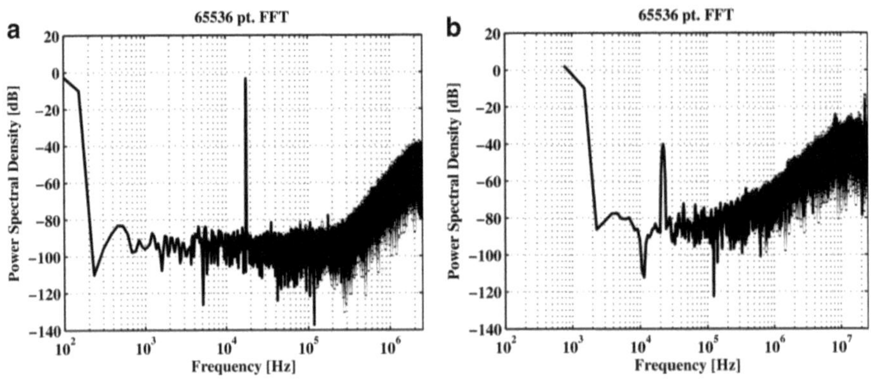

Fig. 12.15 Measured PSD with (**a**) 18 kHz–3 dBFS and (**b**) 22 kHz–40 dBFS input

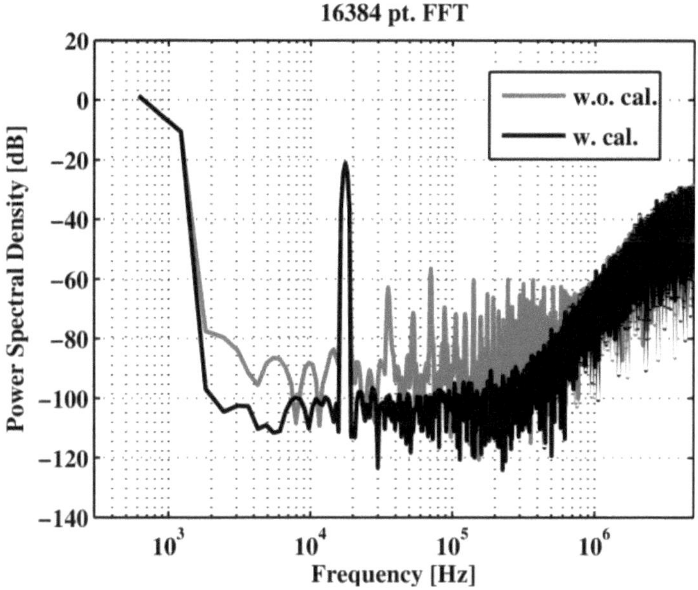

Fig. 12.16 Measured PSD with 18 kHz–20 dBFS input

Table 12.1 Comparison of state-of-the-art TDCs with similar specifications

	[6]	[9]	[15]	[10]	[4]	[12]	CHIP-I	TDC-CAL
Technique	TA	LPI	GRO	SAR	Vernier	Phase-domain ΔΣ	Time-domain ΔΣ	Time-domain ΔΣ
Sample rate (MS/s)	10	180	50	100	15	156	50	10
Resolution (ps)	1.25	4.7	6	1.22	8	2.4	5.6	6
Bits	9	7	11	15	12	10	11	13
Meas. range (ns)	0.64	0.6	12	40	32	3.2	20	100
Power (mW)	3	3.6	21	33	7.5	2.1	1.7	0.7
Core area (mm^2)	0.6	0.02	0.04	1.2	0.26	0.12	0.11	0.08
CMOS	90 nm	90 nm	0.13 μm	0.35 μm	0.13 μm	90 nm	0.13 μm	0.13 μm

The performances of two MASH TDCs are summarized in Table 12.1, and a comparison with state-of-the-art TDCs is also shown. Some report higher performances if only small (± 1 LSB) signals are applied, but in order to obtain a correct comparison only the full scale data is used. Thanks to the MASH ΔΣ technique, the TDC presented in this work achieves the lowest power consumption in the present state of the art with low chip area and high resolution. With the delay-line assisted calibration technique, a wide input range of 100 ns has also been achieved.

12.7 Gamma-Dose Radiation Assessment

Radiation assessments have been performed at the Belgian Nuclear Research Centre. The low dose rate irradiation experiment was carried out from the "RITA" facility, where five TDC samples were irradiated with a ^{60}Co gamma source. The measured dose rate was 1.2 kGy/h. The frequencies of the relaxation oscillators in five TDC samples are monitored continuously during the whole 130 h irradiation period. From 0 to 160 kGy total ionizing dose (TID), no degradation on the frequencies has been found. Only minor ripples are observed, which are mainly caused by the jitter noise coupling into the cable (10 m long). This approves the radiation hardness of the RC relaxation oscillator.

An on-line dynamic measurement under high dose rate radiation has also been performed, proving the TDC's robustness. The experiment was carried out from the "Brigitte" facility, at a dose rate of 30 kGy/h. Figure 12.17 shows the setup of the experiment. The ceramic substrate with the bonded TDC is carried by a container, which is placed in the under water gamma irradiation facility. The substrate is connected to all measurement equipment by a cable with 10 m length. The on-line measurement results are shown in Fig. 12.18. When the system is working in the high conversion rate (50 MHz) mode, the ENOB of the TDC drops only 1 bit after a total dose of 3.4 MGy. This means that a resolution of 10.5 ps can

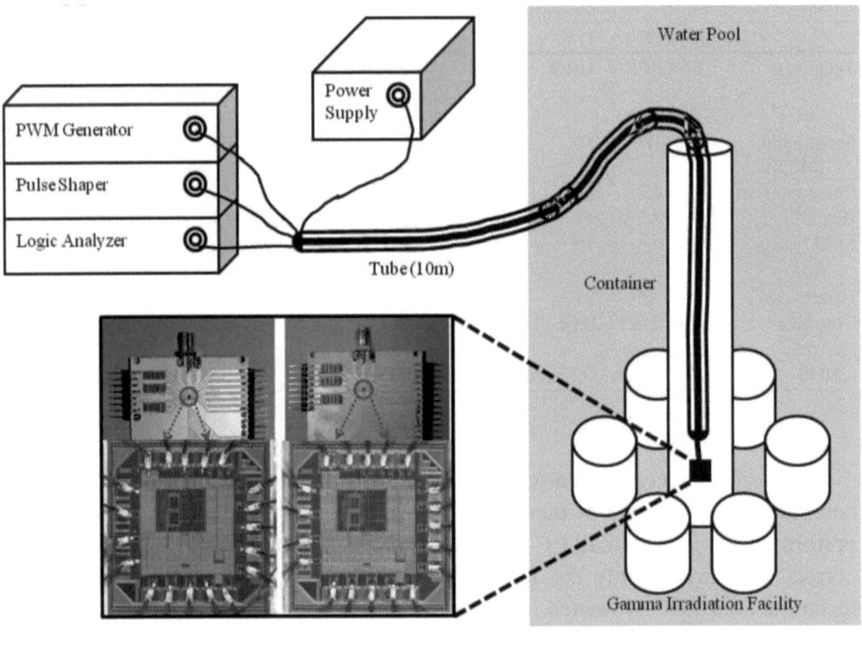

Fig. 12.17 Setup of the high dose rate gamma irradiation experiment. Photos of the substrate and the die before (*left*) and after (*right*) are shown

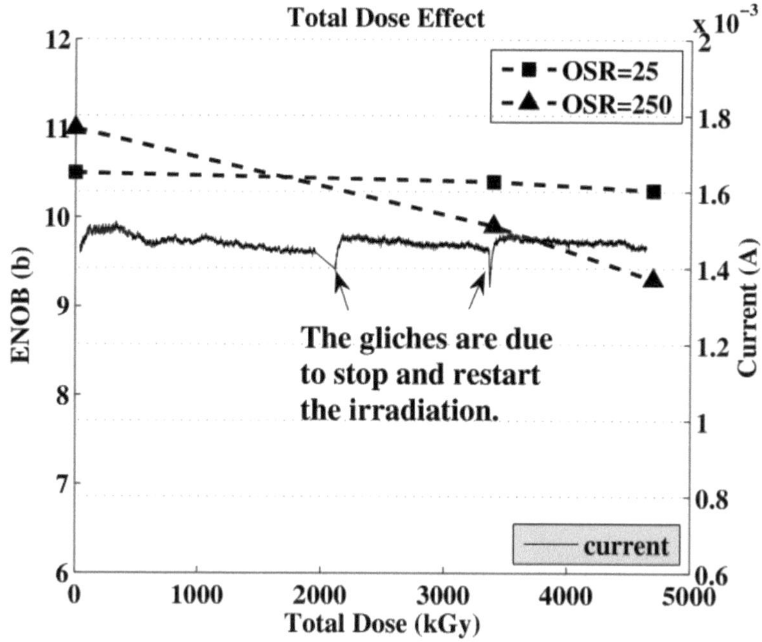

Fig. 12.18 Measured ENOB and total current consumption of the MASH TDC, obtained from the high dose rate experiment

still be achieved. The TDC works functionally till at least 5 MGy. For the TDC working in the low conversion rate (5 MHz) mode, it shows less drop in performance, which can be clearly seen from Fig. 12.18. This is because the timing noise converted from the radiation introduced noise voltage is relatively small compared to the quantization noise in the low conversion rate mode. The current consumption of the TDC system is nearly unaffected during the whole irradiation period. Photos of the substrate and the die before and after 5 MGy total dose irradiation are also shown in Fig. 12.17.

12.8 Conclusions

This paper addresses several practical issues regarding the application of TDCs in harsh environments. Based on a review of some state-of-the-art TDCs, we conclude the noise-shaping TDC as a preferable choice for nuclear instrumentation. We have also presented a third-order MASH $\Delta\Sigma$TDC with 6-ps time resolution. Unlike conventional Nyquist TDCs, its resolution is not limited by the matching property of the technology. PVT variation tolerance is made inherent to the design, which guarantees the TDC's robustness. The first demonstrated MASH TDC is implemented in 0.13 μm CMOS. It consumes only 1.7 mW from a 1.2 V supply, and occupies an area of 0.11 mm^2. The time resolution of this TDC is mainly limited by the comparator delay, which cause skew error when turning on/off the TDC. Fortunately, this skew error can be calibrated by using an on-line calibration method. The large comparator delay can be digitized first by using a coarse delay-line, which delivers a real-time compensation to the skew error. Owing to the power saving in the threshold-detection comparators, the TDC with calibration consumes only 0.7 mW, which is the lowest in the present state-of-the-art and achieves an ENOB of 13 bits. A wide input range of 100 ns has also been achieved.

Gamma radiation assessments with both a low dose rate of 1.2 kGy/h and a high dose rate of 30 kGy/h have been performed, proving the TDC's radiation hardness. The system power consumption is almost not affected and even after an extremely high radiation dose of 3.4 MGy, the ENOB drops only 1 bit and, for an OSR of 250, a 10.5 ps time resolution is still achieved.

References

1. Giraud A (2004) Radiation tolerance assessment of standard electronic components for remote handling. Annual Report of the Association EURATOM-CEA 2004, pp 77–82, available online: http://www-fusion-magnetique.cea.fr/actualites/RA04/eur-cea-technology-2004-full.pdf
2. Staszewski RB, Vemulapalli S, Vallur P, Wallberg J, Balsara PT (2006) 1.3 V 20 ps time-to-digital converter for frequency synthesis in 90-nm CMOS. IEEE Trans Circ Syst II 53 (3):220–224
3. Dudek P, Szczepanski S, Hatfield JV (2000) A high-resolution CMOS time-to-digital converter utilizing a Vernier delay line. IEEE J Solid-St Circ 35(2):240–247
4. Yu J, Dai FF, Jaeger RC (2010) A 12-bit Vernier ring time-to-digital converter in 0.13 μm CMOS technology. IEEE J Solid-St Circ 45(4):830–842
5. Hashimoto T, Yamazaki H, Muramatsu A, Sato T, Inoue A (2008) Time-to-digital converter with Vernier delay mismatch compensation for high resolution on-die clock jitter measurement. In: IEEE symposium on VLSI circuits, Digest of technical papers, pp 166–167, June 2008 Honolulu, Hawaii USA
6. Lee M, Abidi AA (2008) A 9b, 1.25 ps resolution coarse-fine time-to-digital converter in 90 nm CMOS that amplifies a time residue. IEEE J Solid-St Circ 43(4):769–777
7. Seo YH, Kim JS, Park HJ, Sim JY (2011) A 0.63 ps resolution, 11b pipeline TDC in 0.13 μm CMOS. In: IEEE symposium on VLSI circuits, Digest of technical papers, pp 152–153, June 2011 Kyoto, Japan
8. Jansson JP, Mantyniemi A, Kostamovaara J (2006) A CMOS time-to-digital converter with better than 10 ps single-shot precision. IEEE J Solid-St Circ 41(6):1286–1296
9. Henzler S, Koeppe S, Kamp W, Mulatz H, Schmitt-Landsiedel D (2008) 90 nm 4.7 ps-resolution 0.7-LSB single-shot precision and 19pJ-per-shot local passive interpolation time-to-digital converter with on-chip characterization. In: IEEE international solid state circuits conference, Digest of technical papers, pp 548–549, Feb 2008 San Francisco, CA, USA
10. Mäntyniemi A, Rahkonen T, Kostamovaara J (2009) A CMOS time-to-digital converter (TDC) based on a cyclic time domain successive approximation interpolation method. IEEE J Solid-St Circ 44(11):3067–3078

11. van Vroonhoven CPL, Makinwa KAA (2008) A CMOS temperature-to-digital converter with an inaccuracy of ±0.5 °C (3σ) from −55 to 125 °C. In: IEEE international solid state circuits conference, Digest of technical papers, pp 576–577, Feb 2008 San Francisco, CA, USA
12. Young B, Kwon S, Elshazly A, Hanumolu PK (2010) A 2.4 ps resolution 2.1 mW second-order noise-shaped time-to-digital converter with 3.2 ns range in 1 MHz bandwidth. In: Proceedings of IEEE custom integrated circuit conference, Sept 2010 San Jose, CA, USA
13. Ruotsalainen ER, Rahkonen T, Kostamovaara J (2000) An integrated time-to-digital converter with 30-ps single-shot precision. IEEE J Solid-St Circ 35(10):1507–1510
14. Swann BK, Blalock BJ, Clonts LG et al (2004) A 100-ps time-resolution CMOS time-to-digital converter for positron emission tomography imaging applications. IEEE J Solid-St Circ 39(11):1839–1852
15. Straayer MZ, Perrott MH (2009) A multi-path gated ring oscillator TDC with first-order noise shaping. IEEE J Solid-St Circ 44(4):1089–1098
16. Cao Y, Leroux P, De Cock W, Steyaert M (2011) A 1.7 mW 11b 1-1-1 MASH ΔΣ time-to-digital converter. In: IEEE international solid state circuits conference, Digest of technical papers, pp 480–481, Feb 2011 San Francisco, CA, USA
17. Norsworthy SR, Schreier R, Temes GC (1997) Delta-sigma data converters: theory, design, and simulation, IEEE Press New York, USA
18. Abidi AA, Meyer RG (1983) Noise in relaxation oscillators. IEEE J Solid-St Circ 18 (6):794–802
19. Cao Y, Leroux P, De Cock W, Steyaert M (2011) A 0.7 mW 13b temperature-stable MASH ΔΣ TDC with delay-line assisted calibration. In: Proceedings of IEEE Asian solid-state circuits conference (A-SSCC), Jeju, pp 361–364

Chapter 13
A Designer's View on Mismatch

Marcel Pelgrom, Hans Tuinhout, and Maarten Vertregt

Abstract Variability consists of systematic and random components. Many systematic effects can be minimized or circumvented by proper lay-out and dedicated design guide lines. Random variations or "mismatch" between equally designed components are often inherent to the device construction. An overview of the mechanisms and mitigation options is presented as viewed by a designer.

13.1 Introduction

Circuit design greatly depends on the ability to control and reproduce transistor and process parameters. Variation in processing was in the past accommodated by defining process corners. With the improved control over processing, this batch-to-batch variation is largely under control. Statistical variations between otherwise identical components are generally described by "mis-match" parameters. Analog ICs with differential operation were already heavily affected by mismatch. In today's advanced technologies every circuit from SRAM cell to an I-Q mixer must be designed taking statistical variations into account. This paper will review some of the statistical effects and discuss fundamentals, and elaborate on the way analog designers can deal with these issues. Understanding and mitigating these effects requires more and more statistical means. Lessons from the analog domain provide a starting point for the application in the digital domain, and are also helpful in defining targets for process development and test methodology. Designing with knowledge of the statistical margins facilitates achieving the desired yield for a product at minimum power and cost.

M. Pelgrom (✉) • H. Tuinhout • M. Vertregt
NXP Semiconductors Research HTC-32, room 2.22, Eindhoven 5656 AE, The Netherlands
e-mail: marcel.pelgrom@nxp.com; hans.tuinhout@nxp.com; maarten.vertregt@nxp.com

A.H.M. van Roermund et al. (eds.), *Nyquist AD Converters, Sensor Interfaces, and Robustness: Advances in Analog Circuit Design, 2012*,
DOI 10.1007/978-1-4614-4587-6_13, © Springer Science+Business Media New York 2013

13.2 Variability

Variability is generally interpreted as a collection of phenomena characterized by not a-priori modeled parameter variation between individual transistors or components in a circuit. This collection is populated with a large number of effects ranging from offset mechanisms to reliability aspects.

Variability effects can be subdivided along three main axes:

- Time independent versus time variant effects.
- Global variations versus local variations.
- Deterministic versus statistical effects.

Every axis has specific properties that must be considered before defining a useful model and choosing the appropriate design solutions.Table 13.1 subdivides the variability effects in the time domain. Static effects allow a one-time correction of the associated devices in the circuit. These corrections can be implemented during final test of the product by means of non-volatile memories, laser trim or fuses. Slow time-dependent mechanisms must be corrected during operation and can be addressed via circuit compensation methods, such as auto-zeroing and on-chip calibration. However, variations with a time span comparable to the maximum speed of the process cannot be handled in this way. Stretched margins or more power are required.

Table 13.2 lists variability effects along the physical dimension axis from global to local variations. This list shows some well-known examples; new process generations will add new phenomena. The global-to-local axis is subdivided into four levels of granularity. Variations on the level of an integrated circuit are normally considered as standard design space variables. Designers are used to incorporate these variations in their Process-Voltage-Temperature "PVT" analysis. Extensive models of components and well-characterized parameter sets cover these aspects. Differential design, replica biasing etc. are commonly used methods to battle these variations.

Table 13.1 Time dependencies of some variability effects. The lower row lists the methodologies the designer has at his disposal to mitigate these effects

Static	Seconds	Micro-seconds	Pico-seconds
Process corners	Supply voltage	Wiring IR drop	Dyn. IR drop
	Temperature	Temp gradient	Jitter
Lithography	Hot carrier		Substrate noise
Line edge roughness	NBTI, Drift		
Dopant fluctuation	1/f noise	1/f noise	kT noise
WPE, STI Stress	Soft breakdown		
Differential design, calibration	Compensation circuitry	Wire width, chopping, common-centroid placement	Decoupling, shielding, clock cleaning

Table 13.2 From global to local variability

	IC	Sub circuit	Transistor	Atomic
Electrical	Supply voltage	Substrate noise	Wiring IR drop, jitter, cross-talk	
Thermal	Temperature	Temperature gradient	Local heating	Thermal noise
Technology	Process corner	CMP density	Well-proximity	Mechanical stress
Lithography	Line width	Layer density	Lithographic proximity effects	Line-edge roughness
Physics		Wafer gradient	NBTI, soft-breakdown, hot-carrier, stress	Random dopant, mobility variations

Table 13.3 Deterministic and statistical effects

Deterministic	Pseudo-statistical	Statistical
P-V-T	Substrate noise	Line edge roughness
Electrical offsets	stress	Dopant fluctuation
WPE, Stress, STI		Mobility fluctuation
Proximity effects		

The variations that affect each sub-circuit differently are less commonly incorporated in models and design software. Often the only mechanism available to a designer is a set of design rule checks (DRC), electrical rule checks (ERC) or heuristics. Violations are often found only after silicon validation.

Most effects at transistor level are well-modeled as the transistor is the focal point of circuit modeling. Various forms of reliability and hot-carrier effects can be predicted based on the physics involved. Electrical effects on this level of granularity are well understood, but the problem is to judge their relevance within the complexity of the entire circuit. The current variation in a transistor due to substrate noise is trivial if the magnitude of the substrate noise is known. However, establishing that magnitude for a multi-million devices sub-circuit is practically impossible.

In nanometer technologies an increasing number of transistor-related and atomic-scale effects become relevant. Most of these effects are well-described and modeled, and the associated statistical effects are part of design flows. Applying these statistical means requires knowledge of statistical data processing. In a practical design the static local effects are addressed with trimming, auto-zeroing, data-weighted averaging, calibration, re-configuration, error-correction, etc.

From a statistics point of view the time-independent effects can be subdivided in two classes: deterministic and statistical. In designer's terms: offsets and random matching. As simple as this division seems, there is a complication: a number of phenomena is from a physics point of view deterministic, but due to circuit complexity, a pseudo-statistical approach is used to serve as a (temporary) fix. An example is wiring stress, where the complexity of concise modeling is too cumbersome.

13.3 Deterministic Effects

Deterministic or systematic offsets between pairs of resistors, CMOS transistors or capacitors are due to electrical biasing differences, mechanical stress, lithographic and technological effects in the fabrication process [1].

It may seem trivial, but good matching requires in the first place that matched structures are built from the same material and are of equal size. This implies that a 2:1 ratio is constructed with three identical elements. The required 2:1 ratio applies for every aspect of the combination: area, perimeter, coupling, etc. A pitfall can occur when an existing lay-out is scaled and the resulting dimensions rounded off to fit the new lay-out grid. The rounding operation can result in unexpected size deviations in originally perfectly matched devices.

13.3.1 Offset Caused by Electrical Differences

For electrical matching, the voltages on all the elements must be identical. Node voltages are affected by voltage drops in power lines, leakage currents in diodes, substrate coupling, parasitic components, etc. Electrically derived effects must also be considered, e.g. heat gradients due to power dissipation and temporary charging effects due to phenomena as Negative Bias Temperature Instability [2].

13.3.2 Offset Caused by Lithography/Patterning

During the lithography process there are many crucial details that will affect the quality of the patterning. Most of these effects are global variations and become part of the tolerance budget in the definition of the parameter corners of a process.

Lithography critically depends on the flatness of the wafer surface. Damascene technology has been developed to avoid height differences. In a design with less regular wiring patterns, the design rules will prescribe filling patterns: "tiling". If the design tool is allowed to automatically generate tiling patterns, undesired side effects may occur. It is clear that the proximity of a tiling structure will affect capacitor ratios. Also stress patterns and thickness variations can occur. A safe approach is to define and position the tiling patterns during the lay-out phase of critical blocks by hand.

In 65-nm technologies and beyond it is practically impossible to define minimum width lines and spaces with acceptable tolerance at arbitrary positions. In order to create such minimum width patterns pre-distortion is applied to the mask in the form of optical proximity correction (OPC). These tools optimize on patterning tolerances, not on the mutual equality of patterns.

13.3.3 *Proximity Effects*

Variations due to proximity are caused by the presence or absence of neighboring structures. These structures scatter diffused light during the lithography, deplete etch liquids or gases etc. The line width in the open field becomes narrower, while large neighboring structures cause lines to expand. In a lay-out aiming at accuracies below 1% dummy structures are placed at distances up to 20–40 μm.

The well-proximity effect (WPE) is believed to be caused during the implantation of the well in the substrate. The implanted ions interact with the photo-resist boundary and cause a horizontal gradient in the well implantation dose. Variations ranging from 1 μm [3] to 2 μm [4] have been reported. It is advisable to use an overlap of the well mask larger than the minimum lay-out design rule.

13.3.4 *Temperature Gradients*

Gradients can exist in doping, resistivity, and layer thickness. Although structures tend to decrease in dimensions, situations may occur where equality is required in a distance in the order of 1 mm. In older processes CMOS thresholds were observed to deviate up to 5 mV over this distance. Resistivity gradients can reach a relative error of several percent. In advanced processes ($<$ 0.18 μm) process control is much better and technology gradients are hardly present ($<$ 1 mV/cm).

The temperature distribution across a circuit in operation can be a reason for parameter gradients. In a System-on-Chip the memories show a relatively low power density. Output drivers, transmitters, high-speed processors, power regulators (LDO) and input stages (LNA) may consume much more. In larger chips (50–100 mm^2) with 2–5 W power dissipation temperature differences up to 20°C can occur. Local temperature gradients of 2–5 °C/mm are possible. With threshold-voltage and diode temperature coefficients in the order of $-$ 2 mV/°C, an offset of several milliVolts is well possible. It is therefore important to consider the power distribution when temperature sensitive circuits and heat sources are placed on the same die. The circuit designer can position the critical devices on equal-temperature lines or use cross-coupling.

13.3.5 *Offset Caused by Stress*

During the fabrication of a circuit, layers are deposited on the substrate. These layers are built of different materials with different thermal expansion coefficients. After the devices have cooled to room temperature the difference in thermal expansion coefficients leads to mechanical stress. This stress can result in a positive or negative change of the local parameter value. A secondary effect is that the global stress pattern is locally affected by neighboring lay-out features, causing stress modulation in the surrounding components. In high-precision analog design this will lead to undesired systematic offsets.

In resistors and transistors stress predominantly impacts the mobility of carriers. Tensile stress increases the electron mobility and reduces the hole mobility. Compressive stress works opposite and is used to enhance mobility in PMOS transistors. Some major causes for stress are:

- The presence of the die boundary close to sensitive devices. Typically a distance of several hundreds of microns is safe to avoid die-edge related stress effects.
- Plastic packages are molded around the die. After cooling these packages create rather severe mechanical stress. Special gels or polyimide coatings on top of the die relieve this problem. A simple way to detect plastic package related stress is to heat the package with a hot air flow.
- In modern isolation techniques a trench is etched in the substrate and silicon dioxide is deposited and planarized: Shallow Trench Isolation (STI). The different thermal coefficients of silicon dioxide and the substrate cause mechanical stress. A transistor in the substrate with its diffusion areas surrounded by STI experiences this stress. This effect is called "STI-stress" or "LOD-stress" (Length Of Diffusion).
- In a densely packed circuit, there will be many active-to-STI edges. Unrelated edges that are close to a device will influence the stress pattern, this effect is known as "OD-to-OD" stress or "OD-spacing" effect. The Oxide-Definition (OD) mask defines the active area.
- In advanced processes an additional layer can be used to create stress to increase the current in a high-performance MOS transistor. The proximity of other structures will influence this effect, called "PS-to-PS" stress or "poly-space" effect.
- Aluminum has a different thermal expansion coefficient from the dielectric that surrounds it. Asymmetries in wiring will result in stress related offset.

Stress related to Shallow Trench Isolation has been subject of various studies [4–8].In [8] an experiment is reported where the STI-to-active edge of the source and drain is varied, see Fig. 13.1. This paper reports current factor deviations up to 12% and threshold voltage variations of 10 mV. These observations are technology specific but similar effects are reported in [4, 6]. Next to the effect of the STI-to-active edge of the transistor itself, also neighboring edges will modulate the mechanical stress pattern. This may be less relevant in a digital circuit, however, in precision analog design these effects must be taken into account by designing equal surroundings in the lay-out.Also the wiring pattern causes transistors to show offset, see Fig. 13.2.

The coverage of transistors with metal layers can lead to mobility reduction (of 17%) due to incomplete annealing of interface states [9] and to variations in the stress pattern. When a wiring pattern is placed at different spacings on one side of a bipolar pair [10], the resulting current variation is in the order of 1–2%. Even wiring on top level (e.g. tiling patterns) can cause stress [11].

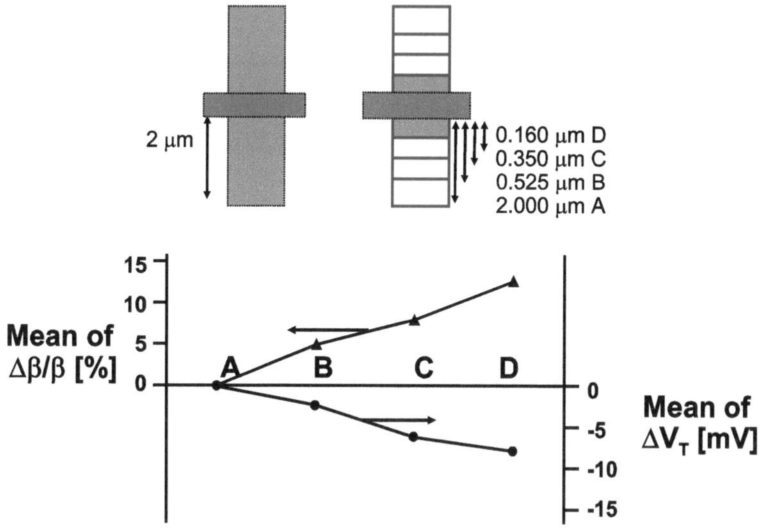

Fig. 13.1 An experiment shows the influence of the STI edge on the drain current and threshold voltage, [8]. *Top*: a 65-nm 2/0.5 μm NMOS reference transistor is designed with the STI edge of source and drain at 2.0 μm. A second device has a similar STI distance (A) or at 0.525 (B), 0.35 (C), and 0.16 μm (D) Bottom: Measured current factor deviation and threshold voltage shift

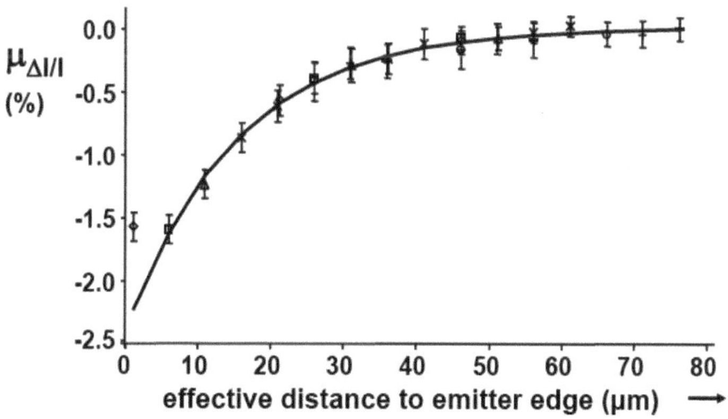

Fig. 13.2 An aluminum wire placed at a certain distance of the emitter causes current deviations, that can be measured up to 40 μm[10]

The measured impact of the wiring pattern halves for every 10 μm distance up to 40 μm. This example shows that a regular, symmetrical and consistent lay-out is required for analog circuits that should yield offsets below 1%. Table 13.4 lists a number of effects.

Table 13.4 Potential magnitudes of deterministic variability effects

Effect	Magnitude
Power supply voltage drop	Voltage shifts up to 100 mV in poorly designed power grids.
Lithography, etch depletion etc	Dimensional errors up to 100 nm, in > 0.25 μm CMOS
Proximity effects	Dimensions errors up to 20 nm in advanced mask making
Temperature gradient	1°C over 20–50 μm close to strong local heat source. With 2 mV/ °C threshold sensitivity.
LOD or STI effec [4, 6, 8]	Mobility 12% between minimum drain extension and large drain
Well-proximity[3, 4]	Threshold shift if mask edge closer than 2 μm
Metallization[10]	Mobility 2% between close and 40 μm far metal track
Metal coverage of gates[9]	Current factors may deviate up to 10–20%

Table 13.5 Guide lines for the design of equal components

1	Equal components are of the same material, have the same form, dimensions and orientation
2	The potentials, temperatures, pressures and other environmental factors are identical
3	Currents in components run in parallel, not anti-parallel or perpendicular
4	Only use cross-coupled structures if there is a clear reason for that (e.g. temperature gradient)
5	Keep wiring away from the components
6	Use star-connected wiring for power, clock and signal
7	Apply symmetrical (dummy) structures up to 20 μm away from sensitive structures
8	Keep supply and ground wiring together and take care that no other circuits dump their return current in a ground line
9	Check on voltage drops in power lines
10	Stay 200 μm away from the die edges to reduce stress from packaging
11	Tiling patterns are automatically inserted and can lead to unpredictable coupling, isolation thickness variations, stress. Do not switch off the tiling pattern generation, but define a symmetrically placed tiling pattern

13.3.6 Offset Mitigation

Despite the complex nature of some variations, a number of guidelines can be formulated to minimize the effect of these offset causes, see Table 13.5 [12].

Common centroid structures are used to reduce the gradient effects [13]. Applying a common centroid geometry, see Fig. 13.3, is not trivial. An asymmetry in the wiring scheme can easily cause more problems than are solved. On the left side is a standard cross-coupled differential pair with common source. The right side shows in-line common centroid structures. The lower structure is exactly common centroid with the disadvantage that the outer devices need dummy structures to compensate for their lack of neighbors. The upper-right structure needs no dummy structures, at the cost of a small spacing between the common centroid points. Other methods for offset reduction are applied on a design level: calibration, auto-zeroing etc.

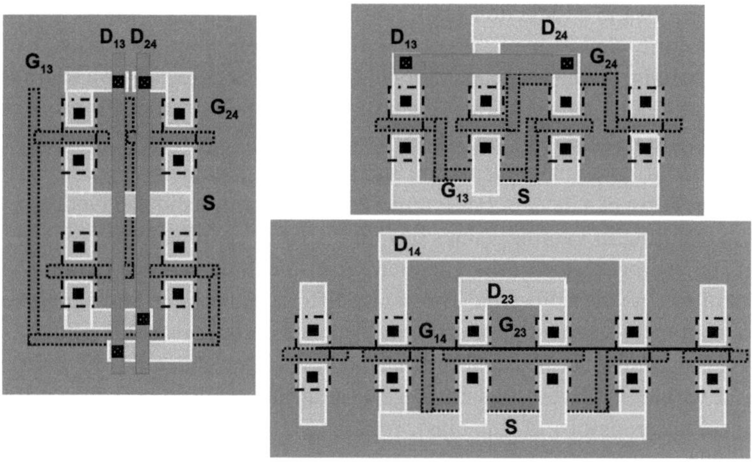

Fig. 13.3 Some common centroid arrangements

By following these design guidelines the effects of systematic errors can be significantly reduced. The obtainable limits in a production environment differ per component and can be summarized as:

- Resistors: The absolute value can change 5–10%. Yet, the relative accuracy of matched resistors is in the order of 10^{-3}–10^{-4} depending on type (diffused is better than polysilicon), size, and environment. Sensitive to substrate coupling.
- Capacitors: The absolute value is usually well-defined in a double poly-silicon or MIM process. Also horizontally arranged capacitors such as fringe capacitors reach excellent performance. The relative accuracy of capacitors is in the order of 10^{-4} for > 1 pF sizes. Minimum usable sizes in design are limited by parasitic elements, relative accuracy or the kT/C noise floor.
- Transistors: The current is sensitive to temperature, process spread and variability effects. The relative accuracy in current is in the order of 10^{-3}.
- Time: with a more or less fixed timing variation or jitter (1–5 ps_{rms}), the best accuracy is achieved for low signal bandwidths (10^{-5}–10^{-6} for signals below 1 kHz).

13.4 Random Matching

13.4.1 Random Fluctuations in Devices

The behavior of devices is the result of the combination of a large number of microscopic processes. The conductivity of resistors and transistors and the capacitance of capacitors is built up of a large number of single events: e.g. the presence of fixed charges in the conduction path, the local distances between the polysilicon grains that form capacitor plates, etc. Various authors have investigated random effects in specific structures: capacitors [14–18], resistors [5, 19], MOS transistors [20–25] and bipolar devices [26]. Accurate measurement methods for this class of phenomena are reported in [27].

In a general parameter fluctuation model [21] a parameter P describes some physical property of a device. P is composed of a deterministic and random varying function resulting in varying values of P at different coordinate pairs (x, y) on the wafer. The average value of the parameter over any area is given by the weighted integral of $P(x, y)$ over this area. The actual difference between two parameters ΔP of two identically sized areas at different depends on the mismatch cause.

One class of distinct physical mismatch causes is considered here as examples for local variations. Every mismatch-generating physical process that fulfills the mathematical properties of this class, results in a similar behavior at the level of mismatching transistor parameters.

Atomic variations are described by a random process in parameter P characterized by:

- The total mismatch of parameter P is composed of mutually independent events of the random process.
- The effects on the parameter are so small that the contributions to the parameter are linear.
- The correlation distance between the events is small compared to the size of the device (basically saying that boundary effects can be ignored).

Statistically the random process is described by a Poisson process that converges for a large number of events to a Gaussian distribution with zero mean. The analysis results in a description of the variance of the difference ΔP between the two instances [21]:

$$\sigma_{\Delta P}^2 = \frac{A_P^2}{WL} \tag{13.1}$$

A_P is the area proportionality constant for parameter P. The proportionality constant is characterized and used to predict the mismatch variance of a circuit.

Equation 13.1 describes the statistical properties of area *averaged* or *relative* values of parameter P. The *absolute* number of events (like the charge in a MOS channel) is proportional to the area of the device WL. Therefore differences in the sums of atomic effects obey a Gaussian distribution with zero mean and

$$\sigma_{\Delta P} = A_P\sqrt{WL} \tag{13.2}$$

In analyzing statistical effects it is important to consider whether the parameter is an absolute quantity (e.g. total amount of ions) or is relative (averaged) to the device area (e.g. threshold voltage).

Many technological processes that cause mismatching parameters fulfill in first order the above-mentioned mathematical constraints: distribution of ion-implanted, diffused or substrate ions (Random Dopant Fluctuations), local mobility fluctuations, polysilicon and oxide granularity, oxide charges, etc.

Apart from theoretical derivations and measurements, 3-D device simulations are applied to analyze the impact of random dopants, line-edge roughness, metal gate and polysilicon granularity in advanced processes [28].

13.4.2 MOS Threshold Mismatch

The threshold voltage is determined by the flat-band voltage, implantation steps, etc. The contribution of a constant substrate doping is given by:

$$V_T - V_{FB} = \frac{Q_B}{C_{gate}} = \frac{qN_x z_d WL}{C_{ox}WL} = \frac{\sqrt{2q\epsilon N_x \phi_b}}{C_{ox}} \tag{13.3}$$

where ε is the permittivity, N_x the dope concentration and ϕ_b the Fermi potential. In the depletion region there are $z_d WLN_x$ doping ions. As matching usually occurs between pairs of transistors, the variance of the *difference between two transistors* is in first order [21, 23]:

$$\sigma_{\Delta VT} = \sqrt{2}\sigma_{singleVT} = \frac{qd_{ox}\sqrt{2N_x z_d WL}}{\epsilon_{ox}WL} = \frac{A_{VT}}{\sqrt{WL}} \propto \frac{d_{ox}\sqrt[4]{N_x}}{\sqrt{WL}} \tag{13.4}$$

Equation 13.4 is the basis for Random Dopant Fluctuation. However, more mismatch mechanisms contribute to the threshold mismatch, such as: interface states, threshold implants, etc.

The function of Eq. 13.4 is commonly depicted as a linear relation between $\sigma_{\Delta VT}$ and $1/\sqrt{area}$. Figure 13.4 shows an example of the measured dependence for $\sigma_{\Delta VT}$ versus $1/\sqrt{area}$. The slope of the line equals the parameter A_{VT}. For the smallest sizes the effective gate area is smaller due to underdiffusion and channel encroachment.This basic threshold-voltage mismatch model has been extended by various authors. More geometry dependence factors can be included to address nanometer CMOS effects [29]. A fundamental limit for dopant fluctuation related mismatch including the distribution in depth, was derived in [30]. Andricciola and Tuinhout [31] shows that the threshold-voltage mismatch coefficients are not affected by temperature. Work reported in [8] indicates that there is no relation between deterministic variations and random dopant fluctuations. The operating regime where the mismatch parameters of the transistor are extracted, has a small effect on the accuracy of the prediction in other regimes [24, 25]. In a practical design this difference can be ignored. In [32] the horizontal axis of Fig. 13.4 is

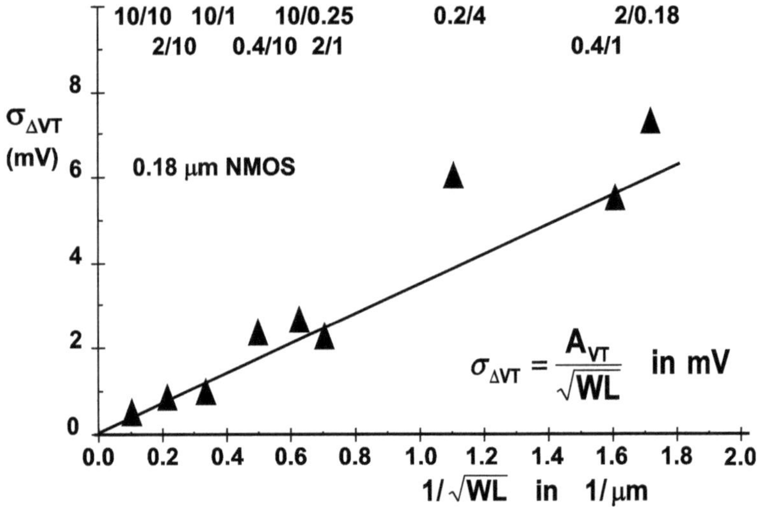

Fig. 13.4 The standard deviation of the NMOS threshold and the relative current factor versus the inverse square root of the area, for a 0.18 μm CMOS process. Device dimensions are as drawn

normalized to the oxide thickness and threshold voltage, called the Takeuchi plot. This approach is useful when the random dopant fluctuation mechanism is dominant.

In nanometer processes the short-channel effects in the channel are controlled by means of "halo" or "pocket" implants. These implants are self-aligned with the gate stack and can introduce some significant variations in the local doping profiles. Next to their own variation, the self-aligned feature prints any line-edge roughness in the doping profile. The pocket implants defy the uniform dopant hypothesis for the calculation of the threshold mismatch of Eq. 13.1. An additional term can be included for the variation due to the pocket implant:

$$\sigma^2_{\Delta VT} = \frac{A^2_{VT}}{WL} + \frac{B^2_{VT}}{f(W,L)} \tag{13.5}$$

where the function $f(W, L)$ still needs to be established.

13.4.3 Current Mismatch in Strong and Weak Inversion

The matching properties of the current factor are derived by examining the mutually independent components W, L, μ and C_{ox}:

$$\frac{\sigma^2_{\Delta\beta}}{\beta^2} = \frac{\sigma^2_{\Delta W}}{W^2} + \frac{\sigma^2_{\Delta L}}{L^2} + \frac{\sigma^2_{\Delta C_{ox}}}{C^2_{ox}} + \frac{\sigma^2_{\Delta \mu n}}{\mu^2_n} \approx \frac{A^2_\beta}{WL} \tag{13.6}$$

The mismatch-generating processes for the gate oxide and the mobility are treated in accordance with Eq. 13.1. The variations in W and L originate from line-edge roughness. At gate-lengths below 65-nm, simulations [28] indicates some role for edge roughness. This role is in measurements hard to identify in the presence of large threshold mismatch. In [21] it is argued that the matching of the current factor is determined by local variations of the mobility. Many experiments show that mobility affecting measures (e.g. placing the devices under an angle) indeed lead to a strong increase of the current factor mismatch. The relative mismatch in the current factor can be approximated by the inverse-area description.

Considering only the threshold and current factor variations, the variance of the difference in drain currents ΔI between two equally sized MOS devices can be calculated. Using the statistical method described in [33]:

$$\sigma_{\Delta I}^2 = \left(\frac{dI}{dV_T}\right)^2 \sigma_{\Delta VT}^2 + \left(\frac{dI}{d\beta}\right)^2 \sigma_{\Delta\beta}^2 \tag{13.7}$$

For strong inversion the simple square-law current model yields:

$$\left(\frac{\sigma_{\Delta I}}{I}\right)^2 = \left(\frac{2\sigma_{\Delta VT}}{V_{GS} - V_T}\right)^2 + \left(\frac{\sigma_{\Delta\beta}}{\beta}\right)^2 \tag{13.8}$$

In weak inversion the current is modeled as an exponential function. Due to the low current level, the current factor mismatch is of less importance. Applying the Taylor approximation[33] gives the mean current and the variance of the resulting log-normal distribution for the current:

$$\mu_I = I(\sigma_{\Delta VT} = 0)\left(1 + \frac{1}{2}\left(\frac{q\sigma_{\Delta VT}}{mkT}\right)^2\right)$$

$$\left(\frac{\sigma_{\Delta I}}{I}\right)^2 = \left(\frac{q\sigma_{\Delta VT}}{mkT}\right)^2$$

Note that the mean value is larger than the nominal current without mismatch. Although the threshold mismatch shows no temperature dependence, the current mismatch certainly will vary due to the denominator term.

Figure 13.5 shows an example of the current mismatch relative to the drain current. At high gate-source voltages the current factor mismatch in Eq. 13.8 dominates. At lower gate-source voltages the threshold-related term in this equation gains importance.

Equation 13.10 predicts in a 65-nm process with $A_{VT} = 3.5$ mV µm a relative current mismatch of 40%, which is confirmed by the measurements in Fig. 13.5.

The standard deviation of the current difference between the 0.12/1 µm transistors reaches $\approx 40\%$ of the mean current in the sub-threshold regime. Obviously this would imply a reverse drain current below -2.5σ for a Gaussian distribution however the current variation follows here a log-normal distribution.

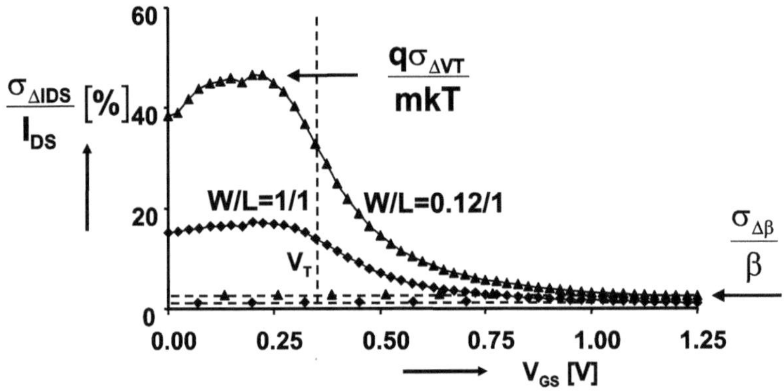

Fig. 13.5 The relative current mismatch for two 65-nm technology transistor geometries swept over the full voltage range. Measurements by N. Wils

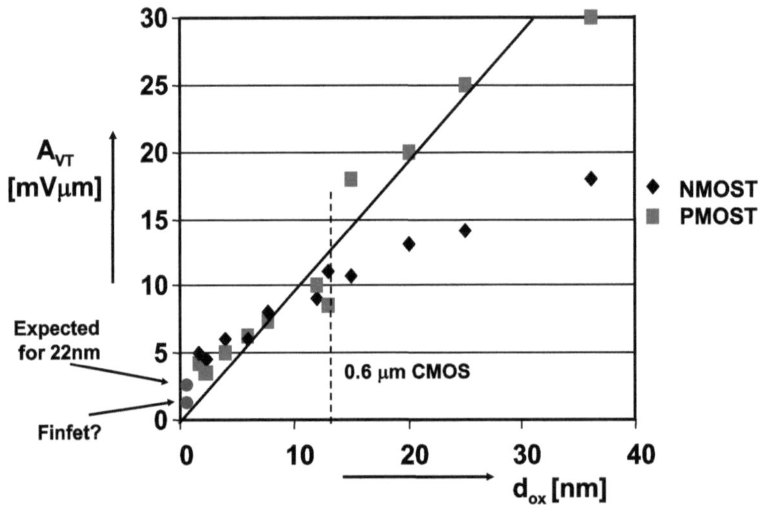

Fig. 13.6 Development of the threshold mismatch factor A_{VT} for NMOS and PMOS transistors as a function of the nominal oxide thickness of the process. The measured data processes covers processes from 45-nm up to 1.6 μm CMOS. The 22-nm and Finfet *dots* are expectations from foundry presentations

13.4.4 Mismatch for Various Processes

In Fig. 13.6 the threshold mismatch coefficient A_{VT} is plotted as a function of the nominal oxide thickness. As predicted by Eq. 13.4 the mismatch coefficient reduces for thinner gate-oxide thickness. Of course many more changes in the device architecture took place, still the oxide-thickness is the most relevant parameter.

The large PMOS transistor coefficients for > 0.6 μm CMOS generations are caused by compensating implants: the substrate, threshold adjust and n-well implants. The net value ($N_a - N_d$) determines the average threshold of the PMOS transistor, while the quantity $N_x = (N_a + N_d)$ is relevant for matching. Beyond the 0.6 μm node a twin well construction with a dedicated well implant for the PMOS transistor is used, thereby avoiding compensating charges.

At the 0.35 μm CMOS node the maximum electrical fields in intrinsic transistors were reached for both the vertical gate-oxide field and the lateral field controlling the charge transport. For this reason and in order to reduce power consumption, the power supply voltage was lowered. The need to tailor the internal fields in the transistor, has led to less uniform and higher implantation channel dope. As can be expected from the theoretical background, the slower scaling of the gate-oxide thickness made that the threshold matching factor A_{VT} stopped decreasing. This became especially pronounced in 65–32 nm technologies, where pocket implants create an additional mismatch source. Shrinking the area of analog blocks in nanometer processes is clearly an important economical issue, but in combination with a rising mismatch coefficient this will lead to lower performance. The reduction in the signal-to-matching coefficient ratio in nanometer CMOS will necessitate changes in the system, architecture, design or technology. In order to maintain high quality signal processing, some enhancements to the standard processes are needed, such as the use of high-voltage devices or precision resistors and capacitors.

In Fig. 13.6 the diagonal line indicates an A_{VT} factor increase of 1 mV μm for every nm of gate insulator thickness. This line is a first order estimate of what a well-engineered process should bring.

Over the same process range the current mismatch factor A_β varies between 1 and 2% μm.

Figure 13.7 shows a simulation of the threshold voltage of 200 NMOS and PMOS 90-nm transistors in their process corners. Although the corner structure is still visible, it is clear that for small transistors in advanced processes the mismatch is of the same magnitude as the process corner variation. The plot is somewhat misleading as it combines global process corner variation and local mismatch. A simple "root-mean-square" addition of these two variation sources would ignore the fundamental difference between the two.

13.4.5 Drift

In literature the term "drift" refers to various phenomena in electronics circuits. Drift can be a long-term parameter change that is caused by aging effects, mechanical influences and temperature. The most frequent appearance of drift in analog-to-digital conversion is a temperature-dependent shift of the input offset. The magnitude of this drift is in the order of several μVolts/°C measured on differential inputs of circuits. Figure 13.8 shows temperature measurements of the relative current

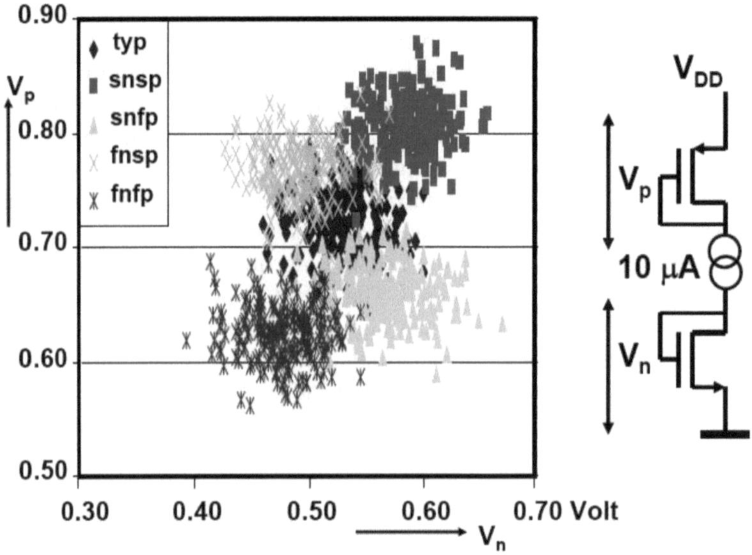

Fig. 13.7 Simulation of 200 0.2/0.1 P- and NMOS transistors in their 90-nm process corners. The notation "snfp" refers to slow NMOS and fast PMOS transistors. The variation due to mismatch is of equal importance as the process variation

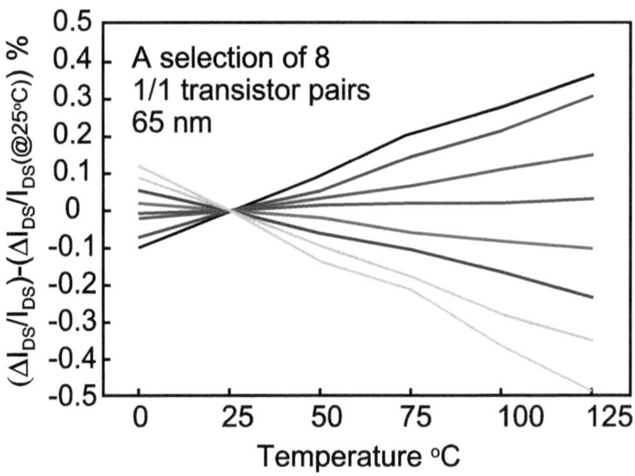

Fig. 13.8 The relative current differences of eight selected pairs of NMOS transistors [31]

differences ($\Delta I_{DS}/I_{DS}$) of eight selected pairs of NMOS transistors. Each curve has been normalized to its 25°C value.

In a completely symmetrical circuit with identical components there is no cause for drift. However, mismatch can result in unbalance that translates in drift. In a differential input pair inequalities will exist due to random threshold mismatch and

Table 13.6 An overview of matching models and value ranges

MOS transistors	$\sigma_{\Delta VT} = \frac{A_{VT}}{\sqrt{WL}}$	$A_{VT} = 1$ mV µm/nm
MOS transistors	$\frac{\sigma_{\Delta\beta}}{\beta} = \frac{A_\beta}{\sqrt{WL}}$	$A_\beta = 1$–2% µm
Bipolar transistors (BJT)	$\sigma_{\Delta Vbe} = \frac{A_{Vbe}}{\sqrt{WL}}$	$A_{Vbe} = 0.3$ mV µm
Diffused/poly resistors	$\frac{\sigma_{\Delta R}}{R} = \frac{A_R}{\sqrt{WL}}$	$A_R = 0.5/5\%$ µm
Plate, fringe capacitors	$\frac{\sigma_{\Delta C}}{C} = \frac{A_R}{\sqrt{C}}$	$A_C = 0.3\ \%\sqrt{fF}$

current factor mismatch. The threshold mismatch is a charge based phenomenon and will create a temperature independent offset. The relative current factor mismatch as formulated in Eq. 13.6 is hardly sensitive to temperature[31]. In contrast the absolute current factor (β) and differences between current factors (Δβ) vary with temperature due to the mobility: $\beta \propto \mu \propto T^{-\alpha}$. As the current factor determines the transconductance, the (constant) relative current factor mismatch is translated into a temperature dependent input referred contribution.

Assume that a current factor difference Δβ exists in a differential pair due to mismatch. This current factor difference must be compensated by an input referred voltage. The temperature dependence of this voltage between the inputs of this differential pair represents the drift of this differential pair.

The offset current is proportional to the current factor difference and can be translated into an input referred offset ΔV_{in} via the transconductance:

$$\Delta V_{in} = \frac{I_{DS}}{g_m}\frac{\Delta\beta}{\beta} \tag{13.11}$$

If a constant bias current is provided the remaining temperature dependence is due to the transconductance. An expression for the temperature dependent drift at the input of a differential pair due to current factor mismatch is found:

$$Drift = \frac{d\Delta V_{in}}{dT} = \frac{\alpha}{T}\frac{I_{DS}}{g_m}\frac{\Delta\beta}{\beta} = \frac{\alpha}{T}\frac{V_{GS}-V_T}{2}\frac{\Delta\beta}{\beta} = \frac{\alpha\Delta V_{in}}{T} \tag{13.12}$$

where α is the temperature exponent in the mobility.

In Table 13.6 matching parameters of various components are listed. In the previous paragraphs the mismatch parameters for the MOS transistor have been extensively discussed.

The behavior of the bipolar transistor is dominated by the number of dopants in the base that are not depleted. The fluctuation of this number, comparable to the fluctuation of the charge in the depletion layer of the MOS transistor, causes the base-emitter voltages between two bipolar devices to mismatch. Therefore a variance can be defined for ΔV_{be}. In [26] various experiments have confirmed the validity of this mismatch model.

Resistors for high precision analog design are formed by diffused n- or p- doped areas. In advanced processes these layers are covered with a silicide layer to

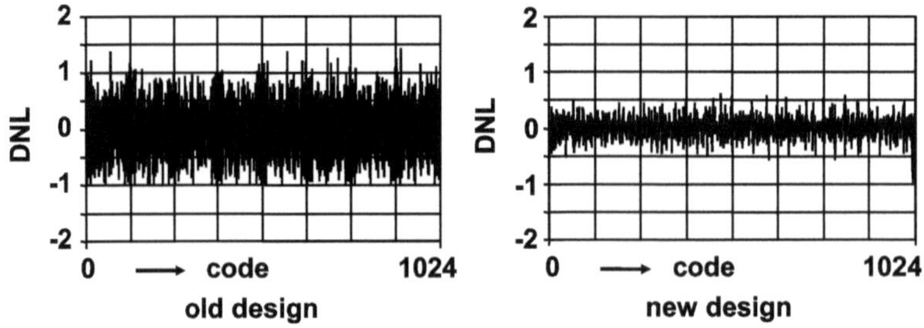

Fig. 13.9 The differential non-linearity is a quality parameter of this 10-bit analog-to-digital converter

lower their impedance to the 2–5 Ω level. A special mask is applied to prevent the deposition of silicide in order to obtain sheet resistances of 50–500 Ω. Polysilicon resistors are enclosed by silicon dioxide acting as a thermal isolator. Dissipated heat will give rise to signal dependant variation or may destroy such a resistor, a small amount of dissipated heat will affect the grain boundary structure and lead to a resistance value shift after cooling. Moreover, the polysilicon grain boundaries give rise to a huge increase in mismatch.

13.4.6 Consequences for Design

A designer has several options to deal with global variations in his circuit. Next to rigorous simulation, the circuit itself can be designed in a way that effects of global variations are minimized. These techniques include differential design and replica-biasing. Unfortunately these methods are ineffective for battling local random variations. Here the designer has to rely on statistical means, including techniques as Monte-Carlo simulation, hand calculation and various other techniques.

The caveat in many analog circuits is in the random offset in differential pairs. An example is shown in Fig. 13.9. The differential non-linearity curve (DNL) is a measure for the error that an analog-to-digital converter makes at every code transition. In high speed converters MOS threshold mismatch of the input pair of the comparators is the dominant contributor to differential non-linearity. The measurement of the first prototype(left) shows significant deviations with excursions down up to 1.6. After analyzing the design with a Monte-Carlo simulation, the major contributors were located and correctly dimensioned, resulting in the measured curve on the right. Monte-Carlo analysis is a standard tool for high-performance circuits in analog design.

Power, speed and accuracy span the design space e.g. [34]. The idea that accuracy must be balanced against power can be easily understood by considering that the voltage uncertainty on the gate capacitance can be described as an energy

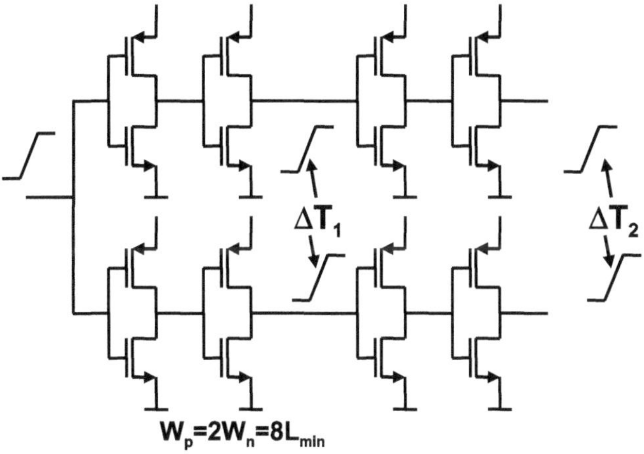

Fig. 13.10 An input pulse is applied to two chains of inverters. Due to mismatches in the transistors there is a time difference at the two outputs. This example mimics a critical timing path

Table 13.7 The simulated standard deviation of the difference in arrival time of the two pulses in the inverter chain of Fig. 13.10

Process node	0.25 μm	0.18 μm	0.13 μm	90 nm	65 nm
Clock period	10 ns	5 ns	2 ns	1 ns	500 ps
$\sigma_{\Delta T2}$, C_{load}=50 fF	16 ps	21 ps	38 ps	68 ps	88 ps
$\sigma_{\Delta T2}$, C_{load}=50..15 fF	16 ps	16 ps	22 ps	33 ps	32 ps

term [35, 36], found by combining the mismatch uncertainty voltage with the associated capacitance:

$$E_{\sigma VT} = C_{gate}\sigma^2_{\Delta VT} = C_{ox}A^2_{VT} = 4.5 \times 10^{-19} \qquad \text{Joule} \tag{13.13}$$

which is independent of the transistor size and corresponds to about 100 kT at room temperature. This energy can be seen as the energy required to toggle a latch pair of transistors in meta-stable condition into a desired position with a one-σ certainty. In circuits with parallel data paths the energy required to overcome component mismatch may hence dominate over kT noise by two orders of magnitude.

Also digital designers experience that for small devices the random component can exceed the process corner variation. An example[37] is shown in Fig. 13.10 and Table 13.7. A pulse is applied to two sets of inverters and it is expected that the outputs will change state simultaneously. Due to mismatch between the transistors of both rows of inverters, a random skew will exist in the difference of arrival time between various samples of this circuit. In Table 13.7 the standard deviation of this random skew is compared to the clock period in five technologies. From an effect in the 0.1% range the random skew can take up to 6% of the clock period in a 65 nm process.

Memory designs are affected by mismatch in several ways. Threshold variations influence the margins for the read and write operations. Moreover low-threshold devices are responsible for larger average leakage currents, see Eq. 13.9. The choice for the size of the transistors in an SRAM cell and in the sense amplifier critically depends on an accurate prediction of the mismatch coefficients. The other problem in memory structures is the large number of cells. This requires simulating statistical distributions up to 7σ. This is not practical in Monte-Carlo simulations. Special statistical acceleration mechanisms (importance sampling, Latin Hypercube sampling) allow to sample tails of statistical distributions [38].

13.5 Limits of Power and Accuracy

One of the main questions in low-power conversion design is the ultimate limit of power consumption. Mathematicians would claim that the mapping of analog values on a discrete amplitude scale should not change the signal and therefore be zero-power.

In physics, however, a lower limit can be derived from the inherent thermal noise. In Fig. 13.11 the maximum bandwidth and resolution in bits are given based on the ratio of the signal power and thermal *kT* noise. This limit is three to four decades away from practical realizations. This is due to the fact that much "overhead" power has to be incorporated in real designs.

In high-speed (conversion) circuits the limits will be based on random variations in the circuit components. This may be the case in multiplexed circuits or in circuits

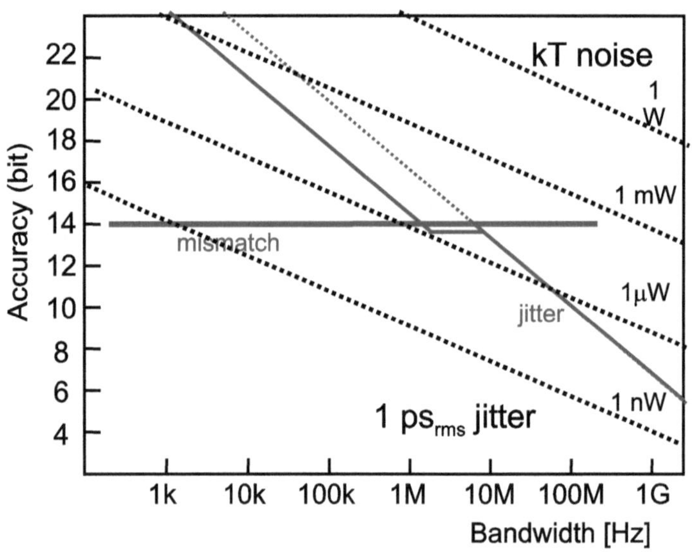

Fig. 13.11 SNR limited by *mismatch* and *jitter* in analog time-discrete blocks

in which the signal path is level-dependent, as in full-flash converters. Table 13.6 sets some limits on how far circuits limited by random variation can be pushed. In Fig. 13.11 this has been summarized to a single level at 14 bit.

Finally Fig. 13.11 shows a limitation due to the jitter of clocks signals as required in any form of sampled data circuit. Here this limit is set to 1 ps_{rms}, a quite aggressive value in a production design. Above the mismatch limit almost all the analog-to-digital techniques are based on sigma-delta modulation, that poses a specific demand on the jitter. Consequently the obtainable performance at 1 ps_{rms} has been reduced to what can be achieved with these techniques.

It is clear that in all practical and fundamental considerations for integrated circuit design variability and specifically its random component will remain an important limit.

References

1. Gregor RW (1992) On the relationship between topography and transistor matching in an analog CMOS technology. IEEE Trans Electron Devices 39:275–282
2. Stathis JH, Zafar S (2006) The negative bias temperature instability in MOS devices: a review. Microelectron Reliab 46:270–286
3. Hook TB, Brown J, Cottrell P, Adler E, Hoyniak D, Johnson J, Mann R (2003) Lateral ion implant straggle and mask proximity effect. IEEE Trans Electron Devices 50:1946–1951
4. Drennan P, Kniffin M, Locascio D (2006) Implications of proximity effects for analog design. In: IEEE custom integrated circuits conference. IEEE, Piscataway, pp 169–176
5. Tuinhout HP, Hoogzaad G, Vertregt M, Roovers R, Erdmann C (2002) Design and characterisation of a high precision resistor ladder test structure. In: IEEE international conference on microelectronic test structures. IEEE, Piscataway, pp 223–228
6. Bianchi RA, Bouche G, Roux-dit-Buisson O (2002) Accurate modeling of trench isolation induced mechanical stress effects on MOSFET electrical performance. In: Technical digest international electron devices meeting. IEEE, Piscataway, pp 117–120
7. Su K-Wi et al (2003) A scaleable model for STI mechanical stress effect on layout dependence of MOS electrical characteristics. In: Proceedings of the IEEE custom integrated circuits conference. IEEE, Piscataway, pp 245–248
8. Wils N, Tuinhout HP, Meijer M (2009) Characterization of STI edge effects on CMOS variability. IEEE Trans Semicond Manuf 22:59–65
9. Tuinhout HP, Pelgrom MJM, Penning de Vries R, Vertregt M (1996) Effects of metal coverage on MOSFET matching. In: Technical digest international electron devices meeting. IEEE, Piscataway/New York, pp 735–739
10. Tuinhout HP, Bretveld A, Peters WCM (2004) Measuring the span of stress asymmetries on high-precision matched devices. In: International conference on microelectronic test structures. IEEE, Piscataway, pp 117–122
11. Tuinhout HP, Vertregt M (2001) Characterization of systematic MOSFET current factor mismatch caused by metal CMP dummy structures. IEEE Trans Semicond Manuf 14:302–310
12. Pelgrom MJM, Vertregt M, Tuinhout HP Matching of MOS transistors. MEAD course material, EPFL/MEAD Lausanne 1998–2012
13. Lam M-F, Tammineedi A, Geiger R (2001) Current mirror layout strategies for enhancing matching performance. Analog Integr Circuits Signal Process 28:9–26
14. McCreary JL (1981) Matching properties, and voltage and temperature dependence of MOS capacitors. IEEE J Solid-State Circuits 16:608–616

15. Shyu J-B, Temes GC, Yao K (1982) Random errors in MOS capacitors. IEEE J Solid-State Circuits 17:1070–1076
16. Tuinhout HP, Elzinga H, Brugman JT, Postma F (1994) Accurate capacitor matching measurements using floating-gate test structures. In: IEEE international conference on microelectronic test structures. IEEE, Piscataway, pp 133–137
17. Tuinhout HP, van Rossem F, Wils N (2009) High-precision on-wafer backend capacitor mismatch measurements using a benchtop semiconductor characterization system. In: IEEE international conference on microelectronic test structures. IEEE, Piscataway, pp 3–8
18. Aparicio R (2002) Capacity limits and matching properties of integrated capacitors. IEEE J Solid-State Circuits 27:384–393
19. Drennan PG (1999) Diffused resistor mismatch modeling and characterization. In: Bipolar/BiCMOS circuits and technology meeting. IEEE, Piscataway/New York, pp 27–30
20. Lakshmikumar KR, Hadaway RA, Copeland MA (1986) Characterization and modeling of mismatch in MOS transistors for precision analog design. IEEE J Solid-State Circuits 21:1057–1066
21. Pelgrom MJM, Duinmaijer ACJ, Welbers APG (1989) Matching properties of MOS transistors. IEEE J Solid-State Circuits 24:1433–1440
22. Michael C, Ismail M (1992) Statistical modeling of device mismatch for analog MOS integrated circuits. IEEE J Solid-State Circuits 27:154–166
23. Mizuno T, Okamura J, Toriumi A (1994) Experimental study of threshold voltage fluctuation due to statistical variation of channel dopant number in MOSFETs. IEEE Trans Electron Devices 41:2216–2221
24. Forti F, Wright ME (1994) Measurement of MOS current mismatch in the weak inversion region. IEEE J Solid-State Circuits 29:138–142
25. Croon JA, Sansen W, Maes HE (2005) Matching properties of deep sub-micron MOS transistors. Springer, Dordrecht. ISBN: 0-387-24314-3
26. Tuinhout HP (2003) Improving BiCMOS technologies using BJT parametric mismatch characterisation. In: Bipolar/BiCMOS circuits and technology meeting. IEEE, Piscataway, pp 163–170
27. Tuinhout HP (2005) Electrical characterisation of matched pairs for evaluation of integrated circuit technologies. Ph.D. Thesis, Delft University of Technology. (http://repository.tudelft.nl/file/82893/025295)
28. Brown AR, Roy G, Asenov A (2007) Poly-Si-Gate-Related variability in decananometer MOSFETs with conventional architecture. IEEE Trans Electron Devices 54:3056–3063
29. Bastos J, Steyaert M, Roovers R, Kinget P, Sansen W, Graindourze B, Pergoot A, Janssens E (1995) Mismatch characterization of small size MOS transistors. In: IEEE international conference on microelectronic test structures. IEEE, Piscataway, pp 271–276
30. Stolk PA, Widdershoven FP, Klaassen DBM (1998) Modeling statistical dopant fluctuations in MOS transistors. IEEE Trans Electron Devices 45:1960–1971
31. Andricciola P, Tuinhout HP (2009) The temperature dependence of mismatch in deep-submicrometer bulk MOSFETs. IEEE Electron Device Lett 30:690–692
32. Takeuchi K, Fukai T, Tsunomura T, Putra AT, Nishida A, Kamohara S, Hiramoto T (2007) Understanding random threshold voltage fluctuation by comparing multiple fabs and technologies. In: IEEE electrron devices meeting, IEDM 2007. IEEE, Piscataway, pp 467–70
33. Papoulis A (2001) Probability, random variables, and stochastic Processes, 4th edn. McGrawHill, New York/London. McGrawHill New York student edition 1965. ISBN: 0-073-66011-6
34. Vertregt M, Scholtens PCS (2004) Assessment of the merits of CMOS technology scaling for analog circuit design. In: 30th European solid-state circuits conference. IEEE, Piscataway, pp 57–64
35. Pelgrom MJM (1994) Low-power high-speed A/D conversion. In: 20th European solid-state circuits conference, low-power workshop. Editions Frontières, Gif-sur-Yvette Cedex

36. Kinget P, Steyaert M (1996) Impact of transistor mismatch on the speed accuracy power trade-off. In: Custom integrated circuits conference. IEEE, Piscataway/New York
37. Pelgrom MJM, Tuinhout HP, Vertregt M (1998) Transistor matching in analog CMOS applications. In: International electron devices meeting. IEEE, Piscataway, pp 915–918
38. Doorn TS, ter Maten EJW, Croon JA, Di Bucchianico A, Wittich O (2008) Importance sampling Monte Carlo simulations for accurate estimation of SRAM yield. In: 34th European solid-state circuits conference. IEEE, Piscataway, pp 230–233

Chapter 14
Analog Circuit Design in Organic Thin-Film Transistor Technologies on Foil: An Overview

Hagen Marien, Michiel Steyaert, Erik van Veenendaal, and Paul Heremans

Abstract In this work an overview is given of the progress which is made in the last few years in the domain of analog organic electronics. Subsequently several building blocks for organic smart sensor systems are brought into focus. The implementations of a two-stage DC-connected opamp, a $\Delta\Sigma$ ADC, a Dickson DC-DC up-converter and a capacitive touch sensor are presented. Special attention is spent to the design techniques applied for embedding the circuits in the given organic electronics technology.

14.1 Introduction

The progress made in the last decade in the domain of organic thin-film transistor technologies has enabled the production of flexible displays. These devices find a market in the sector of handheld consumer devices, such as E-readers and smart phones. Besides for flexible displays, organic electronics has also opened the gates for digital as well as analog and mixed-signal applications. Examples of the latter can be found in flexible and printed RFID tags, flexible lighting and solar panels, and in distributed sensors.

H. Marien (✉) • M. Steyaert
Department of Electrical Engineering, ESAT-MICAS, 91.08, Katholieke Universiteit Leuven, Kasteelpark Arenberg 10 – bus 2443, 3001 Heverlee, Belgium
e-mail: Hagen.Marien@esat.kuleuven.be

E. van Veenendaal
Polymer Vision, Kastanjelaan 1000, 5656AE Eindhoven, The Netherlands

P. Heremans
IMEC, Kapeldreef 75, 3001 Heverlee, Belgium

Department of Electrical Engineering, ESAT-MICAS, 91.08, Katholieke Universiteit Leuven, Kasteelpark Arenberg 10 – bus 2443, 3001 Heverlee, Belgium

A.H.M. van Roermund et al. (eds.), *Nyquist AD Converters, Sensor Interfaces, and Robustness: Advances in Analog Circuit Design, 2012*,
DOI 10.1007/978-1-4614-4587-6_14,

The key benefits of organic electronics are the possibility to deposit circuits on top of various materials for the substrate, e.g. a low-cost plastic foil, the low production temperatures that enable a cost-efficient production. Conversely, organic technology nowadays still suffers from an intrinsically low mobility, a low intrinsic gain and behavioral parameter variations. Furthermore, most of the available technologies are still unipolar, providing p-type transistors only. Nowadays, the research on complementary technologies is proceeding and circuits in complementary technology has been reported [1, 2].

The properties of organic electronics technology makes it a future low-cost medium for flexible large-area and large-scale applications. Nowadays work on digital organic micro-processors [3] and RFID tags [4, 5] has been reported. Analog and mixed-signals circuits have less abundantly appeared in literature. Work on organic ADC's and DAC's [1, 6], an integrated DC-DC converter [7] and a tunable transconductor [8] has been reported. Furthermore, work on various organic sensors [9–11] has been reported. This work gives an overview of today's reported analog organic electronic circuits. A few of these are highlighted and profoundly discussed.

14.2 Technology

The most often used material for organic semiconductors is pentacene ($C_{22}H_{14}$). When sorted in mono-layers, this small molecule obtains semiconductor behavior through the molecule hopping principle. The charge carriers hop from one to the next molecule in the herringbone crystal structure. In practice, pentacene is suitable for making p-type transistors only.

The circuits presented in this work are produced in a unipolar pentacene-based thin-film transistor technology [12, 13] on foil with three metal layers providing p-type bottom-gate transistors and metal-metal overlap capacitors. The minimal length of a transistor is typically 5 μm. The carrier mobility of a transistor is typically around ~0.1–1 cm^2/Vs.

A standard transistor is a three-pin device, since thin-film transistors do not have a bulk contact. The transistor devices can, however, be enhanced to a four-pin device with a backgate. This is a second gate, physically located on top of the bottom-gate transistor. This backgate has a favorable influence on the transistor characteristics, since it influences the threshold voltage in a linear way. As a result, the backgate-source voltage is included in the transistor current equation in saturation region given by Eq. 14.1.

$$I_{sd} = K'.\left((V_{sg} + \xi.V_{sb}) - V_{t,0}\right)^2.\left(1 + \frac{V_{sd}}{V_E.L}\right) \tag{14.1}$$

Where K' is the technology-dependent transistor constant, ξ a factor for the sensitivity of the backgate, V_{sb} the source-backgate voltage, $V_{t,0}$ the initial threshold voltage of the four-pin device when $V_{sb} = 0$ V, V_E the Early voltage and L the length of the transistor. The backgate pin is a powerful tool for designing both analog and digital organic circuits. This is clarified in Sect. 14.4 where the circuit implementations are discussed.

14.3 State-of-the-Art

In terms of digital circuits there is a lot of work reported in literature. A 128-bit RFID tag on foil with a data rate of 1.5 kb/s [5] and a thorough study on the robustness of such tags [14] have appeared. Lately, an 8-bit micro-processor on foil [3] which performs with 40 instructions-per-second is published. The transistor count of this micro-processor with instruction foil is around ~4,000.

The complexity of analog organic circuits is much lower. At the moment, an organic C-2C SAR ADC on glass [1] and a fully integrated $\Delta\Sigma$ ADC on foil [6] have been reported. The C-2C SAR ADC was driven by an external SAR control system and had a 6-bit accuracy. The first-order $\Delta\Sigma$ ADC was integrated on foil and had an accuracy of 26.5 dB. This ADC has a transistor count of ~125. It is further discussed in Sect. 14.4.3. Other reported analog building blocks are DC-DC converters [7] and a tunable transconductance [8].

Organic electronics technology has a lot of opportunities for various kinds of sensors. A nice example of a distributed pressure sensor is found in [9], where an array of 16×16 pressure sensor pixels on a plastic sheet measure the pressure-dependant resistance. This kind of sensor can be used as an artificial skin for robotic arms. In this work an analog touch sensor [15] is highlighted. According to simulations, this 2D capacitive touch pad has an accuracy after interpolation of 0.3 mm.

14.4 Sensor Read-Out

In this work we highlight a set of analog building blocks for the sensor read-out of an organic smart sensor system. The block diagram of the smart system is shown in Fig. 14.1. It is built with a sensor, an amplifier, an ADC, which are all typical building blocks for a sensor read-out path. Furthermore, the system is enhanced with a DC-DC converter. This building block has the very specific function of creating very low and very high bias voltages, which improve the robustness and the performance of the other building blocks, both analog and digital, in the organic electronics technology.

14.4.1 Sensor

The implementation of a 2D capacitive touch sensor on foil is presented in Fig. 14.2. It is built with four row and four column metal plates with a very low

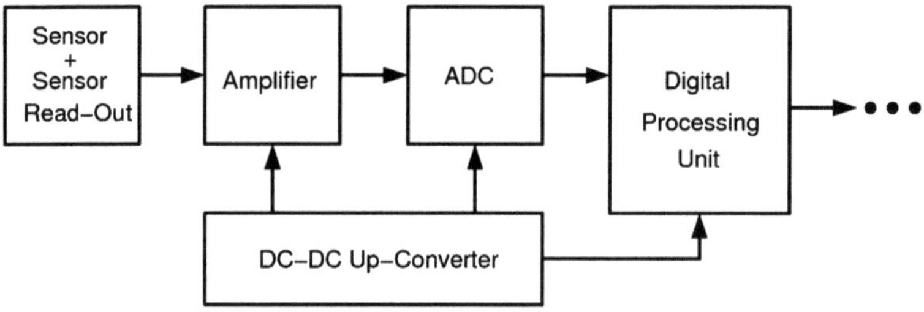

Fig. 14.1 System architecture of an organic smart sensor system

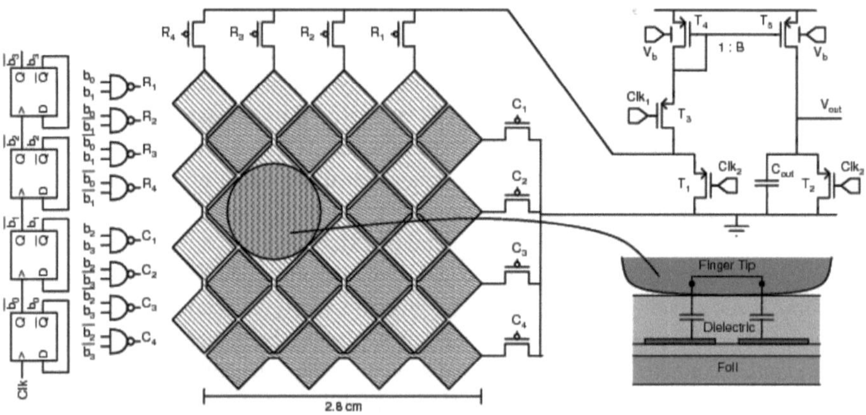

Fig. 14.2 Schematic view of the 2D capacitive touch sensor with a column and row selector. The working principle of the sensor is sketched in the *lower right corner*

mutual overlap capacitance, creating 16 sensor pixels. When a finger is placed on top of one pixel, a series connection of two overlap capacitors is created through the finger tip, as sketched in the cross-section view of the sensor. This capacitance is significantly higher than when no finger is present. A row and column selector circuit selects all pixels in a serial way for measuring the capacitance. This is done with a switched current mirror. In the reset phase φ_1, both the pixel capacitor and the output capacitor are fully discharged. In the sample phase φ_2, the charging current of the pixel capacitor is copied and integrated on the output capacitor. As a result, the output voltage is a linear measure for the capacitance on the pixel.

The measurement results of the sensor are presented in Fig. 14.3. Figure 14.3a represents the case when the finger is exactly on top of one pixel. The sensor is able to detect intermediate positions of the finger. This is demonstrated in Fig. 14.3b where the measured output is given for a finger on positions halfway in between two pixels, and one right in the middle of four pixels. The measured 3 dB-frequency of the sensor is around 1.5 kS/s, which results in a frame rate of 93 Hz. A chip photograph of the presented touch pad is shown in Fig. 14.10a.

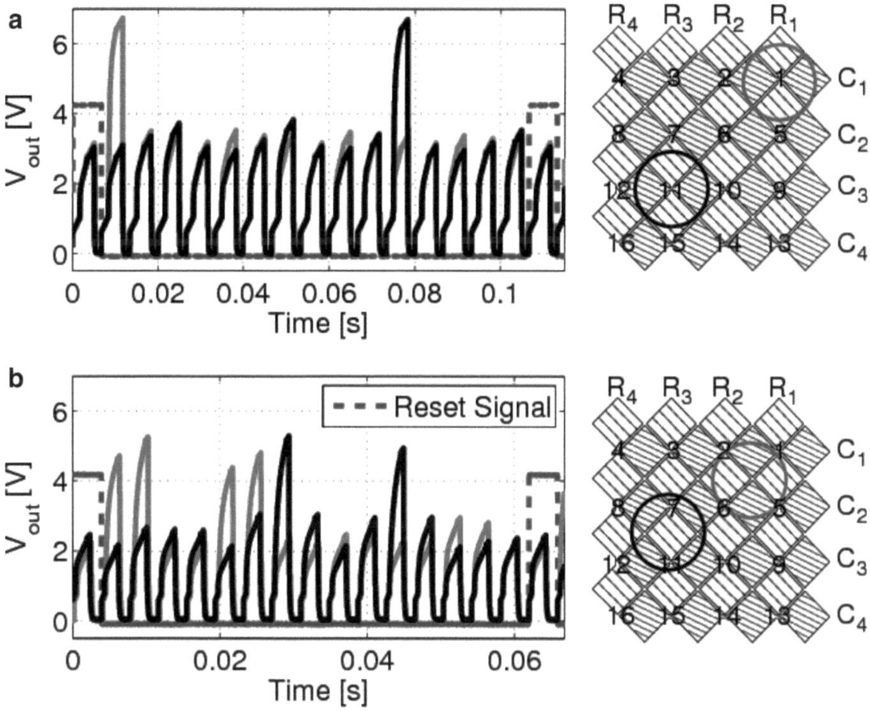

Fig. 14.3 Measurement results of the 2D flexible touch pad. (**a**) Output when the finger is exactly on one pixel, (**b**) output when the finger is on an intermediate position. The measurements are performed with external driving signals for rows and columns and with a shielded conductive finger tip

14.4.2 Opamp

In [6] a three-stage opamp with AC-coupled stages was presented. This circuit was optimized for high gain, but, when employed in larger circuits, it suffered from the high-pass filters in the forward path. The latter create stability issues at very low frequencies around ~0.1 Hz. An improved two-stage opamp design is proposed in Fig. 14.4. Bootstrapped gain enhancement is applied to the p-type load transistors for combining high DC reliability with high gain. Common-mode feedback is applied to reduce the V_T sensitivity of the input pair. The properties of the backgate of a transistor have been employed to bias the single-stage amplifier with identical DC input and output levels. This is obtained by biasing the backgates of the input pair with a very high bias voltage, around 35 V, for a power supply of 15 V. This shifts the V_T of the input pair in a favorable way. As a result, subsequent stages can be connected together. This, of course, solves the stability issues.

The measured bode plot of the two-stage amplifier is presented in Fig. 14.5. The opamp has a gain of 20 dB and a GBW of 2 kHz. The opamp consumes 15 μA from a 15 V power supply voltage. The opamp measures $2 \times 2.4\ mm^2$. The chip photograph of the opamp is included in Fig. 14.10b.

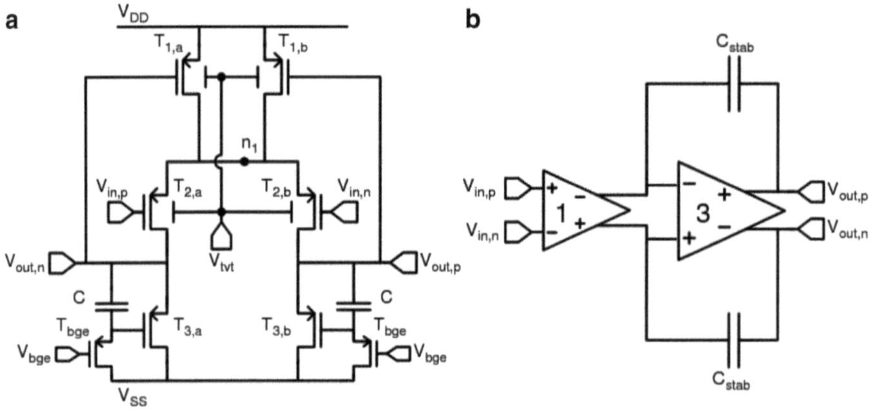

Fig. 14.4 Schematic view of (**a**) the single-stage amplifier and (**b**) the DC-connected two-stage opamp

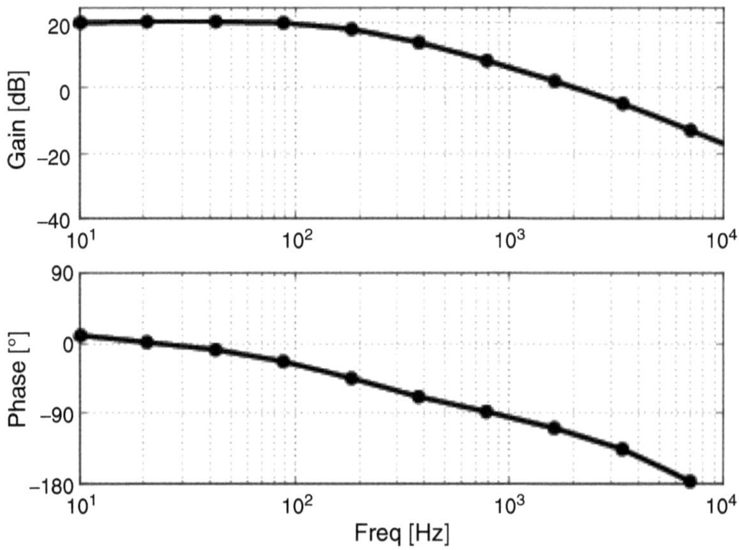

Fig. 14.5 Measured bode plot of the two-stage opamp

14.4.3 ADC

The block diagram of a first-order continuous-time $\Delta\Sigma$ ADC is shown in Fig. 14.6. It is built with a first-order filter, a comparator and a level shifter. The first-order filter is an integrator built with a three-stage AC-coupled opamp, with capacitors and with resistors. The latter are implemented with transistors biased in the linear regime. The comparator is built with single-stage amplifiers and latches, which

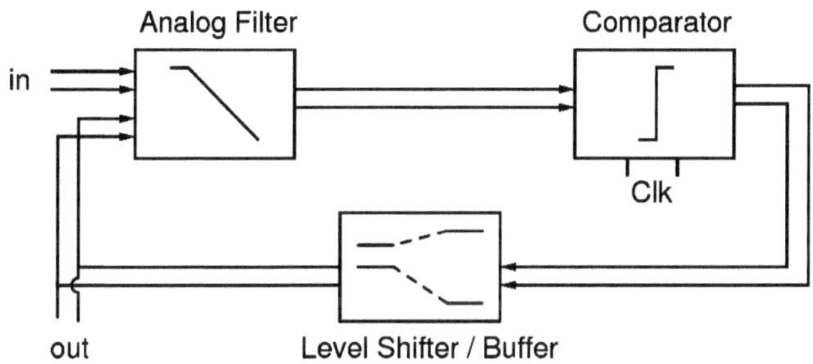

Fig. 14.6 Block diagram of the first-order ΔΣ ADC on foil

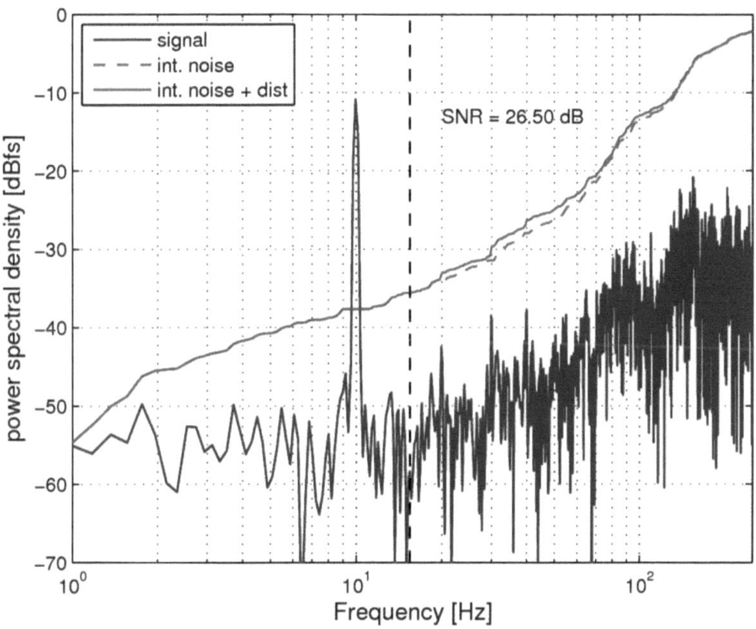

Fig. 14.7 Measured output spectrum of the first-order ΔΣ ADC on foil. The measurement is performed with a 15 V power supply, a 500 Hz clock frequency and an oversampling ratio of 16

digitize and synchronize the data. The function of the level shifter is twofold. It molds the digital signals to the desired DC levels for closing the feedback loop, and it serves as a buffer stage for sending the data to the measurement set-up.

The measured output spectrum of the ADC is shown in Fig. 14.7. It performs with a 26.5 dB precision in a 15.6 Hz bandwidth. The measurement is performed

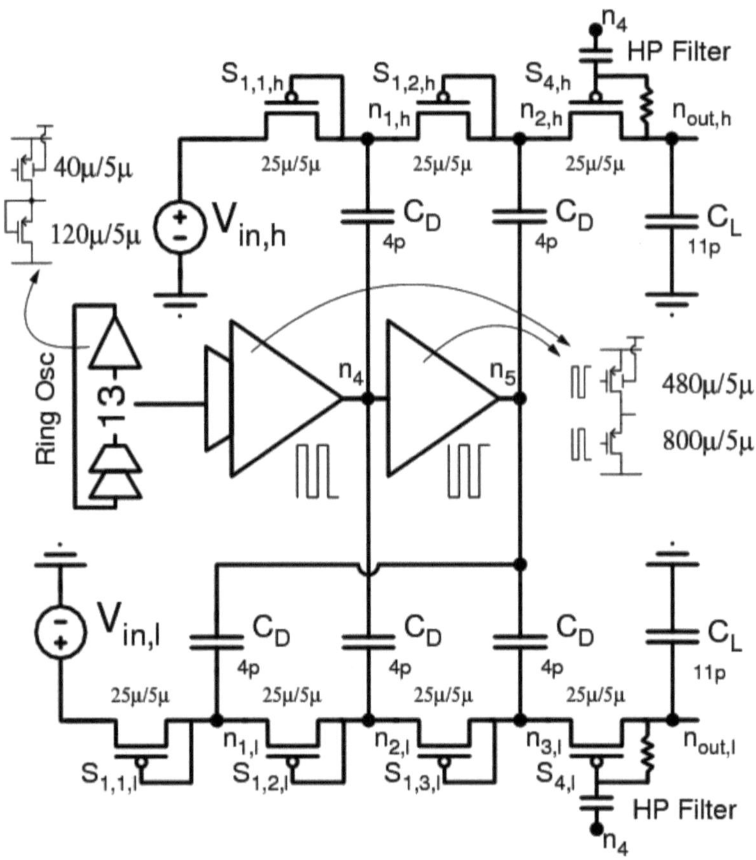

Fig. 14.8 Schematic view of the dual Dickson DC-DC up-converter

with a clock frequency is 500 Hz and for an oversampling ratio of 16 and the ADC consumes 100 μA from a 15 V power supply. The active area of the ADC measures $13 \times 20\ mm^2$. A chip photograph is included in Fig. 14.10c.

14.4.4 *DC-DC Converter*

The last analog building block, proposed for a sensor read-out is a DC-DC converter. It fulfils the demand of the other building blocks for high and, sometimes also, low bias voltages. The proposed architecture of the dual Dickson DC-DC up-converter is shown in Fig. 14.8. It is built with a ring oscillator, a buffer chain and two Dickson converter cores with diode-connected transistors in the forward path. The ring oscillator is constructed with 13 zero-V_{gs} load inverters. The backgates of the pull-up transistors of these inverters is driven by the high-voltage output of the converter.

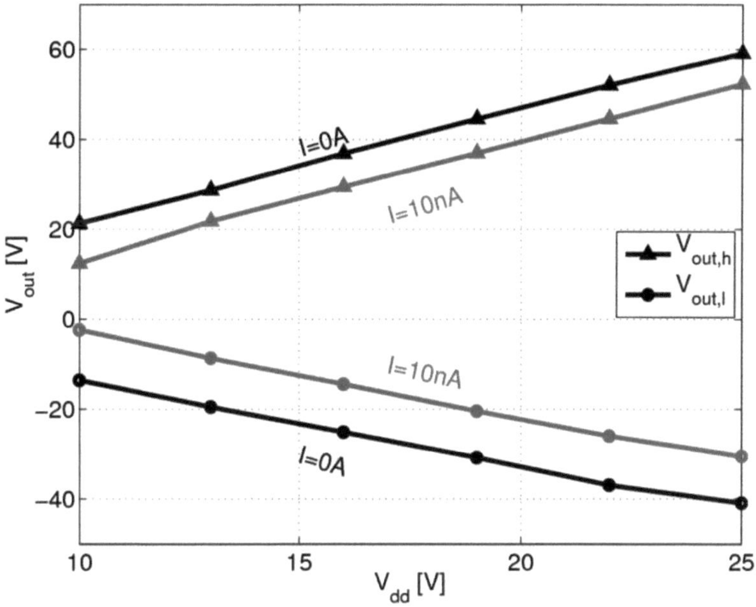

Fig. 14.9 Measured output voltages of the DC-DC converter on foil

This is beneficial for robustness and for the quality of the switching signals. The buffer chain is built with the same type of inverters, except for the last stage of the buffer. In that stage both the pull-up and the pull-down transistor are driven by a clock signal. This improves the pull-down behavior of the signals, at the cost of a 1 V_T loss in the signal swing. Unlike the other diodes in the forward path of the converters, the last diode in the chain is enhanced with a high-pass filter. This filter copies the clock signal on top of the DC-output voltage. Therefore the transistor is biased more off in the off-state and more on in the on-state.

The measured output voltages of the converter are shown in Fig. 14.9. The high output voltage and the low output voltage vary between 20 and 60 V and between −15 and −40 V respectively for a power supply between 10 and 25 V. The current consumption is 1 μA. The converter is only able to draw a few nA at its outputs. This is, however, more than enough to bias gates and backgates of thousands of transistors. A measure for the efficiency of this block is therefore rather found by comparing the power consumption of the converter with that of the building blocks that take advantage of the converter. As an example, the 1 μA of this converter is much lower than the 15 μA consumed in the opamp, presented in Sect. 14.4.2. The DC-DC converter measures $1.2 \times 1.5\ mm^2$. A chip photograph of this converter is included in Fig. 14.10d.

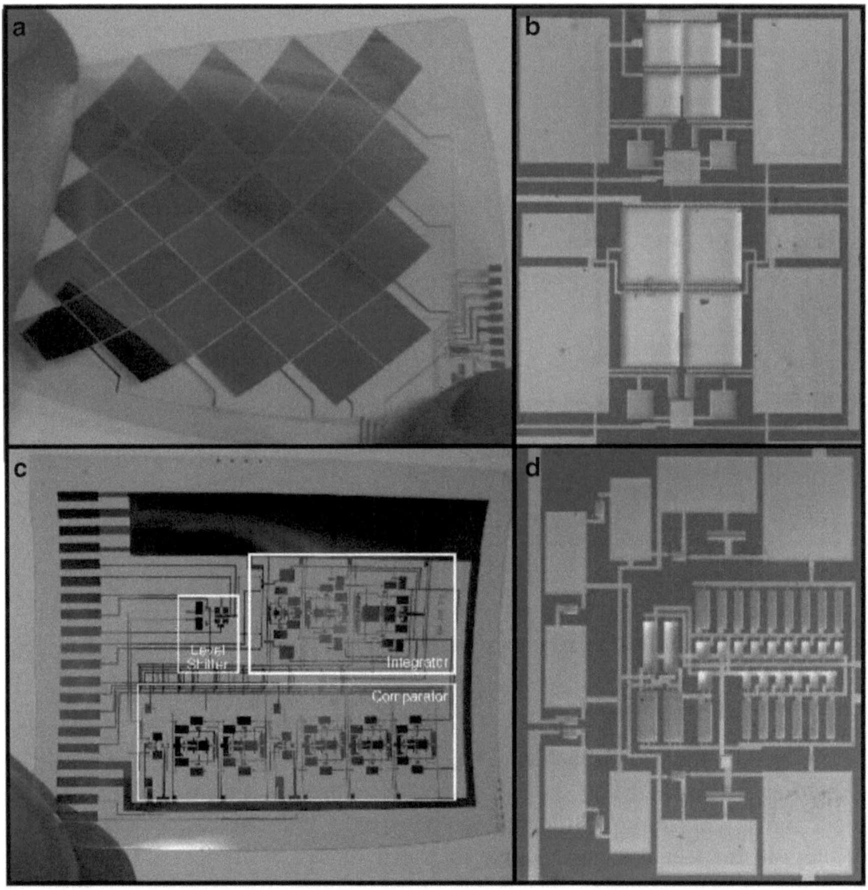

Fig. 14.10 Chip photographs of (**a**) the 2D capacitive touch pad, (**b**) the two-stage DC-coupled opamp, (**c**) the first-order ΔΣ ADC and (**d**) the dual Dickson DC-DC up-converter

14.5 Conclusions

In this work an overview of the development of analog circuits in organic thin-film transistor technologies on foil was given and multiple state-of-the-art circuits were discussed in detail. The architecture of a sensor read-out for smart sensor systems was discussed and implementations of all the building blocks were provided. A 2D capacitive touch pad with an accuracy of 0.3 mm was presented. Next, the implementation of a two-stage opamp with DC-connected stages was proposed. Then a first-order ΔΣ ADC with a measured 26.5 dB precision and a 15.6 Hz bandwidth was discussed. Finally, a Dickson DC-DC converter was proposed. This converter provided the required high and low bias voltages for driving gates and backgates in the other building blocks. The chip photographs of all the presented circuits are shown in Fig. 14.10. The presented circuits pave the way for flexible organic smart sensor system applications, such as food monitoring, health monitoring and logistics.

References

1. Xiong W et al (2010) A 3 V 6b successive-approximation ADC using complementary organic thin-film transistors on glass. In: International solid state circuits conference, Digest of technical papers, San Francisco pp 134–135, Feb 2010
2. Bode D et al (2010) Noise-margin analysis for organic thin-film complementary technology. IEEE Trans Electron Dev 57:201–208
3. Myny K, van Veenendaal E, Gelinck G, Genoe J, Dehaene W, Heremans P (2011) An 8b organic microprocessor on plastic foil. In: IEEE international solid-state circuits conference, Digest of technical papers, San Francisco pp 322–324
4. Cantatore E et al (2007) A 13.56-MHz RFID system based on organic transponders. IEEE J Solid-St Circ 42(1):84–92
5. Myny K et al (2009) A 128b organic RFID transponder chip, including Manchester encoding and ALOHA anti-collision protocol, operating with a data rate of 1529b/s. In: International solid-state circuits conference, Digest of technical papers, San Francisco pp 206–207, Feb 2009
6. Marien H, Steyaert MSJ, van Veenendaal E, Heremans P (2011) A fully integrated ΔΣ ADC in organic thin-film transistor technology on flexible plastic foil. IEEE J Solid-St Circ 46:276–284
7. Marien H et al (2011) DC-DC converter assisted two-stage amplifier in organic thin-film transistor technology on foil. In: 37th European solid state circuits conference, ESSCIRC, pp 411–414, Helsinki Sept 2011
8. Raiteri D et al (2011) A tunable transconductor for analog amplification and filtering based on double-gate organic TFT's. In: 37th European solid state circuits conference, ESSCIRC, Helsinki pp 415–418, Sept 2011
9. Kawaguchi H, Someya T, Sekitani T, Sakurai T (2005) Cut-and-paste customization of organic FET integrated circuit and its application to electronic artificial skin. IEEE J Solid-St Circ 40:177–185
10. Mori T, Kikuzawa Y, Noda K (2009) Improving the sensitivity and selectivity of alcohol sensors based on organic thin-film transistors by using chemically-modified dielectric interfaces. In: IEEE Sensors, pp 1951–1954
11. He D, Nausieda I, Ryu, K, Akinwande A, Bulovic V, Sodini C (2010) An integrated organic circuit array for flexible large-area temperature sensing. In: IEEE international solid-state circuits conference digest of technical papers (ISSCC), San Francisco pp 142–143
12. Gelinck GH et al (2004) Flexible active-matrix displays and shift registers based on solution-processed organic transistors. Nat Mater 3:106–110
13. Gelinck GH et al (2005) Dual-gate organic thin-film transistors. Appl Phys Lett 87:073508
14. Myny K, Beenhakkers MJ, van Aerle NAJM, Gelinck GH, Genoe J, Dehaene W, Heremans P (2011) Unipolar organic transistor circuits made robust by dual-gate technology. IEEE J Solid-St Circ 46:1223–1230
15. Marien et al (2012) 1D and 2D analog 1.5 kHz air-stable organic capacitive touch sensors on plastic foil. In: IEEE international solid-state circuits conference, Digest of technical papers San Francisco (ISSCC)

Chapter 15
Impact of Statistical Variability on FinFET Technology: From Device, Statistical Compact Modelling to Statistical Circuit Simulation

A. Asenov, B. Cheng, A.R. Brown, and X. Wang

Abstract New variability resilient device architectures will be required at the 22 nm CMOS technology node and beyond due to the ever-increasing statistical variability in traditional bulk MOSFETs. A TCAD-based Preliminary Design Kit (PDK) development strategy is present here for a 10 nm SOI FinFET technology, with reliable device statistical variability coming from the comprehensive 3D statistical device simulation and accurate statistical compact modelling. Results from the statistical simulation of a 6T SRAM cell demonstrate the advantages of FinFET technology.

15.1 Introduction

For over three decades, the progressive scaling of bulk CMOS transistors to achieve faster devices and higher circuit density has fuelled the phenomenal success of the semiconductor industry – captured by Moore's famous law [1]. However, with the bulk device feature size reaching decananometre scales, the ever-increasing statistical variability in the device characteristics, introduced by discreteness of charge and granularity of matter, is becoming the major roadblock in the path of traditional bulk CMOS technology scaling. The standard deviation of the threshold

A. Asenov (✉) • B. Cheng
Department of Electronics and Electrical Engineering, Device Modelling Group, School of Engineering, University of Glasgow, Glasgow G12 8LT, UK

Gold Standard Simulations Ltd, Rankine Building, Glasgow G12 8LT, UK
e-mail: Asen.Asenov@glasgow.ac.uk

A.R. Brown
Gold Standard Simulations Ltd, Rankine Building, Glasgow G12 8LT, UK

X. Wang
Department of Electronics and Electrical Engineering, Device Modelling Group, School of Engineering, University of Glasgow, Glasgow G12 8LT, UK

A.H.M. van Roermund et al. (eds.), *Nyquist AD Converters, Sensor Interfaces, and Robustness: Advances in Analog Circuit Design, 2012*,
DOI 10.1007/978-1-4614-4587-6_15,

voltage distribution in minimum-sized square bulk nMOSFETs corresponding to a 45 nm low-power technology generation can exceed 50 mV, with the main contribution coming from random discrete dopants (RDD) in the channel region [2]. New variability-resilient device architectures, such as FinFETs and ultra thin body (UTB) SOI devices, will be required in order to maintain the benefits of technology scaling at the 22 nm node and beyond. Intel has introduced FinFET technology at the 22 nm high-performance technology node [3], and STMicroelectronics has already demonstrated the UTB SOI technology at the 28 nm technology node [4].

From a circuit and system designer's point-of-view, without a reliable preliminary design kit (PDK) at the early stage of technology development, the introduction of new device architectures at the upcoming advanced technology nodes will bring the general design community into unchartered waters. Before the actual silicon data becomes available, a physically-based TCAD simulation can generate the most accurate data regarding the new device characteristics and performance, which will provide solid ground for PDK development. In this paper, we present a comprehensive TCAD simulation study of a 10 nm gate length SOI FinFET technology, and outline a TCAD-based PDK development strategy. The design of template devices is presented in Sect. 15.2, and the impact of statistical variability sources on device characteristics is discussed in Sect. 15.3. The statistical compact modelling of 10 nm SOI FinFET devices, as a vital stage of the PDK development for this technology, is present in Sect. 15.4. The benefits of FinFET technology is demonstrated in Sect. 15.5 by the statistical SRAM simulation enabled by the PDK of the 10 nm template SOI FinFET technology. The final conclusions are drawn in Sect. 15.6.

15.2 The Design of Template FinFET Devices

The TCAD-based PDK development starts with the design of template transistors for the technology node of interest. The device design is based on a generic SOI FinFET structure which has been implemented directly in the GSS 'atomistic' simulator GARAND [5]. Figure 15.1 shows a schematic picture of the FinFET structure, demonstrating the intrinsic 3D nature of the device. Physical and electrical specifications have been taken from the ITRS 2010 update for a 10 nm gate-length high-performance multi-gate device. The devices feature a high-κ dielectric stack with 0.585 nm EOT (3 nm physical oxide thickness with dielectric constant $\varepsilon = 20\varepsilon_0$), and metal gates. Dual metal with gate-last process is assumed in order to eliminate metal gate granularity and the associated work function variability. Relatively high source/drain doping of 3×10^{20} cm^{-3} is used to reduce access resistance. The channel is effectively undoped with a very light p-type doping of 1×10^{15} cm^{-3} for nFinFET, and n-type doping of 1×10^{15} cm^{-3} for pFinFET. The doping profiles for both n- and p-channel devices are shown in Fig. 15.2.

The basic geometrical and electrical parameters are summarised in Table 15.1, in comparison with those specified in the ITRS for this technology. The gate transfer characteristics for the n- and p-channel transistors are shown in Figs. 15.3 and 15.4 respectively, which demonstrate the excellent electrostatic integrity of the template devices.

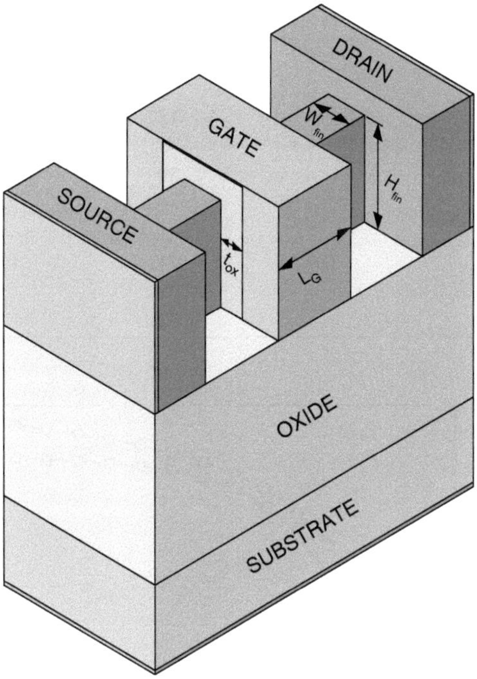

Fig. 15.1 Schematic of the 3D FinFET structure

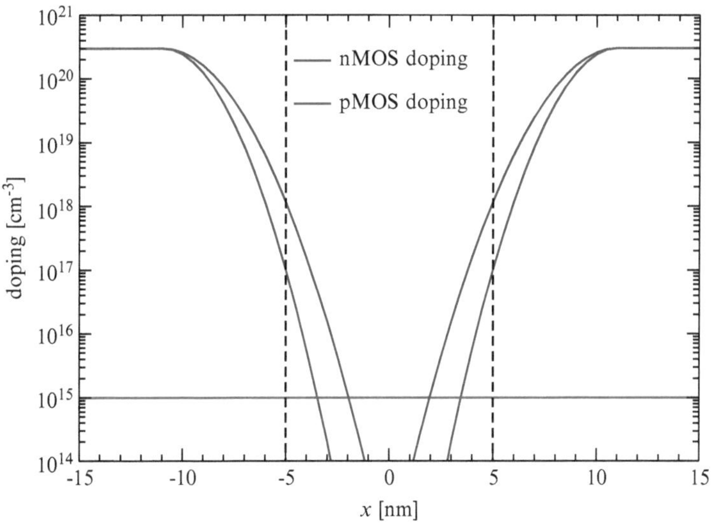

Fig. 15.2 Doping profiles for the nMOS and pMOS FinFET devices

Table 15.1 Structural and electrical parameters for n- and p-FinFETs

	n-channel		p-channel	
	ITRS	TCAD	ITRS	TCAD
L_G, nm	9.7	10	9.7	10
EOT, nm	0.57	0.585	0.57	0.585
H_{fin}, nm	–	12.5	–	12.5
W_{fin}, nm	4.8	5	4.8	5
W_{elec}, ($2H_{fin} + W_{fin}$), nm	–	30	–	30
V_{DD}, V	0.66	0.8	0.66	0.8
I_{DSAT}, mA/μm	2.0	1.9	1.74	1.67
I_{OFF}, nA/μm	100	71	100	70
SS, mV/dec	–	71	–	72
DIBL, mV/V	–	53	–	58

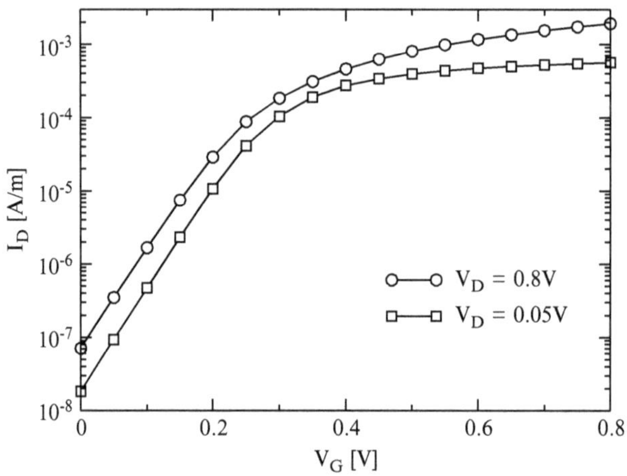

Fig. 15.3 Gate transfer characteristics of the template n-FinFET

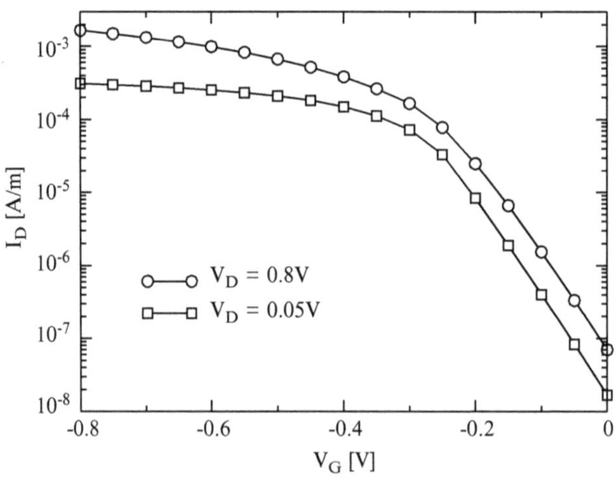

Fig. 15.4 Gate transfer characteristics of the template p-FinFET

15.3 Impact of Statistical Variability on FinFET Devices

Although FinFETscan tolerate very low channel doping, the RDD at S/D region can still play an important role introducing variations in the transistor characteristics. Due to the 3D nature of the FinFET, line edge roughness (LER) in FinFETs not only introduces traditional gate edge roughness (GER) variation, but also introduces fin edge roughness (FER) variation. Previous studies [6] indicate that metal gate granularity (MGG) can be a major source of variability in gate-first technology. However, there is a consensus in the industry that gate-last technology, where the deposited gate metal could remain amorphous, will be the dominant gate technology in sub-20 nm technology nodes, and in this case, MGG is unlikely to be a source of variability in sub-20 nm technology nodes.

The sources of variability considered in the 10 nm SOI FinFET PDK are RDD, GER and FER, as illustrated in Fig, 15.5. An ensemble of 1,000 microscopically different devices is generated and simulated by the GSS 'atomistic' simulator GARAND. Figures 15.6 and 15.7 present the I_D-V_G characteristics of the 10 nm n- FinFETs under the influence of the combined statistical variability sources. The 3σ of the line edge roughness for both GER and FER are 2 nm, and the correlation lengths are 30 nm. Similar Id-Vg characteristics are observed in the p-FinFET.

The impacts of individual and combined statistical variability (SV) sources on important device figures of merit, i.e. threshold voltage (Vth) and on-current (Ion), are presented in Fig. 15.8 for the n-FinFET. As expected, for a fin thickness (width) of 5 nm, FER-induced quantum confinement variation is the dominant variation source in this device. The standard deviation (SD) of Vth and Ion introduced by FER at Vds of 0.8 V are 30 mV and 0.16 mA/μm respectively, which represents 94% and 78% of the SD values introduced by combined sources. The impact of RDD and GER on device characteristics is strongly dependent on the region of operation. RDD have a mild influence on the subthreshold region, but have a

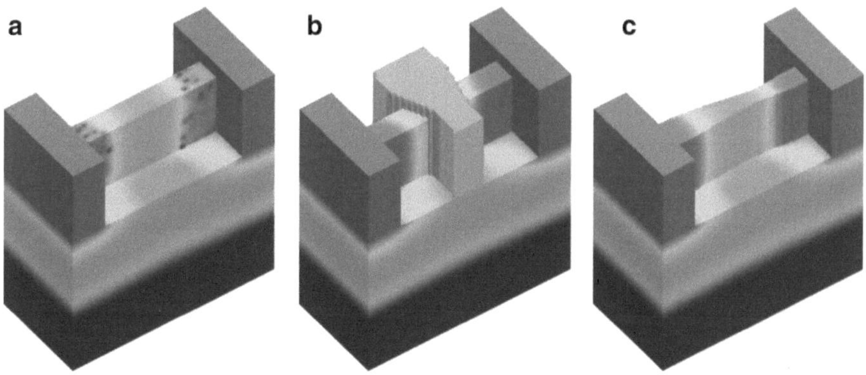

Fig. 15.5 The effect on the structure and electrostatic potential within the device of the different sources of variability: (**a**) random discrete dopants, (**b**) gate edge roughness and (**c**) fin edge roughness

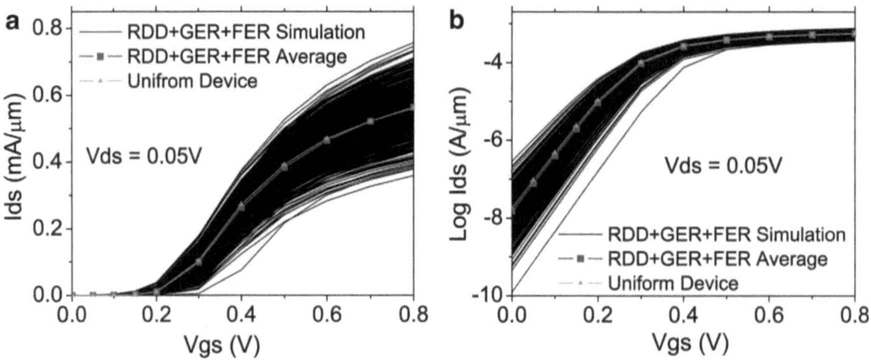

Fig. 15.6 I_D–V_G characteristics at $V_D = 0.05$ V for the nMOS FinFET on (**a**) linear and (**b**) logarithmic scales. The curves for the 1,000 simulated devices are shown, along with the nominal uniform characteristics

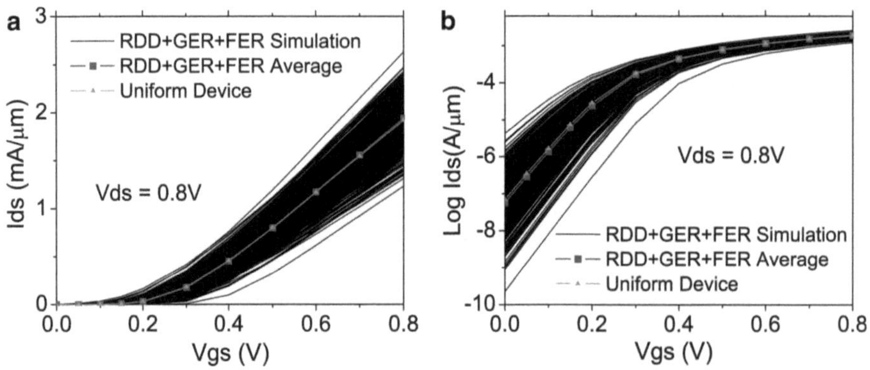

Fig. 15.7 I_D–V_G characteristics at $V_D = 0.8$ V for the nMOS FinFET on (**a**) linear and (**b**) logarithmic scales. The curves for the 1,000 simulated devices are shown, along with the nominal uniform characteristics

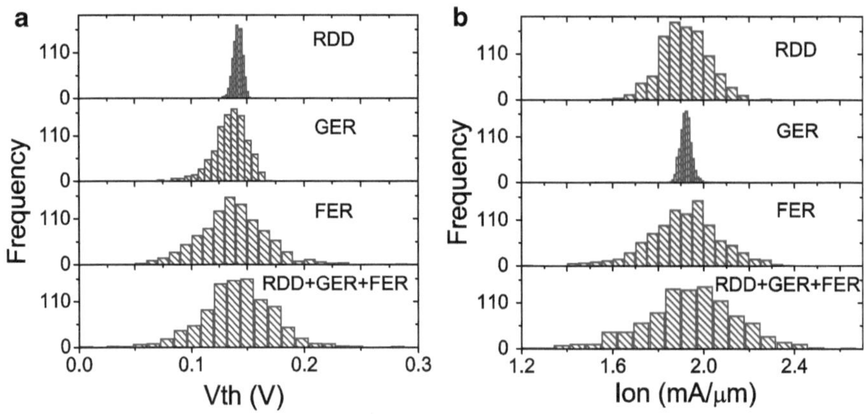

Fig. 15.8 Histogram plots of (**a**) threshold voltage, (**b**) on-current for the nMOS FinFET at $V_D = 0.8$ V

relatively large contribution in the on-current region, while for GER the opposite trend holds.

Similar to n-FinFET, FER-induced quantum confinement variation is the dominant variation source in the p-FinFET. The SD of Vth and Ion introduced by FER at Vds of 0.8 V are 30 mV and 0.15 mA/μm respectively, which represents 90% and 88% of the SD values introduced by combined sources.

15.4 Statistical FinFET Compact Modelling

Compact models are the interface between technology and design. Device compact modelling is a critical part of PDK development work. The precondition to achieve variability-aware circuit design is to have reliable statistical compact models. All the device-variation information present in Sect. 15.3 needs to be captured by the device compact model. The approach adopted here is to generate a different statistical compact model parameter set for each different device that was simulated. The information of distribution and correlations between the statistical parameters extracted from the originally simulated transistors then become part of the PDK. During the statistical circuit simulation stage, a Monte Carlo approach is used where the set of statistical compact model parameters for each of the transistors in the circuit is obtained based on the statistical compact model libraries in the PDK.

A two-stage statistical compact model parameter extraction strategy is implemented [7]. During the first stage, a local optimisation is employed to extract the complete nominal set of BSIM-CMG parameters. This is based on the simulation of the FinFET transistors with *continuous doping concentration and straight edges*. Figures 15.9 and 15.10 presents the nominal BSIM-CMG results of the n- and p-FinFETs against the TCAD simulation results for the template devices. The RMS errors are less than 2%.

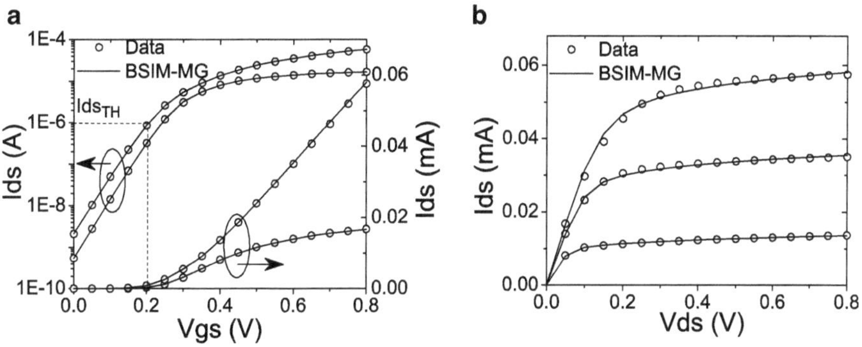

Fig. 15.9 BSIM-CMG results for the 10 nm nMOS FinFET device. (**a**) I_D-V_G characteristics, (**b**) I_D-V_D characteristics

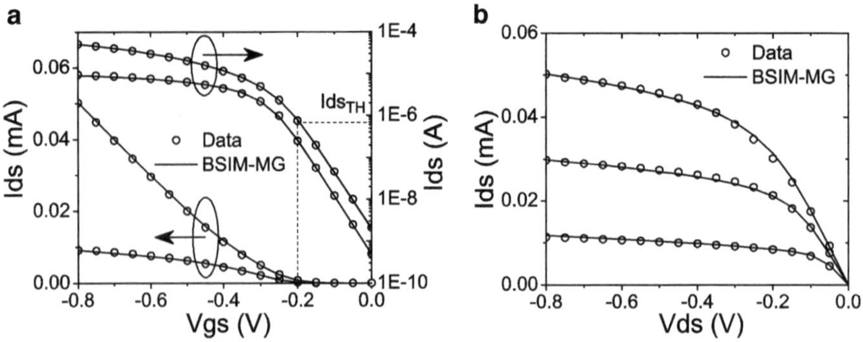

Fig. 15.10 BSIM-CMG results for the 10 nm pMOS FinFET device. (**a**) I_D-V_G characteristics, (**b**) I_D-V_D characteristics

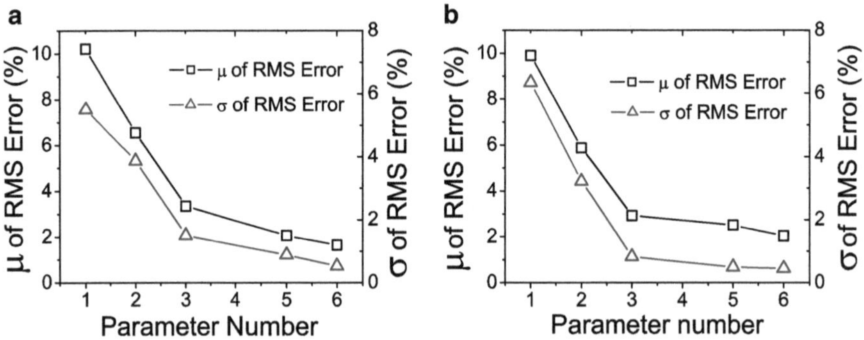

Fig. 15.11 Mean and standard deviation in the RMS error of the extracted statistical compact models of the 10 nm (**a**) n-FinFET and (**b**) p-FinFET devices versus the size of the statistical parameter set

At the second stage statistical parameter extraction, the parameters extracted from the first compact model extraction stage are divided into two groups: those which have been found to be relatively independent of statistical variation that are then fixed during the second stage, and those that are sensitive to the statistical variation, which will be re-extracted for each 'atomistically' different device at the second stage. Six key BSIM-CMG parameters have been identified to represent the effect of statistical variability sources. The accuracy in representing each of the device characteristics and figures of merit under the influence of statistical variation depends on the number of statistical parameters used during the second stage statistical extraction. Statistical parameters are selected, in an order following their statistical significance, to form the parameter sets of different sizes (1–6 parameters). The reduction of mean and standard deviation of RMS error against the increase in the size of the parameter set is illustrated in Fig. 15.11. The worst-case statistical compact modelling results with a 6-parameter set is presented in

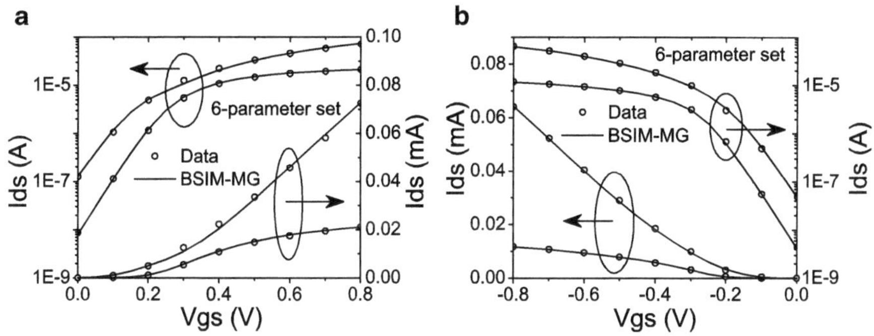

Fig. 15.12 The worst-case statistical compact modelling results of 6-parameter set, (**a**) n-FinFET and (**b**) p-FinFET devices

Fig. 15.12, demonstrating the accuracy of the proposed statistical compact modelling approach.

The above statistical compact modelling results are integrated naturally into the GSS statistical circuit simulation tool RandomSpice through the PDK. With the mean RMS error of statistical compact modelling at less than 2%, RandomSpice can provide the most accurate statistical circuit simulation result for the circuit designer.

15.5 Statistical SRAM Cell Simulation

Among all the circuit components in SOC, SRAM is most sensitive to device statistical variation. The impact of statistical variation on functionality and performance of SRAM is one of the focus fields for variability-aware design [6, 7] and the performance of SRAM reflects the ultimate ability that a technology can offer. A 6T SRAM cell with minimum-size device configuration (Fin number is 1, and cell ratio of 1), which represents a high density SRAM design configuration, is used as a benchmark circuit to demonstrate the benefits that FinFET technology can offer. A much older bulk technology with gate length of 35 nm is used as a reference [7].

Static noise margin (SNM), which is defined as the minimum DC noise voltage needed to flip the memory state during the read operation, is investigated in this study. Figure 15.13 illustrates the bias configuration for SNM simulation, and the butterfly figures used to calculate SNM value are presented in Fig. 15.14. For bulk technology with cell ratio of 1, nearly 10% of the cells will malfunction even under a noise free condition. While for FinFET technology with much smaller device gate length at 10 nm, all cells under investigation are functional. The mean value of SNM is 154 mV and SD of SNM is 13 mV, which indicates that it can pass the "$\mu - 6\sigma$" yield criterion with considerable margin.

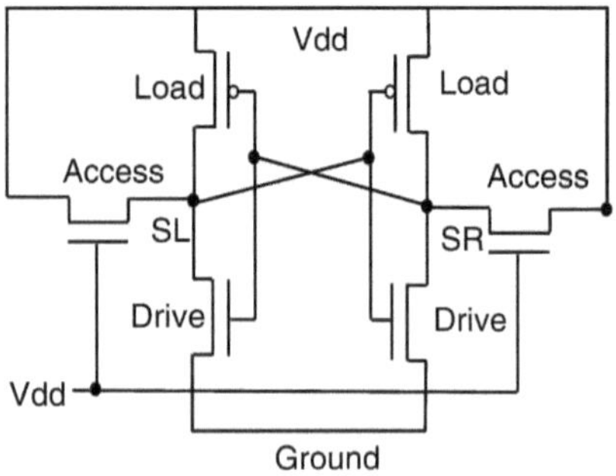

Fig. 15.13 Bias configure-tion for SNM of 6T SRAM cell

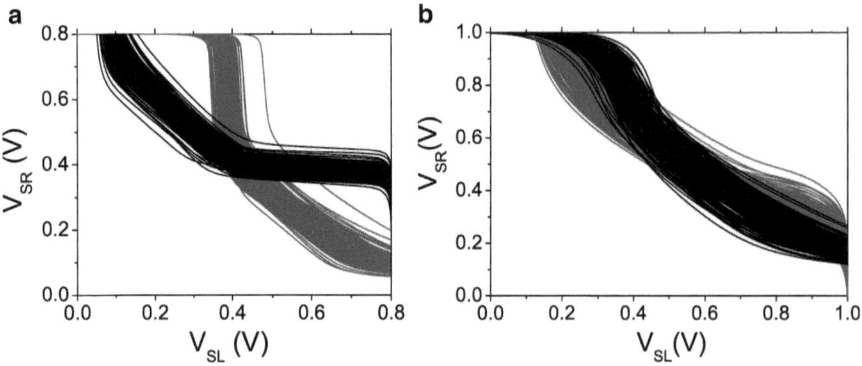

Fig. 15.14 Static transfer characteristics (*butterfly figure*) of SRAM cell with cell ratio of 1, (**a**) 10 nm gate length FinFET SRAM, (**b**) minimal-size 35 nm gate length bulk SRAM

15.6 Conclusions

New variability resilient device architectures, such as FinFET and ultra thin body (UTB) SOI devices, will be introduced at the upcoming advanced technology nodes to tackle the ever-increasing statistical variability challenge. Reliable preliminary design kits (PDKs) at the early stage of technology development will be necessary for the design community to swiftly adapt new technologies. A TCAD-based PDK development strategy is presented for 10 nm SOI FinFET technology, and the PDK-enabled statistical SRAM simulation demonstrates the advantage of FinFET technology.

Acknowledgement This work was supported in part by the European Union through the EP7 Integrated Project Trams.

References

1. Moore GE (1975) Progress in digital electronics. In: Technical digest of the international electron devices meeting. IEEE Press, p 13
2. Cathignol A, Cheng B et al (2008) Quantitative evaluation of statistical variability sources in a 45 nm technological node LP N-MOSFET. IEEE Electron Device Lett 29:609–611
3. http://download.intel.com/newsroom/kits/22nm/pdfs/22nm-Details_Presentation.pdf
4. Arnaud F, Colquhoun S, Mareau AL, Kohler S, Jeannot S, Hasbani F et al (2011) Technology-circuit convergence for full-SOC platform in 28 nm and beyond. In: International electron devices meeting, Digest of technical papers, pp 374–377
5. www.GoldStandardSimulations.com
6. Dadgour H, Endo K, De V, Banerjee K (2008) Modeling and analysis of grain-orientation effects in emerging metal-gate devices and implications for SRAM reliability. In: International electron devices meeting, Digest of technical papers, pp 705–708
7. Cheng B, Roy S, Roy G, Adamu-Lema F, Asenov A (2005) Impact of intrinsic parameter fluctuations in decanano MOSFETs on yield and functionality of SRAM cells. Solid State Electron 49:740

MIX
Papier aus verantwortungsvollen Quellen
Paper from responsible sources
FSC® C105338

If you have any concerns about our products,
you can contact us on
ProductSafety@springernature.com

In case Publisher is established outside the EU,
the EU authorized representative is:
Springer Nature Customer Service Center GmbH
Europaplatz 3, 69115 Heidelberg, Germany

Printed by Libri Plureos GmbH
in Hamburg, Germany